Charles James Blasius Williams, Charles Theodore Williams

Pulmonary Consumption

Its Etiology, Pathology and Treatment

Charles James Blasius Williams, Charles Theodore Williams

Pulmonary Consumption
Its Etiology, Pathology and Treatment

ISBN/EAN: 9783337814229

Printed in Europe, USA, Canada, Australia, Japan

Cover: Foto ©berggeist007 / pixelio.de

More available books at **www.hansebooks.com**

PULMONARY CONSUMPTION

ITS ETIOLOGY, PATHOLOGY
AND TREATMENT

WITH AN ANALYSIS OF 1000 CASES TO EXEMPLIFY
ITS DURATION AND MODES OF ARREST

BY

C. J. B. WILLIAMS, M.D., LL.D., F.R.S.

FELLOW OF THE ROYAL COLLEGE OF PHYSICIANS; PHYSICIAN-EXTRAORDINARY TO HER MAJESTY
THE QUEEN; SENIOR CONSULTING PHYSICIAN TO THE HOSPITAL FOR CONSUMPTION AND
DISEASES OF THE CHEST, BROMPTON; FORMERLY PROFESSOR OF MEDICINE
AND PHYSICIAN TO THE HOSPITAL, UNIVERSITY COLLEGE, LONDON

AND

CHARLES THEODORE WILLIAMS, M.A., M.D. Oxon.

FELLOW OF THE ROYAL COLLEGE OF PHYSICIANS; PHYSICIAN TO THE HOSPITAL
FOR CONSUMPTION AND DISEASES OF THE CHEST, BROMPTON

SECOND EDITION, ENLARGED AND RE-WRITTEN

BY DR C. THEODORE WILLIAMS

WITH FOUR COLOURED PLATES AND TEN WOODCUTS

LONDON

LONGMANS, GREEN, AND CO.

1887

TO

MY COLLEAGUES AND FELLOW-WORKERS

PAST AND PRESENT

ON THE

MEDICAL STAFF OF THE BROMPTON HOSPITAL

WHOSE FRIENDSHIP, ASSISTANCE, AND ENCOURAGEMENT

I HAVE ENJOYED DURING THE LAST TWENTY YEARS

THIS SECOND EDITION

IS GRATEFULLY DEDICATED

PREFACE.

THE first edition of this work appeared many years ago, and
was founded principally on the large and mature experience
in Pulmonary Consumption of Dr. C. J. B. Williams, whose
extensive records of cases were carefully analysed to determine
what influence the progress in the treatment of the disease by
hygiene, medicine and climate had had on its duration, and it
was proved that this had quadrupled. A second great aim of
Dr. C. J. B. Williams was to demonstrate that many of the
phenomena of phthisis were due to a decline or deficiency in
the vitality of the bioplasm, causing inflammatory or other
processes to result in short-lived productions, and that much
might be done by appropriate treatment to correct this tendency and to raise the standard of tissue formation.

That Consumption might possibly have a septic origin
was shadowed forth in the view that one class of the agents
of causation were 'septic influences, which tend to blight or
corrupt portions of the bioplasm of the blood or of the lymphatics, and thus sow the seeds of decay.' The tubercle
bacillus, the great discovery of Koch, may truly be called the
septic element of Consumption, though others may exist, and
it may be said to exercise a powerfully corrupting influence
on the blood and lymph of the body, after the first infective
centre has been established.

My father's retirement from practice nearly twelve years
ago threw the sole responsibility of preparing a second edition
on my shoulders, and I might well have shrunk from the task
had I not enjoyed the advantage of being first his pupil and

then later his colleague, and thus enjoyed the opportunity of
learning much from his clear judgment and original teaching.
Twenty-three years of practice, in which pulmonary diseases
formed a large proportion of the cases, and twenty years'
service on the staff of the Brompton Hospital, have afforded
me unusual opportunities for studying the phenomena and
treatment of Consumption, and will, I trust, absolve me from
the charge of presumption in undertaking this edition, the
objects of which are—(1) to survey the experimental and other
evidence on which the causation of tuberculosis by the tubercle
bacillus rests, and to determine how much of the pathology
and clinical history of Consumption is due to this organism
and its action, primary and secondary, on the tissues, and
how much to other agencies ; (2) to consider, in addition to
the ordinary type of Consumption, the varieties of the disease,
and to treat in some detail of the principal complications ;
(3) to review the present treatment of Consumption in its
various aspects.

To carry out these objects it has been found necessary to
re-write the pathology, and, in fact, the greater part of the
book. Two chapters by Dr. C. J. B. Williams, viz. Chapters I.
and IX., remain practically unchanged, and his ' mine ' of
cases, as the late Dr. Wilson Fox termed them, have been
utilised as well as largely added to from my Brompton and
private note-books. The chapters on ' Predisposing Causes '
and ' Hæmoptysis ' have been enlarged, and Chapters II. to
VI. (Pathology), VIII., XI. (Clinical Aspects of the Tubercle
Bacillus), XII. (Temperature of Consumption), XIII. (Diarrhœa),
XIV. (Pneumothorax), XV. (Albuminuria), XIX. (Fibroid and
Laryngeal Phthisis), and XXII. (Prophylactic Treatment),
and XXVI. (Antiseptic Treatment), fourteen in all, are new.
The chapter on ' Climate ' has been re-written, and contains
the conclusions which sixteen years' additional experience
has yielded, including the remarkable curative influences of
high altitudes on Consumption.

The rest of the work has been carefully revised, and eight coloured plates drawn principally from Dr. Percy Kidd's and my own preparations introduced, and my best thanks are due to that gentleman for kindly superintending their execution. Ten woodcuts have also been added, and it is hoped that these additions will increase the interest of the work.

My present view of the treatment of Consumption is, that while we enter upon a life and death struggle with our enemy, the tubercle bacillus, to destroy him in the nest he has made for himself and to eject him from the living patient, we must never omit prophylactic and anti-phthisical measures, which may render his attack harmless, or, if he has effected a lodgment, may confine him to the outwork and prevent his entry into the citadel; and as proof that this can be done, we may point to the remarkable success which has attended such measures, and specially the so-called 'mountain cure.'

The outlook of the future of Consumption is decidedly hopeful, for there is little doubt that much of the disease is due to preventable causes, which the coming reign of hygiene will sweep away, and that in many cases the disease will be nipped in the bud by a combination of anti-phthisical and bacillicide treatment, while in more advanced cases life will be prolonged even beyond its present lengthened duration.

My best thanks are due to kind friends who have assisted me with their suggestions, and many apologies to my readers for the defects and shortcomings of the work.

C. THEODORE WILLIAMS.

2 UPPER BROOK STREET, *June* 1887.

CONTENTS.

CHAPTER I.

DEFINITION.

CHAPTER II.

MORBID ANATOMY OF PULMONARY CONSUMPTION.

CHAPTER III.

HISTOLOGY OF CONSUMPTION.

CHAPTER X.

CLINICAL ASPECTS OF TUBERCLE BACILLUS.

CHAPTER XI.

HÆMOPTYSIS AND THE HÆMORRHAGIC VARIETY OF CONSUMPTION.

CHAPTER XII.

THE TEMPERATURE OF CONSUMPTION.

CHAPTER XIII.

THE DIARRHŒA OF CONSUMPTION.

CHAPTER XIV.

PNEUMOTHORAX AND PYO-PNEUMOTHORAX.

CHAPTER XV.

ALBUMINURIA OF CONSUMPTION.

CHAPTER XX.

CHRONIC TUBERCULAR CONSUMPTION.

CHAPTER XXI.

THE DURATION OF PULMONARY CONSUMPTION.

a

PLATES *to come all together at beginning of book.*

PLATE I.

(1.) SPUTUM FROM A PHTHISICAL PATIENT (CAVITY STAGE) SHOWING TUBERCLE BACILLI. STAINED BY THE WEIGERT-EHRLICH PROCESS. BACILLI COLOURED RED. EPITHELIUM AND PUS CELLS BLUE. × 550.

(2.) THE SAME AFTER CULTIVATION IN A SOLUTION OF BEEF JUICE SHOW-ING AN ENORMOUS INCREASE IN THE NUMBERS OF TUBERCLE BACILLI. × 550.

Drawing by M. H. Lapidge, from specimens by Dr. C. Theodore Williams.

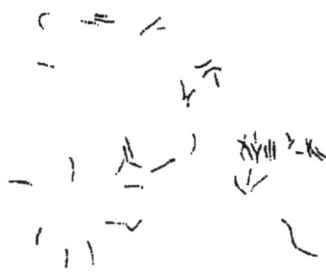

PLATE II.

(1.) ACUTE MILIARY TUBERCULOSIS. LUNG. SECTION OF ALVEOLUS SHOWING DESTRUCTIVE ACTION OF TUBERCLE BACILLI.

> *a, d, e, f.* Alveolar wall, broken through at *d* and *e.*
>
> *b.* A mass of epithelioid cells of various sizes crowded together within alveolus, the result of bacillar irritation. Many are undergoing caseation.
>
> *c.* Tubercle bacilli closely intermingled with the mass, and seen sometimes between, and sometimes within, the cells. At *e* they have destroyed the alveolar wall, and are penetrating it at *f.*   × 550.

Drawn by J. Purkiss, from a Specimen by Dr. Percy Kidd.

(2.) CHRONIC PHTHISIS. LUNG. SECTION OF A CASEOUS MASS SHOWING TRACES OF ALVEOLI. BACILLI ABUNDANT IN ONE PORTION.

> *a.* Cavity extending towards *b.*
>
> *b.* Bacilli swarming in fresh caseating portion of wall of cavity.
>
> *c.* Caseous tract showing traces of alveoli, but devoid of bacilli and cellular structures.   × 75.

Drawn by M. H. Lapidge, from a specimen by Dr. Percy Kidd.

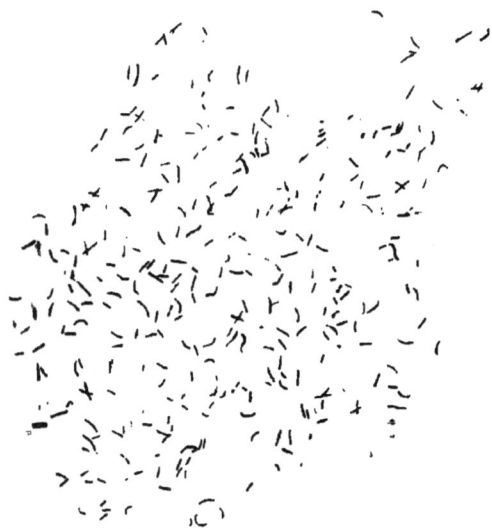

PLATE III.

(1.) ACUTE MILIARY TUBERCULOSIS. SPLEEN. SECTION SHOWING A GIANT CELL UNDERGOING DESTRUCTION BY TUBERCLE BACILLI.

> *a.* Giant cell composed of amorphous protoplasmic material (yellow) seen in its centre and projecting in points along its upper border.
>
> *b, b.* Nuclei (dark blue) distributed principally at the ends.

Tubercle bacilli are seen in close connection with the nuclei, and lying between and across them. One has penetrated between them and reached the centre of the giant cell. × 550.

Drawn by W. Collings from a preparation by Dr. Percy Kidd.

(2.) TUBERCULAR ULCERATION OF THE TONGUE. VERTICAL SECTION SHOWING THE DESTRUCTION OF THE VARIOUS TISSUES BY TUBERCLE BACILLI.

> *a.* Indented surface of ulcer from which the epithelial layer has vanished.
>
> *b.* Large celled growth in sub-epithelial layer.
>
> *c.* Degenerated muscular fibres.

Large masses of tubercle bacilli are seen on the ulcerating surface and penetrating the tissues at various depths below it. × 550.

Drawn by M. H. Lapidge, from a preparation by Mr. H. H. Taylor.

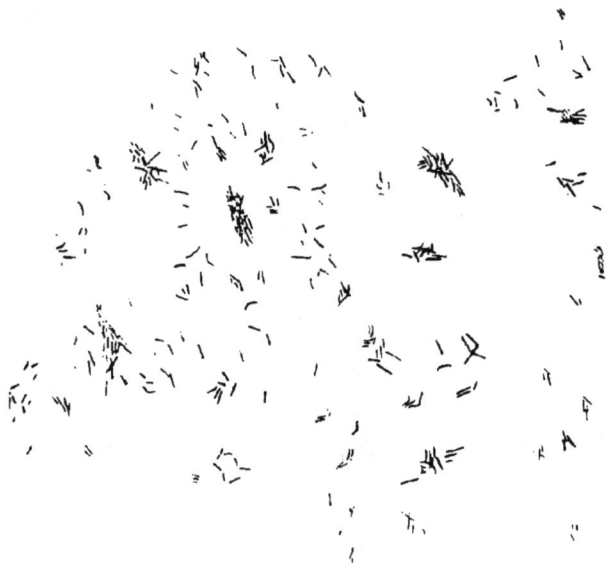

PLATE IV.

(1.) TUBERCULAR MESENTERIC GLAND. SECTION SHOWING ON THE RIGHT THE NORMAL GLAND STRUCTURE WITH A DILATED VENULE FULL OF BLOOD CORPUSCLES TRAVERSING IT, AND ON THE LEFT A ZONE OF PALE INDISTINCT LARGE CELLS COMMENCING TO CASEATE. TUBERCLE BACILLI IN GREAT NUMBERS IN THIS ZONE, WHICH IS EXTENDING TOWARDS THE VENULE.

   a. Normal gland tissue.
   b. Venule.
   c. Blood corpuscles.
   d. Large celled bacillary zone.   × 550.

Drawn by M. H. Lapidge, from a preparation by Dr. Percy Kidd.

(2.) ACUTE MILIARY TUBERCULOSIS. LUNG. ARTERIOLE FROM A TUBERCULOUS NODULE. WALLS OF VESSEL THICKENED AND INFILTRATED WITH EPITHELIOID CELLS SCATTERED THROUGH A FINELY GRANULAR SUBSTANCE (COMMENCING CASEATION). CALIBRE OF VESSEL ENCROACHED UPON BY THE GROWTH. VESSEL STILL PATENT AND CONTAINING BLOOD CORPUSCLES.

   a. Thickened wall of vessel.
   b. Remains of muscular coat.
   c. Cavity of vessel filled with red corpuscles and containing a few leucocytes coloured blue.

Bacilli are seen swarming in the thickened wall of the vessel, and are on the point of penetrating it.   × 200.

Copied from Dr. Percy Kidd's paper, 'Medico-Chir. Trans.,' vol. lxviii.

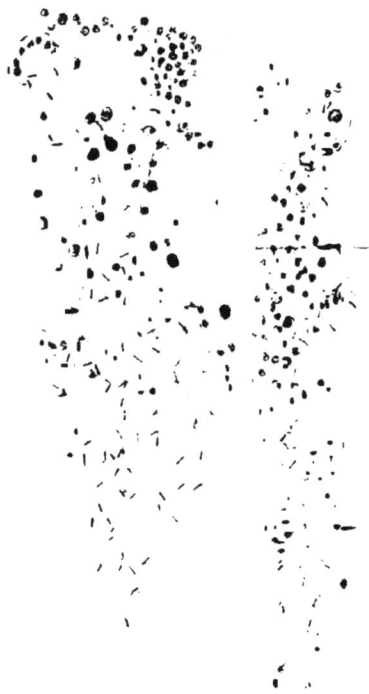

PULMONARY CONSUMPTION.

CHAPTER I.

DEFINITION.

By Dr. C. J. B. Williams.

Illustrations of degrees — Galloping Consumption—Acute Tuberculosis—Scrofulous Pneumonia — More chronic and limited forms — Progress of the disease — Power of Medicine—Unity but not uniformity of Phthisis.

THE DISEASE, too well known to the public, as well as to the medical profession, as PULMONARY CONSUMPTION, is characterised by the symptoms; persistent cough, expectoration of opaque matter, sometimes of blood; a progressive loss of flesh, breath, and strength; often hectic fever, night-sweats, and diarrhœa; and the common tendency of the disease is to a wasting of the body and a decline of its powers, down to its termination in death.

Pathologically considered, pulmonary consumption is characterised by certain changes in the textures of the lungs, consisting chiefly of consolidations, granular or diffused, which irritate their functions and clog their structures, and which proceed to further changes, of degeneration, disintegration, and excavation of some parts, and of induration and contraction of others—all tending to a disorganisation of the lungs, and a wasting away of the flesh and blood of the body.

It is this tendency to degeneration and destruction which stamps the *consuming* character of the disease; and the more strongly this tendency is manifested, the more irresistible and rapid will it be in its fatal course. In certain cases the disease is so acute and extensive as to carry off the patient in

B

a few weeks or months. In others it is more limited and slow, and may not destroy life for five, ten, twenty, or more years. In the former cases medicine has little or no control over the disease; decay and death invade the frame so over-whelmingly, that there is neither sufficient power in nature to resist them, nor time for art to aid that power. One of the most vital organs of the body becomes suddenly invaded by a disease, changing its structure, obstructing its functions, and spreading through it the seeds of further decay, which not only in the organ itself, but by the blood and lymphatics, diffuse its destructive influence through the whole system.

Let us briefly sketch the two most terrible forms of the disease.

A man of middle age is attacked with fever, with pungent heat of the body, cough, viscid expectoration, extreme oppression, and overwhelming weakness, resembling that of continued fever; and the likeness sometimes appears also in the coated or dry brown tongue, sordes on the teeth, and occasional delirium. The vesicular breath-sound is superseded every-where by bronchial rhonchi and mixed crepitation. On per-cussion, the chest is dull nowhere, but less clear in the posterior than in the front parts. This case might be supposed to be one of universal capillary bronchitis, with general pulmonary congestion. So it is; but this is not all. In spite of blisters and other remedies, the breathing remains short and difficult; the pulse becomes more rapid and feeble; the lips, cheeks, and nails become livid; clammy sweats break out, and the patient dies in the third or fourth week from his first attack. The lungs are found congested, and the bronchi loaded with viscid mucus; but more than this, innumerable miliary tubercles are scattered throughout the pulmonary tissue, and these are the obvious cause of the intractability of the case. They break out simultaneously, like the eruption of an exanthem, and by their numbers and bulk induce such an amount of obstruction and congestion in the lungs, as to destroy life before there is time for any considerable degeneration or softening to take place. This *acute tuberculosis* is the worst and most surely and rapidly fatal form of consumption.

The second form of acute consumption begins with pneumonia in one or both lungs. The patient, generally a young subject, is of consumptive family, and may have previously had cough and occasional haemoptysis. The fever attendant on the inflammation may not be very high at first, and the expectoration by no means so viscid and rusty, nor the crepitation so fine and even, as in simple inflammation of the lungs. But the symptoms are more persistent. The pulse and respiration remain frequent. The heat of the body, particularly of the chest, continues remarkably high, almost burning the ear of the auscultator as he examines the back. But this intense heat is alternated with occasional chills and profuse sweats at night. The cough continues distressing, and the expectoration becomes opaque, purulent, clotty, and contains abundant tubercle bacilli and fragments of lung-tissue; the flesh wastes, and the strength ebbs away; and if the appetite does not return, the progress of consumption and decay is rapid. Auscultation reveals the steps of the destructive process in the lung. The affected part, or the whole side, or part of both sides, becomes dull on percussion, only varied with the cracked-pot note from the gurgling within; the loud tubular sounds are replaced by coarse crepitation, in parts amounting to gurgling; and the diffused bronchophony is modified into detached islands of voice, loud and pectoriloquous, or into the snuffling or whispering sounds equally characteristic of a cavity. This form of *galloping consumption* may also prove fatal in a few weeks; and the lungs are found after death in a state of consolidation little more dense than the hepatisation of pneumonia, but their red is mottled with grey and yellow patches of tuberculous matter, and excavated in various parts into numerous small cavities communicating with the bronchial tubes, and containing more or less of the same compound matter which was expectorated during life, consisting of mucus, pus, degenerating epithelium and exudation-matter, with disintegrated fragments of lung-tissue, teeming with tubercle bacilli. This form of acute phthisis, although generally rapidly fatal, is not universally so. When not too extensive, it may sometimes be arrested and brought to a chronic state; and the chance of this result will very much depend on the recovery of the appetite, and the power of the stomach to bear strong nutriment, tonics, and, above all, cod-liver oil.

And in a large majority of the cases of consumption the destructive element is still less extensive and less active, and its progress is much more slow; and we have both time and means to resist its inroads and to fortify the system against its operation to a greater or less extent. In the greater number of instances the disease begins with the symptoms of common cold, often referred to the throat as much as to the chest; and there is, in truth, more or less of bronchial irritation and inflammation attendant on the development of the disease, and recurring with renewed intensity at the time of its increased activity. Often the disorder is mistaken for a common cold, until either its remarkable persistence, or the occurrence of hæmoptysis, of night-sweats, of loss of flesh, or of some other uncommon symptom, gives intimation of its more serious nature. Then it is found that, in addition to the signs of bronchial catarrh, there are some of the signs of consolidation of the lungs, generally near an apex; slight dulness or raised tone of the stroke-sound at or below a clavicle, or at or above a scapula; a tubular character in the breath-sound and voice; an undue intensity or duration of the sounds of expiration, or a weakness or absolute obstruction of the inspiration; and sometimes the various slight degrees of crepitus substituted for the proper breath-sound; and various other signs which it is unnecessary here to detail. These signs, however, are the indications of *incipient pulmonary consumption*—that is, of a disease which tends, sooner or later, to injure and destroy the structure of the lung, and to deteriorate and waste the flesh and blood of the whole body. And the progress of this work of injury and destruction is marked by signs of increased density and diminished motion of parts of the lungs; by more of the moist crepitus, from augmented humidity, in and around the consolidations; and eventually by signs of excavation at one or more points, which announce the removal of the diseased tissue.

The progress of this disease may vary infinitely in time

and in extent. The more extensive the mischief, generally the more rapid will be its progress, which goes on in the worst cases, uninterrupted by any. check or pause, attended by the distressing train of symptoms—harassing cough, opaque clotty expectoration, increasing shortness of breath, burning fever, alternated with profuse sweats and chills, rapid loss of flesh and strength and colour, sometimes diarrhœa, and aphthous mouth ; and terminates in death in a few months.

But in other cases, and these are by far the most common, the destructive lesions are less extensive, and their progress is more slow and intermittent, and often seems in great degree to depend on occasional attacks from cold or other external causes. in the absence of which the disease may be quiescent or stationary, and may not destroy life for years.

Powerless as medicine is in the overwhelming and rapid types of the disease, it has yet considerable influence over these milder forms ; and the following pages will give some evidence that under careful treatment life may be prolonged for many years in comfort and usefulness, and in not very few cases the disease is so permanently arrested, that it may be fairly called cured.

It may well be questioned whether a disease which presents such striking differences in form, intensity, and result, can be truly one and the same disease ; and there was at one time a disposition among pathologists to supersede the comprehensive terms *consumption* and *tubercle* by others more specifically applicable to definite forms of the disease. But the great discovery of the *bacillus tuberculosis* by Koch, and the detection of its presence in all forms of scrofulous and tubercular lesions, has demonstrated conclusively the unity, if not the uniformity, of consumption ; and while admitting considerable varieties in the disease, we may fairly explain them by differences either in the amount of the poison introduced into the system, or in the strength of the constitution to resist the attack, which in the feeble may amount to a general spread, or in the strong may be limited to the invasion of a part of one organ ; or, again, by differences in the predisposing cause and its action in weakening the system, and thus rendering it liable to tubercular attack.

CHAPTER II.

MORBID ANATOMY OF PULMONARY CONSUMPTION.

Various morbid processes and their products—Pathological elements—Grey tubercle — White granulations--Yellow tubercle—Caseous masses—Grey infiltration or catarrhal pneumonia—Red hepatisation—Fibrosis, two kinds—Cretaceous masses—Fibrinous nodules—Vesicular emphysema.

WE will first lay before our readers a brief survey of the principal lesions found after death in cases of consumption, and, having these before our minds, we shall be able to consider the modes of their production and the pathology of the disease ; from thence we can discuss the origin of the principal factors, and later on the clinical features of this common and terrible disease.

The morbid anatomy of phthisis in its various forms, acute and chronic, presents considerable difficulties, partly from the great variety of pathological products found in the lungs, and partly from the complete disorganisation of the normal structure, and even of the invading growths before the death of the individual occurs.   It often happens that several processes have been going on in the lungs simultaneously, and, while all tend to encroach on and abolish the respiratory area, each brings about the work of destruction by a different method and at a different rate, some by obstruction through consolidation, others by caseation and excavation.   On the predominance of one or the other depends the future of the lungs, for we sometimes see one pathological element, which has invaded a large portion of these organs, superseded and gradually destroyed by another of more recent date, but endowed with a higher degree of vitality.   So closely, however, are these processes intermingled, that it is often very difficult to trace the exact part each element plays in the drama of disintegration, and it is only by selecting typical cases, and by closely observing various portions of the same lung in various stages, that we are enabled to recognise the different agencies at work, to trace the steps of each process, and to assign to each its proper share in the pathology of consumption.

In advanced cases the lungs are greatly changed in colour and texture, being for the most part devoid of vesicular tissue and consolidated by various kinds of growths and exudations. They are also occupied by cavities, varying in size from a microscopic point to one of so large a capacity that the lung is converted into a mere bag of thickened pleura; the cavities are of every conceivable form and shape, sometimes oval and well defined, lined with a secreting membrane, at other times irregular, sinuous, anfractuous, and presenting on section either an uneven surface, from which portions of the wall stand out like the columnæ carneæ of the heart, or a very rugged surface, on which ulceration and suppuration appear to have done their worst; but, whatever be their shape or their size, they indicate the destructive character of the retrograde processes by which the disease called pulmonary consumption is characterised.

The consolidations vary in kind, but all partake more or less of a tubercular character. In some cases the lungs are disseminated with miliary tubercles from apex to base, the intervening tissue being free from excavation, and either engorged or consolidated with red hepatisation, or sometimes apparently healthy; in others no trace of miliary tubercle can be found, but the lungs are consolidated throughout by caseous pneumonia, containing cavities of various sizes. Sometimes there are aggregations of the different forms of tubercle— white, grey, and yellow in the same lung—while the opposite organs may be entirely clear; sometimes a lung may be shrunk to the size of a fist, its pleura thickened, and its lobules invaded with white fibrous bands, and its tissue converted into an iron-grey structure by fibroid growth. All these, and many other diverse morbid appearances, are found in the lungs of persons dying of phthisis, and we must endeavour to classify and distinguish them, first describing their naked-eye appearances; secondly, their histological phenomena; thirdly, the changes which take place in other organs of the body;

fourthly, we must indicate the pathological relation these all bear to one another and to the disease generally.

The subjoined are the principal pathological elements :—

1. Grey and dark granulations, or miliary tubercles.
2. White granulations.
3. Yellow granulations, or yellow tubercle.
4. Caseous masses, or yellow infiltration.
5. Grey infiltration, or catarrhal pneumonia.
6. Red hepatisation.
7. Fibrosis.
8. Cretaceous masses.
9. Fibrinous nodules (blood residues).
10. Vesicular emphysema.

1. *Grey granulation, or miliary tubercle*, is seen in the form of small bodies, varying in size from a millet-seed (hence the name miliary) to a hemp-seed, scattered throughout the lung-tissue. When first formed it is greyish white, more or less transparent, and will yield to firm pressure; but, after a while, changes take place, and it either undergoes caseation, becoming converted into the yellow variety, or, losing moisture, it becomes drier and harder, attaining the consistency of cartilage. At the same time it absorbs pigment; the colour passes from a light grey to a dark grey, and to black; the granulations simultaneously drying up and becoming obsolescent. These hard grey granulations are not uncommonly found after death in old persons, and are an evidence of tubercle having appeared at some period of their life, and of its having passed a short and circumscribed existence, and without injury to the individual.

The more common occurrence is for these grey granulations to increase in number, and to form aggregations or clusters, much resembling bunches of berries, standing out on section in bold relief against the healthy or congested lung-tissue, their favourite locality being the upper lobes of the lungs, and especially the posterior portions. In some instances these aggregations spread quickly and extensively, and the whole lungs become so densely packed with miliary tubercle, that it is difficult to find any portion of the respiratory surface free. This rapid formation of tubercle is sometimes sufficient to cause death by asphyxia, but more commonly the intense crowding of the pathological products gives rise to their own decay and destruction. Caseation commences in the centre of the groups, and cavities accordingly form. The discrete form of grey tubercle is generally found in acute miliary tuberculosis, and does not vary much with the different organs attacked by tubercle, as in the peritoneum, pleura, &c. This remarkable identity of form suggests very forcibly the hyperplasia, through irritation, of some normal structure present in all the several organs rather than an adventitious growth.

2. *White granulations.*—These are more opaque, and softer than the grey, and differ from the latter, as we shall hereafter see, in the arrangement of the histological elements, the proportion of epithelial production being greater, and that of reticular growth less, than in the grey variety.

3. *Yellow granulations, or yellow tubercle,* exist in greatly varying sizes, from a pin's head to a pea. They are opaque, soft, granular, amorphous, easily separated from the adjoining tissue, and sometimes surrounded by a circle of pearly, transparent material. Dr. Wilson Fox [1] describes a form of yellow tubercle among children dying of acute tuberculosis, which is not easily separated from the parenchyma of the lungs, but in adults it is generally easily removed, the grey granulations, with which it is so often associated, remaining behind.

Yellow granulation is by far the commonest form of tubercle, and its frequent occurrence in phthisis led Laennec, not unnaturally, to the conclusion that it was a *sui generis* production, essential to the disease. It seldom occurs alone, but is ordinarily associated with the grey and white, sometimes forming with them racemose groups in various parts of the lung, chiefly in the upper lobes. At other times it is the centre of an affected portion, groups of grey granulations apparently radiating from it, thus naturally leading to the supposition that a species of local infection has been set up

[1] *Pathological Transactions,* vol. xxiv.

by the yellow or caseous mass.   These groups, as they increase,
exercise great pressure on the various granulations composing
them and also on the intervening lung-tissue, depriving them of
nutrition, and thus causing death of the part by caseation.
The decayed portion is gradually removed either by absorption
through the lymphatics, or by expectoration; and in the
latter case cavities result.   Careful study of one of these
tubercular groups will demonstrate that the yellow tubercle
is but a later condition of the grey, in which caseation has
commenced, and that the cavities, large or small, in its neighbourhood are the result of the softening and removal of
the yellow tubercle and of whatever lung-tissue happens
to be intermingled with it.   Yellow tubercle is also found in
the various infiltrations, grey and red, which occur in the
lungs of phthisical patients; but here it is more apt to lose
its granular form, and become converted into class 4.   The
caseation of miliary tubercle has generally been considered
due to retrograde changes arising through deficient blood
supply, it being ascertained that tubercle contains no blood-
vessels.   Mr. Watson Cheyne holds that it is caused by a
chemical change in the epithelioid elements, of which tubercle,
according to Mr. Cheyne, is composed, induced by the action
of the specific bacilli, which are found to swarm in fresh
caseating centres.

4. *Caseous masses, or yellow infiltration.*—These are identical in constitution with the yellow tubercle, but differ in
size and form, arising sometimes from the aggregation of a
number of yellow granulations, but oftener from the rapid
caseation of inflammatory exudations, the extensive caseation
being due partly to the action of the tubercle bacillus and
partly to obliteration of the nutrient vessels from pressure;
and in this latter case large tracts become affected with what
is then called yellow infiltration.

5. *Grey infiltration or catarrhal pneumonia,* or the 'gelatinous
infiltration' of Laennec.—The appearance of this varies
much, according to the amount of colouring matter present,
the degree of transparency, and the variety of products.   In
some the grey colour assumes a yellow and in others a reddish

tinge, while in a few a marbling of the texture is manifest, owing to opaque lymph and caseous products being commingled with the pigment of the lung. It is generally recognised that these infiltrations, which involve large portions of the lung in phthisis, are of inflammatory origin, and that the difference in the appearance of the consolidation depends on the relative proportions of epithelial proliferation and lymph-exudation. When a reddish colour prevails it is chiefly exudatory, consisting of lymph and migratory red globules and leucocytes. Where the yellow tint predominates, as is usually the case, extensive proliferation of the alveolar epithelium has taken place. An examination of lungs in this state, which is called catarrhal pneumonia, shows us that the lesion is generally lobular, and commences with bronchial catarrh, which extending downwards causes the choking of certain bronchiæ by epithelial products, and in consequence the collapse of the lobules supplied by them. The pressure on the walls of the alveoli caused by the epithelial aggregations, as well as the inflammatory exudation, gives rise to obliteration of the branches of the pulmonary artery and consequent caseation, and in this way large tracts of grey pneumonia are converted into yellow masses and subsequently become excavations.

Hamilton[1] shows this well, and points out that before softening takes place in the catarrhal aggregation, its structure becomes very dense in the centre, and all traces of alveolar walls cease to be recognisable. Fatty metamorphosis occurs, and caseous centres of varying sizes may be formed, which may possibly become drier or remain in the same condition, and not extend or undergo change for a long period of time. This is the caseous material in which Dr. Percy Kidd[2] and others find few, if any, bacilli. But when liquefaction takes place, the masses break down rapidly, and are converted into the viscid yellow fluid, which abounds in bacilli. Hamilton views this process of liquefaction as a purely chemical one, and quite distinct from the microbiotic ones which led up to it. The

[1] Pathology of Bronchitis, Catarrhal Pneumonia, and Tubercle.
[2] Medico-Chirurgical Transactions, vol. lxviii.

rapid development of bacilli during liquefaction, and their absence from old caseous masses, indicate that the spores must have been present in the caseous masses, and must have been developed quickly, giving rise to those chemical changes, of which breaking down of the masses is the result. Cavities formed in this way present ragged and granular interiors, and are very rapid in their formation, and involve considerable portions of the lungs.

6. *Red Hepatisation*, the result of ordinary croupous pneumonia, is often found associated with one of the above forms of tubercle, but more commonly occurring in the lower lobes than in the upper. It generally accompanies tuberculo-pneumonic phthisis, a form which is intermediate between acute tuberculosis and scrofulous pneumonia or galloping consumption, and a striking feature in the autopsies of these patients are the aggregations of miliary tubercle, grey or caseating, standing out in bold relief against the deep red of the hepatised lung. In some instances the consolidation undergoes resolution, as evidenced by physical signs; but generally it does so only partially, and thickening of the alveolar wall and enlargement of the vessels show a conversion into chronic pneumonia.

7. *Fibrosis* is a pathological process largely present in phthisis, but preponderates in cases accompanied by pleuro-pneumonia, interstitial pneumonia and chronic pleurisy, and in phthisis of long duration. Fibrosis is the great element of the contractile process, whereby the lungs are reduced considerably in size, cavities of large capacity are cicatrised, and caseous masses encapsulated, and not uncommonly grey tubercle is eventually converted into this tissue.

A lung invaded by fibrosis is reduced in size, and presents on section a dense, tough, and very hard structure, resembling cartilage in its resistance to the knife. All traces of the alveoli have disappeared, and nothing remains but a dark grey or black fibrous material, into which run long bands of whitish fibrous tissue, harder than the darker portions. These often consist of the interlobular tissue, largely increased in amount. The pleura is generally thickened, and the septa apparently take their origin from it, and from the connective tissue to be found at the root of the lung, which is also largely increased.

This is what is generally called fibroid phthisis, but fibrosis is also found in limited portions of the lung, in nearly all kinds of phthisis, forming the scars of contracted cavities, or surrounding and thus isolating caseous masses and tubercular aggregations. When miliary tubercle becomes converted into fibroid growth, the resulting tissue is not of long duration, owing to its lacking a proper supply of blood and lymph-vessels; caseation consequently takes place at various points of it, and it thus perishes.

Fibrosis presents under the microscope a fibro-nucleated structure of varying coarseness, and chiefly affecting the alveolar wall of interlobular tissue.   It is generally accompanied in cases of fibroid phthisis by an adherent pleura and some dilatation of the bronchi, the latter lesion being dependent, according to Dr. C. J. B. Williams, on the increased pressure of air on the bronchi from the abolition of so many neighbouring alveoli, and on the traction outwards which an adherent pleura exerts on the bronchial wall.

Tubercle bacilli are seldom or never found in fibroid tissue, and it is generally held that this growth is always secondary and never primary in the lungs, generally supervening on old pneumonia, or more commonly on tubercle, and the statement of the late Dr. Moxon, that fibrosis is the past tense of tubercle, though too sweeping, is not far from the truth, and the clinical features which mark its rapid increase during life are the increasing dyspnœa, and the shrinking of the chest wall overlying the lung affected by it.

8. Cretaceous material is found in chronic cases, lying in small masses in various parts of the lung, chiefly in the apices in the neighbourhood of old cavities or caseous tracts.   Cretaceous material is generally associated with fibroid tissue in which such masses are often encapsulated, and the size and form of these varies greatly in different lungs.   It is commonest in bronchial glands which have been the seat of tubercular disease, and which, by their enlargement and

pressure on the bronchi, cause ulceration of these tubes, and discharge their contents through the openings. Much of the calcareous expectoration, not uncommon in phthisis, has its source in cretaceous bronchial glands, and is not incompatible with active tubercular disease proceeding in the lung itself. Sometimes cretaceous matter is found in cavities; and a marked example will be given where a large cavity of the apex of a lung was filled with it, in addition to smaller accumulations in the opposite lung. The cretaceous material is stated to be composed of phosphate of lime, with the addition (as in urinary calculi) of such dried secretions as may be in the neighbourhood. It may be looked on as the lowest degeneration of tubercular material.

9. *Fibrinous nodules* were noticed by Dr. Reginald Thompson [1] in cases where large hæmoptysis had occurred, and were formerly mistaken for yellow tubercle. They vary greatly in size, consist of inhaled blood, and are situated at portions of the lung where inspiratory action is strongest. When first noted, they appear as white nodules with a zone of red colouring matter, and even in the old specimens some traces of blood in the form of crystals of hæmatine are to be found. Microscopically they are shown to consist of fibrin and red corpuscles, filling the alveoli and even penetrating the alveolar wall. The masses eventually either (1) separate from the surrounding tissue through contraction of the fibrin, leaving an adherent shell; or (2) owing, as Dr. Thompson thinks, to admixture with bronchial secretion or some such septic matter, they soften into a mortar-like material and are got rid of by expectoration; or (3) if the nodule be sufficiently large, and there be exit for its contents, the result is the formation in time of a species of cavity filled with glairy yellow fluid, resembling honey in appearance.

10. *Vesicular Emphysema.*—Two kinds are noted in the lungs of phthisical patients. *Acute Vesicular Emphysema* is found throughout the lungs of those dying of acute tuberculosis. So commonly is this the case that it is not unusual to deduce the presence of the discrete variety from the quan-

[1] *Pulmonary Hæmorrhage.*

tity of acute emphysema present. *Chronic Local Emphysema* occurs in connection with chronic tubercular masses, generally in process of arrest. Specially is it to be found in the neighbourhood of cicatrised cavities. The emphysematous vesicles are few in number and greatly distended, varying in size from a pea to even a pullet's egg; many are the size of hazel-nuts, and these are generally found at the apices of lungs, so puckered by fibroid cicatrices or blocked by obsolescent tubercle and cretaceous masses, that the process of their formation seems to have been due to over-distention of the few alveoli left patent. The commonest site for chronic local emphysema is, after the apex, the anterior margin and the margin at the base of the lung.

A class of cases in which emphysema is a most prominent feature is where extensive fibrosis follows excavation, and especially affects the root of the lung; *i.e.* the cavity contracts towards the root of the lung, and fibrosis is largely present. Here we find emphysema of all the anterior surfaces of one and sometimes of both lungs, and even extending deeper into the organs. Cases are not rare of contracted cavities where, during the last years of life, the patients have suffered from dyspnœa, and even asthmatic attacks, and have lost all appearance of phthisis, eventually dying of the obstructed breathing and circulation, and not from wasting and collapse. After death the lungs have been found chiefly emphysematous, some in the atrophous or flaccid form of the tissue, and having at their roots cicatrices of old cavities with fibrosis and dilated bronchi.

CHAPTER III.

HISTOLOGY OF CONSUMPTION.

Histological elements—Their relation to the Tubercle Bacillus—Exudation of Fibrin and Leucocytes into alveoli—Epithelial elements—Epithelioid and Giant Cells—Their evolution and transformation—Thickening of alveolar wall by lymphoid or adenoid tissue—Opinions of Cornil, Ranvier, Sanderson, and Watson Cheyne—Increase of the interlobular connective tissue—Changes in the bronchi—Rindfleisch's scrofulous inflammatory cells—Infiltration of angles of terminal bronchi—Lesions of the bronchial glands—Pleural adhesions—Their significance—Lesions of other organs.

WE will now proceed to describe the histological elements of phthisis, before attempting any discussion of its pathology; and it may be well to bear in mind that though Koch's discovery of the bacillus tuberculosis has introduced a new and essential element, it has not changed the formerly well-recognised products of the disease, but only somewhat modified the relation of each to its causation. Much of the minute anatomy of phthisis is that of the different inflammatory states of the lung with certain additions of a more or less septic nature, and it is probable that while much may be due to the irritation set up by the bacillus, more may be caused by constitutional weakness or the lowering influence of malnutrition, both of which tend to render the tissues more prone to the attack of the parasite. The microscopic researches of Gulliver, Virchow, Sanderson, Wilson Fox, Schüppel, Friedländer, Rindfleisch, Hering, Klein, Koch, Green, Watson Cheyne, and others, show the principal histological elements of the disease to be, adopting Dr. Green's classification, as follows :—

1. *Exudation of fibrin and leucocytes* into the alveoli, resembling what takes place in croupous pneumonia, the fibrillation not being quite so distinct, nor the coagulation so abundant. Numerous leucocytes, which have emigrated from

the blood-vessels, have been likewise noted entangled in the meshes of the fibrin. In a large number of cases of phthisis, the lung-consolidation consists of exudatory products mingled with epithelial proliferation ; and in some of the most acute instances, these two processes have constituted the only lesion in the lungs.

Dr. Klein[1] states that in some of the most rapid cases of acute tuberculosis no adenoid growth is to be found, but only desquamative or catarrhal pneumonia ; and he remarks that in some specimens examined by him, in the central or earlier portion of a tubercular nodule there was nothing but fibrinous exudation, and at the peripheral, or later formed portion, there was no fibrinous matter, but only spherical multi-nucleated cells, or, again, that the alveolus was filled by one multi-nucleated giant cell.

Klein considers these different appearances represent different stages of the same process, that the fibrinous exudation comes first, and is afterwards absorbed by the surrounding tissue, which becomes infiltrated with fluid, and shows distended blood-vessels. The exudation is replaced by numerous cells, or by giant cells. If the irritation lasts long enough, the small-celled adenoid tissue appears in the alveolar wall.

2. *An accumulation of epithelial cells within the alveoli.*— We have three forms of these. (1) The ordinary epithelial cell, lining the alveoli. (2) Epithelioid cells. (3) Giant cells. The second are generally large spheroidal cells (*see* Plate II. fig. 1) about four or five times the size of a leucocyte, containing granular matter, and a large, and often oval nucleus, and occasionally a nucleolus ; more than one nucleus is occasionally seen. These cells are developed from the epithelium according to Mr. Watson Cheyne,[2] and some of them are transformed later into giant cells. Dr. Hamilton, on the other hand, to judge by his admirable drawings, attributes to apparently similar cells an origin from the tissue of the alveolar wall. In many cases of phthisis the alveoli are stuffed with these epithelioid cells, some perfect, and others towards the centre undergoing

[1] *Anatomy of the Lymphatic System.*
[2] *Practitioner*, April 1883.

caseation : they are of great interest on account of being the
principal haunt of the tubercle bacilli, which are found in these
and in the giant cells, and often in no other lung structure.
It is curious to see how the bacilli are attracted towards them,
and are seen in large numbers both around the cells and
within the cell walls (*see* Plate II. fig. 1).   The third element
is the giant cell, as to the origin of which there has been much
controversy, ending for the most part in the general conclusion
that it is epithelial.

Their actual development from epithelial cells has been
watched by Watson Cheyne ; and the presence of inhaled
particles of carbon in their interior is another proof of their
origin.

Their form varies greatly, being sometimes round, sometimes oval, and often very irregular, with numerous angles
and processes occupying a portion, and often the whole, of the
alveolus.

They appear at first as spheroidal masses of faintly granular protoplasm reaching $\frac{1}{200}$th inch in diameter, containing numerous nuclei, sometimes as many as thirty, and some
bright nucleoli. Around each giant cell, or group of giant cells,
there is a concentric arrangement of delicate fibrous tissue, so
as to give the impression, according to Hamilton, that each
giant cell is the nucleus of a tubercle, or giant-cell system.

As the cell grows older the periphery of the protoplasm
becomes organised, forming a fibrous mantle-like sheath, in
which lie the numerous nuclei, the central portions remaining still granular, and being penetrated occasionally by spaces
or vacuoles. The peripheral portion, or periplast, as it is called,
becoming more fibrous, throws out branched processes, from
which are developed other smaller protoplasmic masses, so
that a branched reticulum is formed round the original giant
cell, connecting it with other giant cells. These branches are
often directly continuous with the lymphoid or adenoid network of the alveolar wall, to be presently alluded to, which
forms a circle round the giant-cell system.

In process of time the fibrous periplast of the giant cell
increasing at the expense of the central protoplasm, the latter

disappears, and the giant cell is converted into a reticular tissue, or a fibrous membrane, the nuclei being found either lying flat on the surface of the reticulum or contained in its meshes. We thus see one of the modes of the transformation of tubercle into fibrosis. System after system of giant cells undergoes fibrosis, and the whole tubercle becomes converted into fibrous tissue.

It may be remarked that giant cells are sometimes not found in the very early stage of tubercle development, and appear generally after the products of exudation have become absorbed. Hamilton has noted them as early as the second week after the appearance of tubercle, but it has been generally observed that the slower the tubercular growth, the larger the number of giant cells. They are devoid of vascular supply, and after the fibrous transformation just described, generally undergo a certain amount of caseation.

Giant cells are regarded by Green as a product of low vitality, incapable of forming organised tissue; where the protoplasm grows the nuclei multiply, but the highest manifestation of cell-life—division of the cell—does not take place.

3. *A thickening of the alveolar wall by a small-celled lymphoid tissue*, consisting of minute cells not exceeding in size a leucocyte, separated from each other by a very delicate reticulum, the existence of which Cornil and Ranvier [1] deny, maintaining that the appearance of the reticulum is due to the action of hardening reagents on the intercellular substance. This growth appears to commence in the walls of the alveoli and terminal bronchiæ, first in the form of a few lymphoid cells, the network appearing later, and has been held by Sanderson to be a hyperplasia of the adenoid tissue already existing in the lungs, for it must be borne in mind that lymphatics and lymphoid tissue are largely present in these organs, and that the alveolar wall is considered one of the densest lymphatic plexuses of the whole body. Watson Cheyne, however, cannot satisfy himself of the existence o any reticulum of tubercle similar to that found in lymphatic glands. He thinks that the reticular appearance is due to

---

[1] *Manuel d'Histologie Pathologique.*

infiltration of the fibrous tissue surrounding the tubercle with leucocytes.

According to Sanderson the small-celled tissue spreads rapidly through the alveoli, invading the walls of the capillaries, the peribronchial and perivascular sheaths, diminishing by pressure the calibre of the vessels, and in time obliterating them, and thus giving rise to necrobiosis by caseation and ulceration of the surrounding tissues. The growth fills up the alveoli, and thus infiltrates whole tracts of the lung, giving them a greyish indurated appearance, and these in time become cut off from both air and blood supply. The future of this adenoid formation is twofold. Either it degenerates by caseation, giving rise to the formation of cavities, or the cells become more spindle-shaped and branched, the reticulum more fibrinated, and then gradual fibrosis of the nuclear tissue takes place. Owing, however, to the disappearance and obliteration of the vessels, this tissue is not properly supplied with nourishment. Caseous and other retrograde changes soon commence, and its life is a short one. Wilson Fox holds that adenoid growth is the basis of tubercular and phthisical changes, and that it is present, at some time or other, in all cases of phthisis, whether the lesion be grey, white, or yellow tubercle, catarrhal pneumonia, or caseous infiltration. According to him the difference between the white and grey tubercles consists in the larger amount of epithelial products mingled with the nuclear growth in the former, the latter containing less epithelium and more nuclear tissue. The appearance of lungs in which the consolidation is due to intra-alveolar accumulations is characterised by a reddish or reddish-yellow colour, by the tissue becoming soft and friable, and by the presence of several cavities ; where the consolidation is due principally to alveolar growth, there is more induration and toughness, the colour is grey, mottled with black pigment, and yellow patches indicative of commencing retrograde changes.

4. *Increase in the interlobular connective tissue* resembling the process prevailing in the liver, kidneys, and other organs during chronic disease, and not necessarily associated with consumption. This feature is most marked in cases

associated with pleurisy or pleuro-pneumonia, or, again, where
the disease is of very long standing, and the result is best
seen in the large fibrous septa often accompanying the bronchi
and great blood-vessels, as is specially exemplified in fibroid
phthisis. Microscopically it is difficult to distinguish between
the interlobular tissue and the alveolar adenoid growth in their
early stages, both being richly cellular, the main differences
being the situation of the former around the lobules, and in
the neighbourhood of the great air and blood-vessels, whereas
the latter is found in the alveolar wall and smaller bronchioles.
Again, the interlobular tissue is not so prone to retrograde
changes, owing to the vascular supply being less liable to
obstruction and obliteration, and again, the alveolar growth
has a more delicate reticulum of fibres. The interlobular
tissue is coarser and tougher, and may, as a rule, be recognised
by its white colour, its position around the lobules, and the
marked contractile effects on the lung and the overlying chest
wall it gives rise to. Whether it appears in connection with
old pleurisy or with fibroid phthisis, it invariably causes a
flattening and contraction of the thorax, and appears to
obliterate large tracts of lung-tissue, not only those affected
with tuberculosis, but some which are nearly healthy, so that,
although this growth has a distinctly limiting and even arresting influence on ulcerative processes, it reduces the amount of
respiratory surface, and thus increases the tendency to dyspnœa.

Changes in the bronchi, pleuræ, and bronchial glands.—
The bronchi become implicated in all cases of phthisis. In
many, catarrh of the mucous membrane, giving rise to a
richly cellular secretion, which forms the greater proportion of
the expectoration of the consumptive, is the principal lesion,
extending in acute cases throughout the whole bronchial tree ;
but in more chronic forms being limited to the bronchi leading
to the affected lobules. A second and more important change
is the infiltration, noted by Rindfleisch,[1] of the sub-epithelial
connective tissue by large cells characteristic of scrofulous
inflammation, and very difficult of absorption. The mucous
membrane appears swollen and opaque, the epithelium may be

[1] Ziemssen's *Cyclopædia of the Practice of Medicine*, vol. v.

shed, and if the sub-epithelial infiltration disintegrate, small
ulcers are formed, which Rindfleisch states are common, but
which are held to be rare by Green.   Perhaps the change of
greatest interest to pathologists is that stated by Rindfleisch
to be the first lesion of phthisis, viz. : a tubercular infiltration
of all the angles and projections of the terminal bronchi
where they become continuous with the alveoli.   Well-marked
greyish nodules of tubercle form on these projecting surfaces,
which afterwards undergo caseation.   Later on, tubercle is
found forming a complete ring around certain bronchi, and
round others a crescent or half-moon, caseation and atrophy
having destroyed the remaining portion of the circle.   The
tubercle lies in the sub-epithelial layer, and is well supplied
with giant cells.   A fourth change is the infiltration of the peribronchial tissue, and the proliferation of lymph-follicles in the
walls of the smaller bronchi, owing to transmission of infective
substances from the branches through the lymphatics.   The
bronchi from these changes become reduced in calibre, and
consequently the adjoining ones, as noticed by Grancher,[1] are
often dilated through the influence of increased air-pressure.

The *bronchial, cervical, mesenteric,* and other *glands* undergo
various changes.   In many, and especially in advanced cases,
the bronchial glands enlarge and become deeply pigmented.
This is, perhaps, more common in very chronic instances, and
specially where arrest of pulmonary disease has taken place.
Pigmentation is rare, if not unknown, in the glands of children,
which generally have a reddish or yellowish tint on section.
In other cases they seem to partake of the changes proceeding
in the lungs ; they become affected with grey tubercle and
caseate, and occasionally cretify, the cretaceous material being,
as a rule, inside the gland at the centre, though I have occasionally witnessed the reverse to be the case, and the calcareous
matter to form a shell over the whole gland.

Caseous bronchial glands often attain a large size, and have
been known, by pressure on the trachea, to cause death by
suffocation, as happened in two cases cited by Dr. Percy Kidd[2]

---

[1] *Nouveau Dictionnaire de Médecine et de Chirurgie Pratiques.*
[2] *Pathological Transactions.*

and Dr. Goodhart.[1] More commonly they soften and open
by ulceration into the bronchi, and then discharge their contents, leaving behind an empty sac, which is often mistaken
for a tubercular cavity. Considering how marked a feature the
bronchial glands are in the autopsies of phthisis, it is curious
how little count is taken of them in the pulmonary diagnosis
during life, though the careful articles of Drs. Gueneau de
Mussy,[2] Baréty,[3] and Quain[4] have supplied us with abundant
information as to the symptoms and physical signs they give
rise to during life. The enlargement of the glands immediately underlying the first portion of the sternum is common
among children, and is the cause of dulness over the sternum
near the clavicles, and sometimes of bronchial breathing in
that region, and similar signs in the interscapular regions
may be traced to a like origin. The tubular sound, without
dulness, which is often heard in children above the scapula,
and frequently assigned to tubercular aggregation, is often
due to enlarged bronchial glands, and disappears under
appropriate treatment.

The pleura undergoes various changes in the progress of
phthisis, and its condition may often be accepted as an index
of the acuteness or chronicity of the pulmonary lesions. In
acute tuberculosis and acute phthisis (scrofulous pneumonia)
adhesions are rare, and in this rarity, and also in the frequently superficial character of the caseating centres, lies the
explanation of the common occurrence of pneumothorax in
acute phthisis, for here large caseous masses appear to extend
to the surface, and to open into the pleura, without any
opposing adhesion.

On the other hand, limited pleural adhesions are common
in all the chronic forms of phthisis, and specially so in
fibroid and chronic tubercular phthisis. In the former it is
not uncommon to find the pleura adherent from apex to base,
and thickened to the extent of three-quarters of an inch or
one inch, the layers near the base being separated, as Dr.

[1] *Pathological Transactions.*
[2] *Nouvelles Etudes sur l'Adénopathie Trachéo-bronchique.*
[3] *L'Adénopathie Trachéo-bronchique.*   [4] *Dictionary of Medicine.*

Douglas Powell [1] has well pointed out, by a gelatinous material consisting of connective tissue, apparently provided to fill up any void that might arise from the contraction of the lung. In the ordinary form of phthisis, limited, dry, or fibrinous pleurisy is found commonly overlying tubercular masses, if they are superficial, and often extending over the whole apex of a lung and forming over it a fibrinous cap. Not uncommonly the friction sound of dry pleurisy in the supra-scapular region is the first sign of the presence of tubercle at the apex, which, later on, is abundantly verified by signs of consolidation in the anterior regions, and in these the existence of adhesions can be traced by the retraction of one or more intercostal spaces during expiration.

The pleura, peritoneum, arachnoid, and even the pericardium, may become the seats of miliary tubercle in the most acute form of phthisis, viz., miliary tuberculosis; but it is generally noted that the lungs are the first organs attacked, and it is extremely rare for tubercle to exist in any organ without being also present in the lungs.

The morbid changes noticed in the larynx will be described under the head of laryngeal phthisis, and the various lesions which give rise to diarrhœa in phthisis will be included in a special chapter.

As regards other organs, the liver is rarely normal, but generally undergoes either fatty or lardaceous degeneration, the spleen is sometimes softened, and very commonly lardaceous. The kidneys are not generally affected, but where albuminuria has prevailed towards the close of the disease, fatty or lardaceous changes are generally detected, and these will be considered in the chapter on phthisis and albuminuria. The heart is usually small, and the muscular tissue pale, and very often in a state of fatty degeneration (Quain). Fatty growths are sometimes found on the external surface.

Having now described the principal elements of the morbid anatomy and histology of phthisis, we can pass on to consider the organism which plays so important a part in the etiology of the disease.

[1] *Pathological Transactions*, vol. xx.

CHAPTER IV.

THE BACILLUS TUBERCULOSIS.

History of the discovery—Inoculation experiments of Laennec, Villemin, Simon, Marcet, Andrew Clark, and Lebert proved the specific nature of tubercle—Demet's human inoculation—Experiments of Sanderson, Wilson Fox, Waldenburg, Cohnheim, Fraenkel, and Schottelius negatived the conclusion of specificity—Klebs' suggestions—Test experiments of Koch, Cohnheim, Fraenkel, Salomonsen, Baumgarten, Watson Cheyne, and Dawson Williams confirm the specific nature of tubercle—Koch's discovery of the Tubercle Bacillus by a series of cultivations—Great success of his inoculations—Results dependent on number of bacilli in fluid, and mode of inoculation—Injections under skin and into veins—Inhalation of tubercular spray—Methods of staining Tubercle Bacillus: Koch's—Baumgarten's—Ehrlich's—Heneage Gibbes'—Double stain—The Weigert-Ehrlich method—Ziehl's modification—Gram's plan.

THE discovery of the Bacillus Tuberculosis was the last link in that important chain of facts by which the specific character of tubercle has been proved, but it will be advisable, before treating of it, to give some account of the various experiments which led up to it, and especially of the interesting series of artificial inoculations which were carried out in France, England, and Germany to determine the real nature of tubercle.

Laennec is stated to have originated the idea of producing tubercle by inoculation from an observation he made on himself. In opening some vertebræ affected with tubercle, the forefinger of his left hand was slightly scratched by the saw. The next day a little redness appeared and there followed gradually afterwards, a little swelling under the skin, of the size of a large cherry-stone. In eight days the skin opened at the scratch, and there appeared a yellowish compact body exactly like crude yellow tubercle. He cauterised it with butter of antimony, and then squeezed out the contents, which, being softened by the liquid caustic, exactly resembled softened tubercle. After repeating the cauterisation to the remaining little cyst, the wound healed without further inconvenience. Laennec died of phthisis some twenty years later, and though we can hardly conclude the disease originated from this accident and from no other cause, without further evidence, the coincidence is striking.

Villemin in 1865 communicated to the French Academy the results of a series of experiments on guinea-pigs and rabbits to determine whether tubercle could be inoculated. The matter of tubercle, grey and yellow, was inserted under the skin, and in the course of from two to six months the animals were killed, when tubercles, both grey and yellow, were found in the lungs, liver, spleen, lymphatic glands, peritoneum, and other organs, the yellow being most manifest in the animals which lived longest. M. Villemin concluded that tubercle was a specific poison capable of being communicated from one animal to another by inoculation, a conclusion which was soon confirmed by a series of experiments carried on by Mr. Simon, Dr. Marcet, Sir Andrew Clark, and Lebert.

In 1874 Demet and Paraskova Zablonus,[1] of Syra in Greece, succeeded in inoculating a man of fifty-five with tuberculosis. The patient was dying of gangrene of the left foot through obliteration of the femoral artery. Phthisical sputum was inserted into the upper part of the right leg, the lungs having been previously examined and pronounced perfectly healthy. Three weeks after the inoculation, signs of commencing induration of the right apex were detected, and seventeen days later (i.e. thirty-eight after inoculation) the patient died of gangrene. The autopsy showed seventeen tubercles, varying in size from a mustard-seed upwards, at the right apex, and a smaller number at the left apex, all evidently of recent formation. This instance proves the possibility of inoculating man with tubercle.

In 1868 Drs. Burdon Sanderson[2] and Wilson Fox[3] succeeded in producing tuberculosis in guinea-pigs, not only by the inser-

[1] *Nouveau Dictionnaire de Médecine et de Chirurgie Pratiques-* Phthisie.
[2] Tenth Report of the Medical Officer of the Privy Council, 1868.
[3] The Artificial Production of Tubercle in the Lower Animals, 1868.

tion of tuberculous material, but also by that of non-tuberculous. Dr. Fox used the following non-tuberculous materials :
putrid muscle, pus of various kinds, pneumonic products,
lardaceous liver, cirrhosed kidney, vaccine matter, pyæmic
abscess of spleen ; and in a large proportion of the cases produced tuberculosis of various organs, from which other guinea-
pigs were successfully inoculated. Drs. Sanderson and Fox
also produced tuberculosis by inserting setons of cotton-thread
under the skin of these animals without inoculating them
with any morbid material, but in reference to these experiments Dr. Sanderson cautiously guarded himself against
hasty deductions by stating, ' With reference to the traumatic
origin of tuberculosis in the guinea-pig, another possibility
claims consideration, namely, that of the influence of the air
and of the organisms which it contains. It has not yet been
proved that injuries which are of such a nature that air is
completely excluded from contact with the injured part are
capable of originating a tuberculous process. The following
experimental results seem, indeed, to suggest that they may
not be so. Setons, steeped in carbolic acid, were inserted in
ten guinea-pigs on September 24, 1868, each animal receiving
two. At the same time, extensive fractures of both scapulæ
were produced in five others, care being taken not to injure
the integument. No tuberculosis or other disease of internal
organs has resulted in either case.'

Waldenburg, Cohnheim, and Fraenkel found that in the
Pathological Institute at Berlin all the guinea-pigs into whose
abdominal cavities they introduced pieces of cork, paper, and
cotton-thread, &c., became tuberculous, and therefore concluded
the non-specific character of tubercle ; but at a later date the
two latter were led to modify their opinions, as will be seen
later on.

Schottelius of Würzburg produced granular pulmonary
tuberculosis in dogs, by making them respire air charged with
pulverised phthisical sputum, but he produced similar results
with air charged with the expectoration of bronchitis, with
Limburg cheese, and with vermilion. Klebs, firmly convinced
of the specific nature of tubercle, had described an actively
moving organism as its cause. Schüller and Toussaint had

pictured a spherical micrococcus in connection with the disease. Aufrecht had found more than one form of organism, and thus paved the way for the next step, which was the discovery of Robert Koch.

Koch, having by means of certain aniline dyes detected the bacillus tuberculosis, succeeded through a series of ingenious cultivations in procuring it pure and simple. He first took tubercle, and after washing it with a solution of corrosive sublimate, removed the outer layers and separated a portion into which he might fairly expect that no bacteria of putrefaction had penetrated. This he spread over a nutrient soil, consisting of the blood plasma of the ox, which had been previously sterilised by boiling in a test tube. The coagulum of this, with the tubercle added, was introduced into a test tube with a cotton-wool plug, and kept in an oven at a temperature of 37° to 38° C. (98·6° to 100·4° F.). Nothing appeared during the period of incubation of the ordinary bacteria of putrefaction, but at the end of ten days there were seen on the dry surface of the coagulum a number of very small points or dry-looking scales, surrounding the pieces of tubercle, spread out in circuits more or less wide according to the distribution of the tubercle fragments. After a few weeks' more exposure these crusts ceased to enlarge, and were then transferred to a fresh test tube, containing blood plasma similarly prepared. After another interval of ten days the scales appeared, became confluent, covering more or less of the surface of the coagulum, as the seed was scattered, and so from test tube to test tube the experiment was carried out under the most rigorous antiseptic conditions as many as a dozen times, and for a period extending over one hundred and fifty days. With the results of these culture experiments two hundred rabbits and guinea-pigs were inoculated, the places selected being under the skin, the peritoneal cavity, or the anterior chamber of the eye. With one exception all these animals acquired tuberculosis of the lungs, liver, spleen, and other organs, the tubercles having the structure of true tubercle, and including giant cells, which latter were found to contain bacilli.

Such being Koch's results, how are we to account for the

production of tubercle by non-specific inoculation, such as Schottelius and others performed? Klebs suggested that they might be ascribed to infection, as the experiments were performed in laboratories at Berlin in the presence either of animals already tuberculous, or where tuberculous materials might remain from previous investigations. Cohnheim and Fraenkel, to test this, repeated their experiments with non-tuberculous materials on similar rodents, Cohnheim in the laboratories of Kiel and Breslau, and Fraenkel in his own home, and with absolutely negative results in both cases, thus demonstrating that the septic atmosphere, and not the simple irritation, was the cause of the tuberculosis. Salomonsen and Baumgarten's experiments with tubercular and non-tubercular materials exactly confirm this conclusion; the former's also showing that caseous material, if boiled, or treated with absolute alcohol, could not produce tuberculosis in rabbits. Cohnheim held, previous to Koch's discovery, that the best test of tubercle was its inoculability, and that it was conveyed by specific organisms to the lungs, and affected thus both pleura and bronchial glands. Mr. Watson Cheyne [1] confirmed Klebs's suspicions by a series of experiments on rodents, which he states he performed under the best hygienic conditions, with complete isolation of the animals from each other, and thorough disinfection by heat of the instruments used. In six cases setons of various kinds were introduced either subcutaneously or into the anterior chamber of the eye. In ten, vaccine lymph was used; in three, pyæmic pus was injected either into the eye or under the skin, or into the abdomen; in six, various materials, such as cork, hardened tubercle which had been soaked for three months in alcohol, or worsted, were introduced into the abdominal cavity. In none of these rodents did tuberculosis result. Mr. Watson Cheyne also made experiments with cultivations of bacilli obtained from Koch. He inoculated twelve animals with these organisms, chiefly into the anterior chamber of the eye; all of them became tuberculous, and more rapidly than after the inoculation with tubercular tissue.

[1] *Practitioner*, April 1883.

Dr. Dawson Williams,[1] at Dr. Wilson Fox's request, repeated some of his earlier experiments with non-tubercular materials. Care was taken to avoid the contamination by tubercular material of the instruments and fluids used, but no antiseptics were employed. The repetition of the experiments with putrid fluids gave entirely negative results. All the animals (guinea-pigs) which survived the primary infective fever, when this occurred, recovered entirely ; and, when killed, after intervals of varying length from the time of inoculation, presented no lesions of either a tubercular or pyæmic character. The seton experiments were repeated, but the animals were unaffected, and when killed found to be perfectly healthy. Dr. Dawson Williams's researches were carried out with the greatest possible care and skill, and quite convinced Dr. Wilson Fox that there must have been some fallacy in the mode in which his original experiments were performed, which might have been due either to insufficient cleansing of instruments, or to associating the guinea-pigs, who had been subjected to some injury, with the tuberculous guinea-pigs.

We may therefore conclude that as far as experiments of inoculation are concerned, the specific character of phthisis is proved, and we will now consider certain points in connection with the production of tuberculosis, on which Koch's elaborate and complete series of experiments has thrown great light. First, as regards the material used. It appears that the inoculations were successful, whether the material used was phthisical sputum, grey tubercle, or caseous matter, or scrofulous matter, or pus from an abscess in connection with caries of the spine : or again, if it were lupus material, (which, he showed, contained tubercle bacilli), or, lastly, if perlsucht, or the tubercle of cattle. In all cases tuberculosis of various organs was produced, but none of these had such a potent effect as the culture liquid produced by cultivation of the bacillus in some congenial medium. This Koch tried largely, and succeeded with it in inoculating not only

[1] *Pathological Transactions*, vol. xxxv. p. 413.

rabbits, guinea-pigs, and hens, but even dogs, which were formerly supposed not to be receptive of the tubercle inoculation.

A second important point is the mode of inoculation used, according to which the results varied. Koch, as has been before stated, injected fluid containing tubercle bacilli into the anterior chamber of the eye of guinea-pigs, and found, if the solution was weak, miliary tuberculosis resulted; if strong, then tubercular infiltration and rapid caseation. In most of the animals the wound was made in the abdominal wall, and the spleen, liver, lymphatics and lungs became studded with miliary tubercle. Now, Koch kept a number of healthy guinea-pigs with the inoculated ones, and it was found that after three months' interval fresh cases of tuberculosis appeared among the former, the number of which diminished in proportion as the number of the inoculated ones was diminished. When the infected guinea-pigs—as we will call them —were killed, the lungs were found to contain masses of tubercle, for the most part undergoing caseation, and considerable-sized cavities; the bronchial glands were enormously swollen and caseous, the other organs remaining free. This striking contrast from the results of artificial inoculation received confirmation in some experiments on inhalation of tubercle.

A solution of tubercle was used in a spray apparatus for half an hour on three successive days, and eight rabbits, ten guinea-pigs, four rats, and four mice were made to inhale the spray. Symptoms of dyspnœa were manifested by some of them in ten days; three rabbits and four guinea-pigs died at periods varying from fourteen to twenty-five days after the last inhalation, the rest were killed on the twenty-eighth day. All the animals had tubercle of the lungs, and those which survived the infection longest had tubercle of the liver and spleen. The state of the organs in these animals resembled that of the guinea-pigs infected by association with the inoculated ones. The lungs were affected with caseous pneumonia, breaking down into cavities. Koch also states that among all the numerous guinea-pigs he kept, no case of spontaneous tubercle occurred; and so we are not surprised at his conclusion that the tuberculosis of the infected animals was exclusively derived from inhalation. It should be mentioned that the rats and mice when killed had tubercular organs, but not in so advanced a condition of disease as the guinea-pigs.

Another mode of inoculation which was tried was the injection of tubercle fluid into the veins of rabbits, that of the ear being selected for the purpose. Twelve animals were experimented on, and all died of phthisical symptoms in from nineteen to thirty-one days after inoculation. Miliary tubercle was found in the lungs, liver, and spleen of all, and in those dying first the tubercles were most numerous, but smaller: whereas in those surviving longer, they were fewer in number, but more developed. From this it appears that inoculation into the veins was by far the most effectual method of producing tuberculosis, but that various channels answer, including the anterior chamber of the eye. The certainty of success of inoculation and the rapidity of its occurrence depended on the number of bacilli contained in the inoculating material, thus pointing to the conclusion that they were the essential cause of the tuberculosis.

We will now turn to the bacillus, and, before describing it, give some account of the processes by which it can be detected. Koch originally used an alkaline solution of methylene blue, prepared thus:—

Distilled water 200 ccm.
Saturated alcoholic solution of methylene blue 1 ccm.

Shake, and, while shaking, add ·2 ccm. of a 10 per cent. solution of caustic potash. The specimens were kept in this solution for twenty-four hours, and then placed in a watery solution of vesuvin, which eliminates the blue from everything except the tubercle bacilli, which thus remain blue against a brown background.

Baumgarten's plan was to treat the specimen with a very weak potash solution and then stain it with aniline blue, which resulted in the surroundings being blue and the bacilli themselves remaining colourless. Ehrlich's original solution was 5 ccm. of pure aniline added to 100 ccm. of distilled water, well shaken and filtered through moist filter-paper; he added to the filtrate an alcoholic solution of fuchsin, methyl violet, or gentian violet. The subsequent process was the same as the Weigert-Ehrlich one, to be described below.

Heneage Gibbes used the following solution for staining them :—

Magenta crystals .	.	2 grammes.
Pure aniline .	.	3 ccm.
Alcohol, spec. grav. ·830	.	20 ccm.
Distilled water .	.	20 ccm.

The specimen remained twenty to thirty minutes in the solution. He then washed it in strong nitric acid (1 to 2 of distilled water), dipped it in a solution of methylene blue or chrysoidin to tint the background, washed again in water and alcohol, and then mounted it in Canada balsam. This process gave excellent and speedy results, the only fault being that the magenta solution was too concentrated, and coloured the bacilli too deeply.

Heneage Gibbes also introduced a double stain to simplify the process of staining with two colours, and also tried to omit the use of the nitric acid. The advantage of this process, according to Gibbes, is that the other bacilli and micrococci are coloured blue, and the tubercle bacilli red. In my hands this last solution has not given satisfactory results, partly from its opacity and partly from the tubercle bacilli not standing out in sufficient contrast to the other organisms. The process, which has yielded the best results in the detection of tubercle bacilli, both in sputum and tissues, is known as the Weigert-Ehrlich, and has been largely used by Koch, the Brompton and Victoria Park Hospitals, and many observers, including myself. It is as follows. The sputum is spread out on a cover-glass, dried, and then passed through the flame of a spirit lamp, or Bunsen burner, to coagulate the albumen and prevent the film being washed off. The cover-glass is then placed, with the film downwards, in a watch-glass containing the following staining fluid :—

Saturated watery solution of aniline .	100 ccm.
Saturated alcoholic solution of fuchsin	11 ccm.

The fluid should be strained, and Koch adds 10 per cent. of absolute alcohol to keep it.

In this solution the cover-glass remains twenty to thirty minutes; but if, as Rindfleisch suggests, heat be applied sufficient to cause vapour to rise from the surface of the fluid, five minutes, or even three or four, will amply suffice to stain the bacilli. It is then transferred to a strong solution of nitric acid (1 to 3 or 4 of water), or to diluted sulphuric acid (5 per cent. of acid). In this it remains long enough to change the colour from red to yellow, is then washed in distilled water or methylated spirit, which partly restores the red colour, and then placed in a saturated watery solution of methylene blue for five minutes; it is again washed in distilled water or methylated spirit, next in absolute alcohol,[1] then dried and mounted in Canada balsam dissolved in benzole. Though the process has many steps, it is not really a long one, and provided the magenta solution be heated, it need not occupy more than ten minutes. The object of using the nitric acid is that it is found to remove the colouring matter from all bacilli or micrococci, except the tubercle bacillus, which still remains red, owing, according to Ehrlich, to its possessing a sheath impenetrable to acids. This does not accord with my experience; for I have often found the putrefactive bacilli and micrococci of various kinds tinted red after the nitric acid and distilled water, though the sharp outline and characteristic shape of the tubercle bacillus render it easy of recognition. The addition of the chrysoidin or methylene blue at once tints all other bacilli, leaving those of tubercle unaffected; and this should always be done in doubtful cases. The omission of the second dye considerably reduces the length of the process, and answers perfectly well for experienced workers.

Gentian violet may be used for the first stain, and in this case vesuvin or Bismark brown is best for a second tint.

The fault of this otherwise admirable process lies in the staining of the bacilli not being permanent, but often fading after a few months, and the solution cannot be depended

---

[1] This last washing in alcohol may be omitted where water has been used if there be no hurry, and plenty of time can be allowed to admit of gradual drying. Such specimens are remarkably well stained.

on for longer than three or four weeks. According to Ziehl it is the property of the tubercle bacillus to take up colouring matters very slowly, and to hold them, when absorbed, in spite of acids and alkalies, and the rate of absorption is quickened by combination with aniline, carbolic acid, resorcin, and pyrogallic acid. Ziehl replaces the aniline in the Weigert-Ehrlich solution by carbolic acid, and arrives at a staining fluid which keeps for months and dyes the tubercle bacilli permanently. The following is the fluid, in which the covers must be immersed without warmth for half an hour, and subsequently treated on the Weigert-Ehrlich plan.

Saturated alcoholic solution of fuchsin . . .	10 ccm.
5 per cent. solution of carbolic acid (watery solution)	100 ccm.

Mix and keep in a stoppered bottle. Sections should be left in it for five to ten minutes; sputum half an hour, and, in doubtful cases, all night or for twelve hours.

This is an excellent method, and is now generally adopted at the Brompton Hospital.

Gram uses the Weigert-Ehrlich formula, substituting gentian violet for fuchsin, and transfers the sections or sputum to water, then to absolute alcohol, in which they are immersed from one to three minutes. They are transferred to a solution of—iodine 1 gramme, iodide of potassium 2 grammes, and water 300 cubic centimetres, for three minutes, then to absolute alcohol and afterwards dried and cleared in oil of cloves, and mounted in Canada balsam. The result is that the bacilli are stained violet and the tissues a faint yellow.

The application of these tests has been accompanied by such wonderful results that I have thought it necessary to enter rather fully into the details of the processes. They are all so easy, and can be made so accessible, that any practitioner or student can speedily master them, and soon become a proficient in their employment.

CHAPTER V.

THE TUBERCLE BACILLUS (*cont.*).

Size, form, external membrane and spores Presence in sputum, blood, urine, fæces of consumptive patients—Its detection in tubercular and scrofulous lesions, in lupus, in milk of tubercular cows, in air of hospitals—Dr. Ransome's and author's inquiries—Baumgarten's experiments on rabbit's eye Veragut's and Ribbert's inoculations—Bacillus in pulmonary lesions Its relation to miliary tuberculosis, caseation, and fibrosis— Tubercle bacilli and giant cells—Four channels of bacillar infection—Ptomaines—The Micrococcus Tetragenus.

THE tubercle bacillus is generally described as varying in length from $\frac{1}{8000}$ to $\frac{1}{13000}$ of an inch, and having a breadth of one-fifth of its length, or from one-quarter to one-half the diameter of a red corpuscle. As regards breadth, it is generally uniform, though I have noted some bacilli double the thickness of others, the thickest ones being invariably beaded and ready to divide. I have, therefore, concluded that these are the older rods, and that the others are of younger growth. As regards length, they vary greatly. Some are only twice as long as they are broad, others extend over a twelfth of the diameter of the microscopic field. In form, most are straight, with somewhat rounded extremities. The bacilli are not always straight, but often have a curve, sometimes forming a portion of the arc of a circle. With reference to the varying thickness above mentioned, I am aware that this might be due to two bacilli lying on each other, as is often the case; but this can be easily detected by adjusting the focus to different parts of them. Bacilli are found singly or in groups. In the latter case they appear like bundles of sticks, numbering three, four, five—a very common position is two bacilli touching at one end, at an angle of 45°. Koch concludes they have an external membrane for three reasons.

1. He has noticed that where stained with methylene blue, the bacilli appear thinner than when stained with violet or fuchsin. 2. That the bacillus dyed with methylene violet does not fade equally, the outer portion disappearing first, leaving the inner, which is of the same size as the bacillus when stained with methylene blue. 3. Bacilli under cultivation appear glued together in masses, and he thinks this to be due to the presence of an external envelope.

Bacilli appear to multiply by spores, and not uncommonly we see one long straight rod, and parallel with it a second one, which is evidently in process of division, i.e. with slight spaces separating three or four segments, all lying in the same line. Often we see a third bacillus in the vicinity, still more broken up into numerous segments, these being separated by wider intervals, but the original connection being still recognisable by their distribution along the same line. Even under a low power they frequently present a beaded appearance, dark spots, which are the spores, alternating with clear spaces, generally of an oval shape. These form beadings, varying in number according to the length of the rod. As a rule five or six are seen, sometimes a larger number, in long rods about to divide. My experience is that the beaded rods may break up into single beads, a spore being contained in each bead. As each of these forms a fresh bacillus, the rate of multiplication is enormous.[1]

The tubercle bacillus has been so frequently detected in the sputum of phthisical patients that failure to do so is generally regarded as more likely to be due to want of practice than to absence of bacilli. The sputum of two hundred consumptive patients was examined at the Brompton Hospital, and in only three did we fail to find tubercle bacilli, and in

[1] Von Schrön, of Naples (*London Medical Record*, January, 1887), after numerous cultivations, concludes that the tubercle bacillus is, in its early stage, a torula chain, the granules of which recede from each other during development and become connected by a band, formed by a secretion from the granules. He holds that by regressive metamorphosis these granules are set free as bacillus spores, and enlarging, they become mother spores with a capsule and granular contents (daughter spores). The latter burst the capsule and issue forth, either as separate spores, or as a torula chain (young bacillus).

these the failure arose either from insufficient opportunities
of investigation, or from some mechanical obstruction to a free
supply of lung secretion, as in the case of a contracted cavity,
where the expectoration was purely bronchitic. Tubercle bacilli
have also been detected in the blood of hæmoptysis, and first
noted, I believe, by Dr. Perez and myself in a patient of Dr.
Tatham's. On the other hand, in the sputum of twenty-three
patients suffering from pulmonary affections other than
phthisis, no bacilli could be found. This is in exact accordance with the observations of Balmer, Fraenkel, D'Espines,
Heneage Gibbes, Heron, West, Dreschfield, Whipham, Hunter
Mackenzie, and others. Tubercle bacilli have been detected
in the various lesions of tubercular disease of the lungs (to
which we will recur more fully presently), also in tubercular
meningitis, in tuberculous ulcers of the tongue (see Plate III.
fig. 2), in tuberculous kidney, bladder, and ureter, and in the
urine in the above cases ; in tuberculosis of the supra-renal
capsules and spleen : in tuberculosis of the uterus and fal-
lopian tubes and in tuberculous testicle, though here, as in
many other lesions, it was confined to the giant cells ; also in
a case of tuberculous abscess of the brain. Scrofulous glands
of the neck, axilla, and groin have been found to contain
them, though not in such abundance as bronchial or mesenteric glands similarly affected, and here, as above, they are
found chiefly in connection with the giant cells. They are
tolerably abundant in the intestinal ulcers of phthisis, and
have been detected in the fæces by Gaffky,[1] and by Mr. H.
Taylor at the Brompton Hospital in a patient of mine. Koch
found them also in scrofulous disease of the various joints,
but failed to detect them in pus discharged from caries of the
spine, though the same pus when inoculated into animals
produced tuberculosis, Koch alleging this to be due to the
presence of spores which are not coloured by the staining
fluid. Schuchardt and Krause[2] examined forty cases of
surgical tuberculosis in which the synovial membranes, the
bones, glands, tongue, testicle, and female genital organs

[1] *Mittheilungen aus dem Kaiserlichen Gesundheitsamte*, vol. i.
[2] Volkman's *Klinik*.

were respectively affected, and found bacilli in all, their seat being generally the giant cells.

Doutrelepont detected them in sections of lupus skin in seven cases, and gives an instance where meningitis supervened on lupus. Koch also detected them in lupus, but, judging by his illustrations, only in small numbers, each giant cell containing one ; but he succeeded in cultivating them in appropriate nutritive media. This is very interesting as regards lupus, as placing this disease at once in the category of tubercular lesions, from which many authorities were inclined to separate it.

Tubercle bacilli abound in perlsucht or bovine tuberculosis, and in the tubercle of horses, and were found in all the tubercular lesions of the various carnivora and rodents experimented on. Bovine tubercle contains bacilli in greater abundance, as a rule, than human tubercle, and is stated to be more effective for inoculation purposes, the experiments of Koch and Watson Cheyne showing that the relative success in inoculating animals depends on the number of bacilli contained in the inoculating material. Bovine tubercle produces more certain results than human, and culture fluid containing the largest number of bacilli being the most successful of all. Bacilli have been detected in the milk of cows affected with bovine tuberculosis : but it appears that this is not the case, unless the milk glands or udders are affected. If these are not affected, the milk of animals severely attacked with tuberculosis does not contain tubercle bacilli. Moreover, the tubercle bacillus has been found in the breath exhaled by advanced cases of phthisis by Dr. Ransome[1] and myself.[2] Dr. Ransome detected them by an apparatus arranged over the patient to receive and test the expired air. The method which I employed was to fill one ward of the Brompton Hospital with cases of advanced phthisis, and to place in another ward, similarly situated, a number of non-consumptive patients (cases of asthma, bronchitis, and heart disease). Each ward was carefully isolated, and glass plates, well cleansed, sterilised and smeared with glycerine, suspended in the extraction shafts of

[1] *Proc. Roy. Soc.*, 1882. [2] *Lancet*, March 1883.

each ward for two or three days at a time. The plates were then carefully tested, and it was found that those from the non-consumptive ward contained no bacilli, those from the *consumptive* ward contained a few well-marked ones.

The true part played by the bacillus in the causation of tuberculosis is well illustrated by the changes in the eye of a rabbit, as noted after inoculation of the anterior chamber with tubercle by Baumgarten,[1] which, from the observations being made during life, and from the protection from other organisms being specially complete, are most interesting. The first change after the wound was made was the formation of a scar and the encapsulation by granulation tissue of any foreign bodies which had entered. After the second day an increase in the number of enclosed bacilli is observed, and the newly formed ones reach the granulation tissue. Increasing in numbers, they press in masses through this, and extend into the iris and sclerotic. On the fifth day isolated bacilli are seen in the above membranes, partly free in the intercellular substance and partly contained in fixed cells, in neither case giving rise to any changes in the normal histology of the part. On the sixth day there are found in connection with masses of bacilli in the cornea and iris, some few cells of an epithelioid character, some being giant cells and others lymphoid, characteristic, according to Virchow, of miliary tubercle. On the tenth and eleventh days the changes become macroscopic.

The different stages of the production of tuberculosis are here well defined : viz.—the multiplication of bacilli, their extension to neighbouring parts, where, if only present in small numbers, they appear not to cause any disturbance, but if in masses they give rise to the formation of giant cells and to cell proliferation, and at a later date to disorganisation of the whole tissue. We can hardly imagine a more complete proof of the distinct action of bacilli as causes of tuberculosis.

Veragut's[2] experiments showed the early stages of the bacillar attack on the lungs. He rubbed down tubercular sputum with water, and then diffused it in the form of a spray,

[1] *Centralblatt für die Medicinischen Wissenschaften*, 1883, No. 42.

[2] *Archiv für Experimentale Pathologie und Pharmakologie.* Zürich, 1883.

which he made three goats and eighteen rabbits inhale for an hour every day for fourteen or twenty days. He also wounded the vocal cords and performed laryngotomy on some of the animals, to produce, if possible, primary laryngeal tuberculosis. In this latter attempt he was unsuccessful. In the first rabbit, killed fourteen days after the inhalations, from one to three tubercle-bacilli were noted in each field, in sections taken from the pleura overlying the upper border of the lung. The capillaries were infected, and the infundibula and alveoli contained brittle masses of bell-shaped epithelial cells, partly dissolved and partly adherent to the alveolar wall. These contained bacilli. The second rabbit, killed sixteen days after the inhalations commenced, showed to the naked eye two nodules in the lung, and the alveoli full of epithelium, abounding in bacilli.

In the third rabbit, killed after twenty-two days, the vocal cords had been wounded. Caseous masses were found in the neighbourhood of the vocal cord and scattered throughout the lungs. These contained abundant bacilli.

The fifth rabbit was killed after forty-two days, and giant cells, containing bacilli, were found in the lungs.

The sixth animal was a goat, killed sixty-six days after the inhalation; the lungs contained cheesy masses as large as beans, and some calcareous material.

In a goat killed one hundred and fifty days after inhalation, the lungs contained caseous and cretaceous masses, and cavities the size of hazel-nuts. The caseous material showed few bacilli.

These experiments demonstrated that the inhaled bacilli entered the alveolar epithelium, causing it to swell and desquamate, the epithelial inflammation being accompanied by hyperæmia of the capillaries and exudation of lymphoid cells into the infected alveoli. The walls of the epithelial cells were then destroyed, and masses of bacilli thus let loose into the exudation. Around these masses was developed a reacting connective tissue, probably fibroid, which tended to encapsule and isolate the caseous material, necrosis proceeding in the centre of the tissue, and the bacilli disappearing. Single ones, or perhaps spores, passed through the encapsulating tissue and entered the lymphatics, producing secondary growths, which appeared to be miliary tubercles with typical giant cells, containing bacilli. In time, all these structures, including the giant cells, caseated, and some cretified, and the bacilli disappeared. In the animals killed seven days after inhalation, Veragut found no bacilli.

Ribbert's observations on tubercle in poultry, and its minute distribution in the liver, though pertaining to a different class of beings, throw considerable light on the mode by which the tubercle bacilli spread through an organ. The tubercle was present in the form of very small nodules, which, under the microscope, showed the following arrangement :—1. A central caseous zone. 2. A zone of tubercular cells free from bacilli. 3. A bacilliferous zone. 4. An outer zone of inflammatory tissue. According to Ribbert it appears that with the growth of the tubercle the bacilliferous zone continued to extend its area, and that caseation occurred, not from direct action of the bacilli, but from shutting off of the blood supply. He found that the indurated vascularised areas surrounding the caseating tubercle showed the following. The parallel bundles of connective tissue forming the walls of veins were seen to be thickly invaded by the bacillus, which occurred in irregular-shaped masses, in places encroaching upon the lining membrane of the vessel. The bulgings in this membrane produced the appearance of minute tubercles, and by constricting the calibre of the vessel rendered thrombosis probable. The bacilli were specially massed within the mesh-like spaces of the connective tissue forming the vascular wall, and caseation was not to be found in these structures. The arterial walls were likewise similarly invaded by the bacillus. A like invasion of blood-vessels in the human lung would explain early hæmoptysis; for the blood-vessel, weakened by invasion of organisms, would yield before the blood stream, and thus give rise to hæmorrhage.

We will now consider the tubercle bacillus in its relation to the pulmonary lesions of phthisis, and see what conclusions may be drawn from the researches of Koch,[1] Mr. Watson Cheyne,[2]

[1] *Op. cit.*　　　[2] *Practitioner*, April 1883.

and Dr. Percy Kidd[1] in this department. Tubercle bacilli are invariably found in freshly formed tubercle; but in grey or miliary nodules of some date they are often absent, probably from their having undergone fibrosis. They are very abundant in the nodules of acute miliary tuberculosis. In caseous masses the distribution is irregular, depending again, probably, on the age of the material. Where the caseation is uniform and firm, and therefore old, according to Dr. Kidd no bacilli, as a rule, can be found; but where the process is commencing, or where softening is going on, and microscopical cavities are present, bacilli are often seen in great numbers. In fibrosis they are entirely absent. They are found in the greatest abundance on the walls of cavities in process of formation, both in the cavity contents and the wall itself. The more rapidly cavities form, the more abundant are the bacilli in their walls. They therefore abound in cavities formed in infiltrated tissue, as in caseous pneumonia or acute phthisis, and are scanty in excavations where the walls are fibrous, or where the numerous ribbed processes show that excavation has taken place slowly, or is limited by the interlobular tissue, as is so common in chronic tubercular phthisis undergoing fibrosis, or in fibroid phthisis. The fibroid element seems to oppose a direct barrier to the growth and multiplication of these organisms, and in large tracts of lung-tissue which have been converted into this material, often not a bacillus can be detected. They are found in the infiltrated walls of the small bronchi in acute phthisis.

Recurring to the four groups of histological elements of phthisical lesions, we find that the bacilli are not largely present in the fibrinous exudation, but they have been traced into leucocytes in the lower animals. Neither are they found in the interlobular tissue; and not abundantly in adenoid or lymphoid tissue. Their principal haunt is the alveolar epithelium, and its products, the epithelioid cell and the giant cell. Reaching the alveolus through the breath, the bacillus enters the epithelial cell, and causes proliferation by irritation. The alveolus becomes stuffed with cells, and if the irritation

[1] *Medico-Chirurgical Transactions*, vol. lxviii.

be a gradual process, it gives rise to a highly nucleated product, the giant cell. If the irritation be great, caseation is rapidly produced, though whether this be due to necrobiosis from overcrowded proliferation, or, as Mr. Watson Cheyne thinks, to a chemical change in the cells, brought about by the action of the bacilli, is not yet determined. The epithelioid cells previously described are found containing bacilli, and surrounded by them, and Mr. Cheyne advises these cells to be the first object of search in the bacilli hunt. Koch found them more particularly in the giant cells, and in some of these he counted as many as fifty bacilli. The gradual destruction of the giant cells, which is well represented in Plate III. fig. 1, has been traced step by step, and may be thus described :—The bacilli appear on the side of the cell where the nuclei are fewest, but after a while increase in number and invade them, separating the nuclei and breaking down their walls. The bacilli always take up such a position that their axes are perpendicular to the giant cell. This in time disappears entirely, and its exact site is occupied by a circle of the destroying bacilli. Many such circles have been noted in lung sections. Bacilli have also been seen in the wandering cells, which, according to Koch, are converted into giant cells ; and in the guinea-pig and rabbit leucocytes have been seen to contain them. A curious feature is the way in which bacilli spread themselves. Cells appear to take them up and to distribute them among other cells, so that where at first one giant cell was seen to contain a large number of bacilli, later on they were found distributed among a number of cells. The after-progress of the bacilli depends on the rate of their growth, and on the fitness of the soil for their development. If they multiply quickly, caseous pneumonia and rapid excavation of the lung will ensue ; but if slowly, the irritation of their presence and of the resulting caseation may cause thickening of the alveolar wall with inflammatory tissue, and a fibrous barrier may be opposed to further infection.

The spread of the bacilli in the lung may proceed through various channels—(1) By continuity, the bacilli passing through the alveolar wall, having first destroyed the epithelium (see Plate II. fig. 1). (2) By reinhalation of bacilli-laden sputum. Where the power of expectoration is feeble, the sputum, having been raised by coughing to a point where two or more bronchi join, is drawn by deep inspiration into an adjoining one, and thus a fresh set of alveoli are infected. (3) Through the lymphatics. This is probably the way in which the pleura, which is network of lymphatics communicating with lung, becomes involved; nor is it unlikely that the distribution of miliary tubercle around a cavity or caseous centre may originate in lymphatic infection. Ponfick demonstrated that the system may become infected through tuberculosis of the thoracic duct, especially in children. (4) Through the arteries and veins. The fact of masses of bacilli appearing in fresh parts of the lung, Koch explains by their having entered the general circulation by penetrating the walls of vessels, and being distributed secondarily in the lungs. How the arteries may become involved was shown by Koch in a case where acute tuberculosis was produced through tubercular disease of a bronchial gland, the walls of the artery supplying it containing masses of bacilli extending into its interior. A beautiful example of this is seen in Plate IV. fig. 2, taken from a lung section of Dr. Percy Kidd's, a case of acute miliary tuberculosis, which displays an arteriole from one of the nodules with a minute tubercle containing bacilli, projecting into the lumen of the vessel and encroaching upon its calibre. The vessel is still patent and contains blood corpuscles, but the entry into it of the bacilli is evidently at hand.

It may be remembered that Weigert and Mügge discovered in cases of acute tuberculosis tubercular masses in the walls of the pulmonary veins, and on one occasion, infection of the innominate veins from the bronchial glands. The inferior vena cava has also been found infected in a horse from a tubercular abscess of the peritoneum, and acute tuberculosis thus set up.

Ribbert's evidence on tubercle in the fowl affords further proof of the vascular system being penetrated by bacilli, and the fact that in tubercular meningitis the bacilli are found most abundantly in the neighbourhood of the smaller arteries, where Koch has seen them forming plugs, indicates that their channel of communication has been the blood-vessels, and principally the arteries.

It is not certain that in all cases of tuberculosis the bacilli themselves are the carriers of the virus in the circulation, and in many instances of well-marked general tubercular infection all attempts to detect bacilli in the blood have failed, and this too when they have been detected abundantly in the sputum. It is quite possible that they produce a chemical poison (ptomaine), which may give rise to the irritative fever both of tubercular formation and of secondary absorption, which are characteristic features of the disease. The process of caseation is held by some to be due to such a chemical process; and when we consider the light Brieger's researches[1] have thrown on the nature of 'ptomaines,' we must not deny the possibility of their acting an important part in tuberculosis.

We must not omit all mention of another organism, the *micrococcus tetragenus*, which has been found by Koch and others in the walls of tubercular cavities and in phthisical sputum. Under the microscope it appears in groups of four micrococci, each group having a diameter of one-third of a red blood-corpuscle, and in arrangement much resembling the micrococci of sarcina ventriculi. When cultivated in peptone gelatine and inoculated into mice, guinea-pigs, and rabbits, they caused death in ten days in the two former, but had no effect on the rabbits. *Post-mortem* examination of the mice and guinea-pigs showed zoogloea masses of this micrococcus, even visible to the naked eye, in all the organs, and especially in the blood-vessels. From the spleen they had penetrated to the abdomen, producing peritonitis. What part this organism plays in connection with phthisis is at present unknown, but its absence from the early lesions (of grey and miliary tubercle) render it improbable that it is not as important as the tubercle bacillus.

[1] *Ueber Ptomaine.*

CHAPTER VI.

PATHOLOGY OF CONSUMPTION.

Relation of Tubercle Bacillus to the four histological elements—Part due to irritation of organism—Protecting influence of fibrosis—Entry of bacilli through inhalation—Amount of lesion dependent on their number and fitness of soil—How the bacilli perish—Cavities—Reasons for apex excavation—Frequency of cavities in dorso-axillary and mammary regions—Secondary caverns, mode of formation—Contraction of cavities, and resulting changes—Expansion of healthy lung—Displacement of neighbouring organs—Of heart, liver, stomach, spleen—Changes in pleura—Collapse of chest wall—Spinal curvature—Cause of vagrancy of cavernous sounds.

The discovery of Koch has undoubtedly changed our views of the pathology of phthisis, and has made us regard much of the morbid anatomy and histology of the disease from a new standpoint. One great gain has been to establish the unity of phthisis on a firm basis, and to break down all distinctions between phthisis, tubercle, and scrofula, which, judging from the numerous connecting links furnished by clinical experience, we have long held to be artificial. Such differences can now be shown to depend either on the channels through which the organism is distributed, or on the tissues or structures first attacked, or again on the strength of the individual to withstand such attack.

We must now regard many of the histological elements of phthisis, on which stress was formerly laid, not as essential factors of the disease, but simply as results of the irritation to the tissues, caused by the presence of the bacillus, their number and variety depending on the rapidity of the onset and the length of time permitted to nature to organise defences against the invader.

Taking the four principal histological changes in order, we may consider the exudation of fibrin and leucocytes and the epithelial proliferation as the first effect of the entry of the

bacillus into the alveoli. If the bacilli are present in large
numbers, or if the irritation they set up be very considerable,
the exudation and epithelial proliferation may be sufficiently
abundant to distend the alveolus with products, which would
rapidly caseate ; if, however, the irritation be less, the process
is a slower one, and giant cells are evolved which, according
to some authorities, are an attempt on the part of the organ
to protect itself from the bacillus. Others do not hold this
opinion. Whether it be so or not, we know that giant cells
accompany the bacillus in all chronic cases, and the steps by
which these remarkable growths are destroyed by the bacillus,
have been already described.

The inter-alveolar lymphoid growth is probably the result
of irritation to the alveolar wall, when this has been penetrated
by the bacillus or its virus, and may also be another attempt
on the part of the tissues to repel the invader. Unfortunately,
this growth is of rapid but not lasting character ; so that
while it causes wholesale destruction to the alveoli, from the
walls of which it originates, either through the action of the
bacilli, or through want of blood supply, it soon undergoes
caseation, and appears to add only to the general disorganisation of the lung-tissue. On the other hand, the fourth histological element—growth of the interlobular tissue, though it
also may arise from bacillar irritation—has a distinctly
limiting influence on the inroads of that organism, and
appears to present an effectual barrier to its progress. It is
rare to find bacilli in this growth, or in any form, indeed, of
fibroid growth.

The characteristics of the action of the bacillus in an
organ are *irritation*, exemplified as we have seen in various
ways, and *infection*, as shown after it has entered one of the
channels of communication and has diffused itself more
widely, either in the same or in other organs.

It is probable that in most cases of pulmonary consumption the bacilli reach the alveoli through inhalation : but the
explanation why out of a number of persons placed under
conditions apparently similar, and necessitating the inhalation
of tubercle bacilli, in but few they increase and multiply, is

not easy.   A predisposing cause to fit the organs for the
reception of the bacillus, seems to be, in phthisis, as necessary
as the bacillus itself; and this, according to Koch, is to be
found in the denudation of the bronchial mucous membrane,
occurring after measles, whereby it loses the protection of its
epithelium, and especially of its ciliated cells.   This would
account for the main bronchi being exempt from tubercular
lesions in the early stages of phthisis, when their ciliary and
epithelial layers are complete, and for the alveoli, where the
epithelium consists of but one thin layer of pavement cells,
being the first to succumb.   In the same way denudation of
the intestinal mucous membrane is said to predispose to
intestinal ulceration arising from swallowing bacilli-laden
sputum.   Chronic pleurisy, by crippling the movements of
the lungs, and thus promoting congestions and exudations,
prevents the proper expansion of the alveoli, and thus affords
a nidus for the bacillus.

The limited lesions of early phthisis are explained by the
inhalation of bacilli into only a few alveoli, and the fact of
the apex of the lung being the most liable to attack, by the
presence of inflammatory exudation, the result of more or less
injury to the pulmonary vessels through stagnation of the
blood, the whole being more or less due to the imperfect
exercise of the upper lobes during respiration.   The entry of
the bacilli causes proliferation of the epithelium; the alveolus
becomes stuffed with epithelioid cells; other alveoli are infected partly by coughing and re-inhalation of bacilli, and
partly by continuity, that is, through the alveolar wall being
attacked and destroyed; caseous masses are thus formed and
expectorated, and a cavity is the result, the walls of which
appear to present the conditions most favourable to the growth
and multiplication of the bacillus, and here they are found
swarming.

The further spread of the organisms in the lungs proceeds, partly, through re-inhalation into other lobules, notably
those communicating with the same bronchus as the cavity,
and thus is formed what Dr. Ewart [1] calls the secondary

[1] *British Medical Journal*, 1882.

cavity, or the cavity of infection. By re-inhalation the opposite lung may become infected, and this occurrence can often be foretold during life if the expectoration be not free, while the number of bacilli contained is large. In these cases I have often noticed that the physical signs appear in the mammary regions, and not in the sub-clavicular or supra-scapular; showing that the opposite lung is infected through channels different from those which the poison follows in the ordinary course of phthisis.

The extension of the centres of caseation leads to the lymphatics and blood-vessels becoming involved, and in the former case we get a formation of grey tubercle immediately in the neighbourhood of the caseous mass; in the latter, acute tuberculosis of all the organs, or at any rate tuberculisation of the opposite lung takes place. If the bacilli enter the pulmonary vessels, infection of the whole system may occur, and we may have tuberculosis of the membranes of the brain and spinal cord, of the peritoneum, liver, spleen, and other organs. It is therefore the infection of the pulmonary blood-vessels which may convert a case of chronic phthisis into one of general tuberculosis. On the other hand, where the physical signs and general condition point to undoubted arrest of disease, we may conclude that a great proportion of the caseous centres, the chief haunt of the bacilli, have been expectorated, leaving a cavity, which the rapid growth of the fibroid elements tends to limit, and afterwards to contract. After a while, bacilli cease to be found in the expectoration, and while this may be due to the bronchus leading to the cavity being blocked, it may generally be ascribed to a closing of the secreting surface of the cavity. The flattening of the chest wall demonstrates further progress in the contraction of the cavity, and thus the bacilli become isolated and in time perish.

Cavities play so considerable a part in the pathology of phthisis that a few words further may be said on their formation, varieties, and position in the lungs. We learn from Dr. William Ewart's valuable *post-mortem* statistics [1] of the

---

[1] 'Gulstonian Lectures,' 1882, *British Medical Journal*.

Brompton Hospital, that the relative liability of the two lungs to excavation does not differ much, but that the left lung is rather more liable than the right to extensive excavation, such as that of a whole lobe or of an entire lung.

The relative frequency of excavation in the different portions of the lung in three hundred and four lungs was thus shown :—

Excavation at apices	. . . .	in	282
,, in the dorso-axillary region	.	,,	227
,, ,, mammary	,, .	,,	189
,, ,, sternal	,, .	,,	61
,, at the base	. . .	,,	32

Here we see well brought out the great liability of the apex, and the very slight tendency of the base, to excavation. The sternal region is also comparatively exempt, its liability being less than a quarter that of the apex, and this is more marked in the lower portion of the sternal region than in the upper, where there appears to be some tendency for the disease to spread from the apex.

The greater liability of the apex to excavation, as well as to tuberculisation, and the comparative immunity of the base, are undoubtedly due to the same causes. First, to defective movement of the upper parts of the thorax, and the consequent imperfect ventilation of the apex. Second, to the frequent presence in the upper lobe of inflammatory exudations, arising, according to Dr. Green, from more or less injury to the vessels through stagnation of blood. Thus are formed the necessary conditions for bacillar growth and multiplication and the destructive changes resulting from them.

On the other hand, the base of the lung is, through the action of the diaphragm and lower ribs, the subject of very complete ventilation, the respiratory movements being remarkably free in this region, and even if pleuritic adhesions have taken place, the action of the diaphragm is still sufficient to cause fair inflation of the lower lobe.

The frequency of cavities in the dorso-axillary and mammary regions arises, according to Dr. Ewart, from inhalation of secretions from apex cavities, due partly to the distribution

of the bronchi, and partly to the suction, promoted by the
inspiratory movements of the middle of the lung, being
stronger than that of the apex.

Dr. Ewart calls attention to large dense masses of tubercle,
which differ from the usual racemose formation in having a
cuboidal rather than a spherical outline, and are always found
in connection with cavities, and on their distal side.   Their
seats of election are—(1) the upper axillary border, external to
the cavity, very usually found in the outer part of the upper lobe;
(2) the middle third of the axillary region, commonly occupying
the wedge-shaped summit of the lower lobe, and bordering
the cavity of the dorso-axillary region; and (3) a smaller

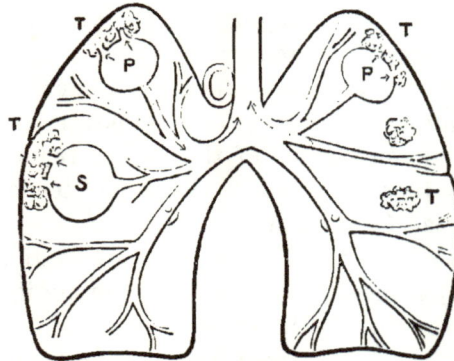

Fig. 1.— Mode of Origin of Secondary Cavities in *Dorso-axillary* Region (adapted from
Ewart).   P. Primary, s. Secondary Cavity, T. Secondary Tubercle.

mass is sometimes found at the outer apex, apparently the
result of a cavity which did not reach the summit of the lung.
All these masses follow the distribution of certain cavities,
which have formed at the terminations of bronchi in parts
where the inspiratory power is strong.

The formation of the dorso-axillary cavity, on which much
stress has been laid, appears to arise from the secretions from
the apex cavity being imperfectly propelled by the deficient
expiratory power of the diseased apex, and then sucked by the
deep inspiration after cough into the neighbouring bronchus
which supplies the axillary portions of the lung; the first
appearance of the axillary nodule thus produced being trefoil-

like in form, showing the terminal divisions of a bronchiole. The mass gradually enlarges, and softening takes place, and thus a cavity secondary to the apex one is formed, which is peculiar in its rapid formation and in having no trabeculæ. Like the other cavities due to inhalation, it gives rise to tubercles on its distal side. The accompanying woodcut, adapted with modifications from Dr. Ewart's Lectures, shows well the mode of formation of secondary cavities.

Cavities contract, and in some cases become completely cicatrised. This last occurrence is somewhat rare, and has been denied by Andral, Louis, and Walshe, though maintained by Laennec, Roger, and Boudet Heitler,[1] Pollock, and the writer. More common is a shrinking of the cavity and partial obliteration, with a blocking of the bronchus leading to it, causing cessation of the cavernous sounds during life.

The great cause of the retraction of cavities is fibrosis, which, starting from an adherent pleura, or from the interlobular tissue, or from obsolescent tubercle, rapidly spreads round cavities and causes their shrinking and contraction.

When this takes place the laws of atmospheric pressure necessitate certain changes in the adjoining structures to fill in the vacuum which would otherwise occur. This is accomplished in various ways :—

1. By an expansion of the lung-tissue round the contracting cavern. This rarely occurs alone, but as there is generally adhesion of the pleura, it is accompanied by collapse and flattening of the chest wall overlying it. The dilatation of alveoli is often irregular, some being largely dilated, and some not at all so, and thus is formed a localised emphysema, the well-known accompaniment of lung cicatrisation.

When the upper lobe of the *right* lung is the seat of a contracting cavity, it is not rare for the middle and lower lobes to be drawn up and rendered emphysematous by traction, and this accounts for the less frequent displacement of other organs on the right than on the left side.

When adhesion of the pleura is limited to the portion over-

[1] *Anzeiger der K. K. Gesellschaft der Aerzte in Wien*, June 1880.

lying the cavity, band-like adhesions of the pleura, of some
length, pass into the involution of the lung surface, which
are caused by the retracting cavity [1] being surrounded on either
side by dilated spongy lung-tissue. These pleuritic bands
generally indicate the presence of a contracting cavity. I have
known a cavity, sufficiently large to give rise to tinkling sounds,
completely contract and cause at first no displacement of
organs, and only slight flattening of the chest wall, owing to
the large expansion of lung-tissue round the cicatrix.[2] When
the pleura is adherent over a large space, the contraction very
persistent, and the thoracic wall very rigid, either the pleura
becomes thickened or appears to develop meshes,[3] into
which fluid is effused, or else fat is developed between the ribs
and parietal pleura to fill in the void. When the dia-
phragmatic pleura is adherent, the contraction of the lung
draws up the diaphragm, producing invagination of a portion
of that organ.

2. By a drawing over of the opposite lung, which becomes
hypertrophied, and often increases to nearly double its normal
size. This is the case where complete contraction of the
lung containing the cavity takes place, and the viscus is
found lying at the back of the thorax, occupying about one-
third of its normal area.

We then find the healthy lung extending across the median
line, far into the usual limits of its neighbour.

As a rule, however, the anterior and upper portion of the
healthy lung is drawn across, and the upper lobe alone is
involved. Dr. Ewart draws attention to the various involutions of the healthy or hypertrophied lung, which appear to be
an attempt on the part of Nature to gain an increase of
breathing surface. Their direction depends apparently on the
shape of the thorax; in long narrow chests, they take a vertical
direction, and in short and wide ones, a horizontal one. When
there is a contracting cavity in the upper lobe of a lung, the
lower generally possesses one or more involutions, produced by

[1] Ewart, 'Gulstonian Lectures,' 1882
[2] *Clinical Transactions*, vol. v. p. 27.
[3] R. D. Powell *Pulmonary Consumption*.

its being in a state of compensatory hypertrophy, and therefore unable to be contained within its former limits.

3. By displacement of the neighbouring organs; this is rarely absent if the cavity be of any size. The organs liable to be thus displaced are the heart, liver, stomach, and spleen, but whether one or more of them is displaced, and the extent of such displacement, depends upon which lung contains the cavity, and on what portion it is situated in.

Where the contracting cavity is in the right lung the displacement generally consists of the liver being drawn up, and the upper edge of this organ may be raised as high as the fourth rib. The left lung and heart may be drawn across the median line, and the pulsation of the latter may be visible in the second and third interspaces, the apex being drawn across, and its beat detected as far as the right nipple. This is, however, a rare occurrence, for the heart's position in these cases is more generally beneath the sternum, and slightly to the right of it. I can, however, call to mind a few instances of contracting cavity in the right lung, where the heart seemed entirely shifted to the right of the sternum, and the impulse was felt between the fourth and fifth ribs, outside the right nipple.

When the cavity is in the left lung the displacement of organs is, as a rule, more marked. The right lung may be drawn considerably over to the left side; the stomach may mount up to the fifth rib, but does not generally rise higher than the sixth. The heart is uncovered and drawn upwards through retraction of the left lung, and comes into close contact with a considerable surface of the chest wall, and on this account its action is sometimes visible in the first and second interspaces. In this change of the heart's position the course of the apex seems to be upwards, outwards, and towards the axilla, often describing the arc of a circle, of which some point in the neighbourhood of the aortic valves is the centre. First, the apex is felt beating between the fourth and fifth ribs, immediately below the nipple; then between the third and fourth, rather to the outside of the nipple; and at last it reaches the axilla. There is also sometimes a twisting of the heart on itself, and this, owing probably to its causing a slight obstruction in the aorta, sometimes gives rise to a systolic murmur; but, according to Dr. Pollock,[1] no lesion is found after death in these cases.

4. The other mode of filling up the void is by collapse of the chest wall. This generally takes place sooner or later in all cases of contracted cavity, but not always at first, as often the displacement of organs, particularly of the abdominal viscera, the stomach, and liver, is sufficient to fill the vacancy. The flattening generally commences under the clavicles, but overlies the cavity; and when the collapse is considerable the circumference of the affected side by measurement through the nipple line shows a diminution of from one to two inches.

The falling-in of the ribs leads to an increase in the convexity of the costal arch, and this secondary bulging, as Dr. Ewart well remarks, is usually seen in the axilla, and is a good confirmation of the occurrence of excavation. Curvature of the spine may be produced by a contracting cavity, as I showed in a remarkable case,[2] where a cavity large enough to give rise to tinkling sounds, contracted enormously and caused spinal curvature and marked lowering of the shoulder. In addition to the external changes which accompany and promote the contraction of cavities, there are internal ones, which are (1) the retraction of the trabeculæ of the cavity. Owing to their derivation from collapsed alveolar substance, and to their enclosing the branches of pulmonary veins and arteries, elastic tissue enters largely into their composition, and this forms a contractile element, the force of which may be seen in the recoil of the ends, when the trabecula is cut through. It is possible that when these undergo fibrosis, as is usually the case in chronic phthisis, their tendency to contraction is increased. (2) The shortening of the bronchi connected with the cavity. The bronchi increase in thickness, the thickening being mainly due to an abundant growth

[1] *Elements of Prognosis in Consumption*, p. 200.
[2] *Clinical Transactions*, vol. x. p. 113.

of fibro-cellular material within the peribronchial sheath. This is the beginning of fibrosis, and the bronchus, terminating in a vomica, becomes distinctly shortened in proportion to its thickening, and thus the cavity is drawn more and more towards the root of the lung. It is probably this shortening and contraction of the bronchi, combined with the pressure of the expanding pulmonary tissue, which draws small cavities, situated originally at the periphery of the lung, towards the root, and gives rise to that extraordinary vagrancy of the cavernous sounds which often forms such a remarkable feature of the contraction of cavities, and will be described later on.

The periods occupied by cavities in contracting differ greatly—and vary from one month to several years. I have had instances of a cavity contracting in the course of one month, while in other cases the same process has occupied years. The part of the lung involved in the cavity greatly influences its chance of contraction, for the length of bronchus and the amount of movement which its contents are likely to be subject to, as also the freedom, for expansion, of the adjoining lung-tissue, all tell on the probabilities of a cavity healing.

For these reasons, the sub-clavicular region is found to be the most favourable for contraction, and especially the inner portion of it. Primary cavities in the axilla heal; secondary ones, according to Dr. Ewart, never: the sternal region is not favourable, and the base the most unfavourable of all.

CHAPTER VII.

PREDISPOSING CAUSES OF CONSUMPTION.

Predisposing causes general and local—Family predisposition—Consumption
proved to be hereditary—Opinions of Niemeyer, Virchow, and Waldenburg
—Hereditary Tuberculosis in animals—Offspring of gouty, syphilitic,
asthmatic, and aged parents often consumptive—Prevalence of family
predisposition—Evidence of Louis, Copland, Cotton, Fuller, Pollock, Briquet,
and the Author—Its more frequent occurrence among females than males—
Dr. R. Thompson's results—Paternal and maternal transmission—Influence on age of attack—Heredity and acute phthisis—Double heredity—
Influence on symptoms—Cases—Influence on duration—Relations affected
—Age at death—Deductions—Impure air and improper food—Continued
fevers—Scarlatina and measles—Cessation of discharges—Unfavourable
confinements and over-lactation—Mental depression—Damp—Buchanan
and Bowditch's researches—Local causes—Dusty occupations—Syphilis—
Alcoholism—Intermittent fevers—Infection—Evidence of Brompton and
Victoria Park Hospitals—Musgrave Clay's Cases—The Collective Investigation Report—Conclusions.

WE have now to consider what agencies influence the human
frame to such an extent as to render it more or less liable to
successful invasion by the tubercle bacillus, in other words,
what are the predisposing causes of phthisis.

These may be divided into (1) *general* predisposing causes,
which act chiefly by weakening the constitution, and thus
rendering it more liable than formerly to attack. Some of
these are family predisposition, want of pure air and good
food, typhoid fever, scarlatina, miscarriages, and bad confinements, over-lactation, damp soil, and mental depression.
(2) *Local* predisposing causes, which offer obstacles to the
free play of the lungs and chest wall during inspiration—
such as adhesive pleurisy, tight lacing, occupations involving
much stooping, or again pulmonary lesions tending to congestion of the lungs, and epithelial proliferation, *e.g.* bronchitis,
whooping-cough, croup, trades and occupations in a gritty
atmosphere, and injuries to the chest wall.

Family predisposition has by general consent held a very

prominent place, but the value of its influence in the causation of phthisis has been modified of late years by the fuller recognition of other causes which had been to some extent overlooked—such as damp, inflammatory attacks, &c. These and other direct sources of phthisis must exercise in our calculations a depreciatory influence on the amount we assign to hereditary transmission, and numerous cases of this disease which have hitherto been held to originate in a consumptive ancestry, will now be traced to a nearer and more direct cause. Nevertheless, no small number of cases owe their origin to hereditary predisposition, though it is not always easy to demonstrate their hereditary character. Its exact value as a predisposing agent, its mode of transmission, the varieties of the disease in which its influence is most apparent, all these and other points of interest are by no means settled questions, but still open to further inquiry.

One of the most striking proofs of the hereditary character of phthisis is the presence of tubercles, often demonstrated, in the lungs of a fœtus or of a young infant[1] of consumptive parents; another, though less striking one, is to be found in instances where a consumptive and healthy person marry, and the children become consumptive; but, on the death of the affected parent, the sound one marries again and the offspring of the second marriage is healthy.

Niemeyer only admits that the tendency to consumption is inherited if the parents were consumptive at the time of begetting the offspring. ' But it is not,' he says, ' the malady which causes inheritance, but the weakness and vulnerability of constitution, which had already laid the foundation of the

---

[1] Waldenburg and Virchow deny this as regards miliary tubercle. We would draw their attention and that of our readers to the following passage in Sir Charles Scudamore's work on *Pulmonary Consumption*, p. 55:—' I examined the body of an infant which died of extreme emaciation at the age of four months, the mother having been in the last stage of tubercular phthisis when she gave birth to it. I never witnessed so remarkable and extensive a display of tubercles, both miliary and of a larger size, the former *semi-transparent*, the latter *grey* in colour. The lungs on each side, both upper and lower lobes, the liver and spleen, the mesentery and peritoneum, were universally studded with tubercles.' The italics are our own, but the description needs no comment.

consumption in the parents, or which had arisen in them in consequence of that disease.' [1] Waldenburg denies the direct hereditariness of consumption by means of a specific contagion, and only allows, in any case, the transmission of the phthisical *habitus* and innate disposition of the parents to phthisis. Both these authors, regarding tuberculosis as a secondary product, and never a primary one, hold that tubercular consumption can never be directly inherited, but the caseous deposits from which it arises may be produced by scrofulous or phthisical *habitus*. [2]

We propose now to consider some of the principal features of the subject, and to give, as concisely as possible, the conclusions which our researches and those of others warrant us in adopting; and first, we would explain why we prefix as the heading, 'family predisposition,' instead of hereditary predisposition. As it was our object to work on a broad basis, it was thought desirable to estimate indirect as well as direct influence, and for this purpose the existence of consumption was noted in uncles, aunts, brothers, and sisters, as well as in parents and grand-parents, the principle being to accept as instances of disease in the family all those derivable from a common stock. Several phthisical uncles and aunts throw a suspicion of consumptive taint on the grand-parents, while several brothers and sisters thus affected would lead us to suspect, though not necessarily to conclude, that the parent's health was not quite sound. [3]

It may, however, be objected, that the fact of several of

[1] *Text-book of Practical Medicine*, vol. i. p. 213.

[2] *Die Tuberculose, die Lungenschwindsucht und Scrofulose*, p. 524.

[3] Waldenburg cites an instance of six brothers and sisters, who, when they first came under his notice were strong and blooming, but five out of six died of phthisis between the ages of 24 and 34, the disease beginning in many of them with hæmoptysis. The father died of a different complaint, and the mother, up to the age of 53, enjoyed uncommonly good health, being quite free from any symptoms of consumption. She was then suddenly attacked with hæmoptysis, and expired the same day, her death being subsequent to the commencement of the disease in the children. Waldenburg says the children's ill-health could not be traced to endemic causes, as they had lived in different localities, and separated from each other, and had been brought up under favourable circumstances. He concludes, 'This is a very remarkable instance where the mother, without having phthisis herself, had the disposition to consumption, and transmitted it to the children, who died of phthisis.'

the same generation being affected may be explained by a
local cause; for instance, a healthy married couple may take
up their residence on a clay soil, and their children born and
bred in this unhealthy locality may be attacked one after
another with consumption. In this instance it would be
clearly incorrect to attribute the disease to family predisposition, as the origin is probably endemic; but we must bear in
mind, that this objection applies, to some extent, in nearly all
cases of family predisposition, as when the son of a consumptive father is attacked with the disease, it is often difficult
to say for a certainty that the son's disease is hereditary and
has not been acquired, especially if, as may be the case, his
brothers are quite exempt.

We must, therefore, take into consideration all degrees of
family predisposition, and try to estimate each at its proper
worth, and we shall find a certain number of undoubted
hereditary cases.

It may be remarked in passing, that man is not the only
animal cursed with hereditary consumptive disease. Delafond
states that a phthisical ram in a flock of merinoes transmitted
his disease to sixteen or twenty of his progeny. Dr. Sanderson told me that at the great cattle breeding establishment
at Lyons, the offspring of a Scotch bull whose lungs after
death were found to be stuffed with tubercles, were also
infected with the disease, and several of the guinea-pigs who
were tuberculised in his and Dr. Wilson Fox's experiments
produced tuberculous offspring.

We must not overlook the fact, that it is not necessary for
parents to have consumptive disease in order to produce
tubercular predisposition in the offspring. The children of
very aged parents, of syphilitic, gouty, or asthmatic [1] parents,

---

[1] The tendency of asthmatic parents to have phthisical children is hardly
sufficiently recognised. Among my out-patients at Brompton I found many
instances, and our tables include several cases.

The reason of this probably is, that many cases of asthma were at their
commencement cases of limited phthisis, which have been arrested. In the
course of this process induration at the root of the lung has taken place
causing contraction of the bronchi, and giving rise to asthmatic symptoms. The
children of these patients, if phthisical at all, are generally more decidedly so,
than the parents.

or of those whose constitutions have been greatly weakened
by drink, sexual indulgence, or other debilitating causes, are
prone to phthisis. We cannot agree with Niemeyer, that
these parents are as liable as consumptive parents to beget
children who come into the world with a predisposition to
consumption ; on the contrary, the proclivity exists to a far
less extent than in those with a strong tubercular family
history.

The extent in which hereditary predisposition exists among
cases of phthisis, is a subject on which authorities differ ; but
these differences, as Dr. Fuller remarks, are attributable to
the various degrees of relationship included. The 1,010 cases
of the Brompton Hospital Report included only parents, and
gave an average of 24·4 per cent. Dr. Fuller's [1] 385 cases
embraced grand-parents, uncles and aunts, and furnished 59
per cent. Dr. Cotton's [2] 1,000 cases included parents, brothers,
and sisters, giving 36·7 per cent., and Dr. Pollock's [3] 1,200,
similarly estimated, showed 30 per cent.

Dr. Copland [4] gives 47 per cent., but does not state his
number of cases or the list of relations included. Louis,
Briquet, and Rufz base their calculations on so small a number
of patients as hardly to furnish fair evidence, but, as far as it
goes, it is confirmatory of the above estimates.

Our 1,000 cases give 484 instances, the particulars of which may here
be seen :—

    10 had grand-parents affected.
    43 „ father affected.
    67 „ mother affected.
    10 „ both parents affected.
    48 „ uncles and aunts affected.
    72 „ father's or mother's family affected (particulars unknown).
    224 „ brothers and sisters affected.
    10 „ cousins affected.
    ——
    484

Where more than one relation was affected, as was the
case in 60 of these patients, it has been the rule not to make
a double entry, but to record the nearest relative of the

[1] *Diseases of the Chest.*   [2] *Consumption*, 2nd edition.
[3] *Op. cit.*   [4] *Dictionary of Medicine.*

preceding generation, *e.g.*: 'mother and brother' affected is entered under 'mother,' 'father and sister' under 'father,' and so on. The greater number of duplicates occurred in those of the same generation as, brothers, sisters, and cousins.

Our percentage of *family* predisposition was 18·4, but, as will be seen by the above table, the number of purely hereditary cases, that is, with parents alone affected, was only 12 per cent., thus differing greatly from the percentages of Drs. Cotton[1] and Fuller[2] of the first Brompton Hospital Report, which was about 25. The only explanation we can offer of the discrepancy, is a difference in the *class* of the patients, on which our statistics are based, from those on which the above authorities found their estimate. Their calculations were based on hospital practice (except Dr. Fuller's, which included some private patients): ours entirely on the results of private practice.

Great pains were taken to arrive at an accurate result. The patients were all closely questioned as to their dead and living relations. In many instances the existence of consumption in the family was at first denied; but after cross-questioning, not only was it admitted, but undoubted cases of death from that disease were traced among their relatives.

On the whole, we think that an average of 12 per cent. for direct hereditary predisposition, and of 18 per cent. for family predisposition, are not unfair estimates for the upper classes, and that the average of the above authorities is probably correct for the lower classes; and we think it also likely that our smaller percentage in a class, which from its wealth is able to banish many of the most fertile causes of phthisis, gives a more just estimate of the influence that hereditary predisposition, unaided by poverty and exposure to divers pernicious influences, exercises on the causation of phthisis. Having discussed the prevalence of family predisposition, we now proceed to consider the following questions regarding its transmission.

[1] 24·1.

[2] 25·7.

First, the influence of *sex*.

1. Family predisposition is much more common among women than among men. This was clearly proved, as far as *hereditary* influence is concerned, by the first Brompton Hospital Report, which showed that, in this particular, the relative proportions of the sexes were as two to one. Our own cases of *family* predisposition show similar, though, as might be expected from the wider list of relations included, less striking evidence. Fifty-seven per cent. of the females, and only forty-three per cent. of the males, were thus affected. Dr. Reginald Thompson [1] has collected some important evidence on the relative proclivity of the sexes to develop hereditary phthisis from the life histories of 80 families, the parents of which appear to have succeeded in transmitting the disease. In these 80 cases paternal inheritance was present in 24, maternal in 30, double heredity in 14, and atavism or more distant inheritance in 12. The 80 families consisted of 385 children—203 male, and 182 female, and the result after they had attained puberty was as follows :—

	Males.	Females.
Phthisis developed in	98	96
Died in childhood .	21	16
Non-phthisical . .	84	70
	203	182

This certainly makes the development of hereditary predisposition the same for both sexes ; but we hesitate to accept a conclusion which is opposed to that arrived at by others, including ourselves.

The question is one of great difficulty, as it is possible that the children of phthisical parents, though not developing the disease themselves, may be the silent carriers of it to the next generation ; and, on the other hand, they may have the disease, but in a markedly acquired form. We prefer on the whole to accept the conclusion which has been arrived at by authorities investigating the matter from a broad basis in the upper and lower classes of society, and whose statistics were

[1] *Family Phthisis*, p. 42.

not collected specially to determine the sexual question, which is that family predisposition is more common among females than males, a conclusion which, apart from statistics, every-day experience abundantly confirms.   In considering the hereditary tubercular influence on the two sexes, we must bear in mind that in acquired phthisis all evidence goes to show that the male has a power of resistance to attack superior to that possessed by the female, and that early susceptibility and feeble resistance are the characteristic of, what has been justly termed, the weaker sex.   As we shall see later on, this results in a larger number of acute cases, in a greater fatality, and in a shorter duration of life generally, for female consumptives.

2. The transmission of phthisis is more common through the mother than through the father.[1]   Mr. Ancell states that, as far as limited statistics go, the evidence on this point is strong.   We are glad to be able to confirm this statement, from our own statistics, which, in one thousand cases, record sixty-seven consumptive mothers against forty-three consumptive fathers.   Mr. Ancell remarks: 'Regarding it [phthisis] as a disease of the blood, this result might have been predicated, since there is only one period at which the father's influence could be exercised, viz., that of conception; whereas the influence of the mother is exercised at that period, and also through uterine gestation.'

3. Fathers transmit more frequently to sons, and mothers to daughters, than the converse.   This important fact was first established by the first Brompton Hospital Report, and having been amply confirmed by numerous competent observers, may now be said to be matter of every-day experience.

How does family predisposition influence *age* in consumption?   This is a very important question, bearing intimately on the prognosis of the disease, and deserves a fuller discussion than is to be found in most works.   Our researches[2] have, we trust, cleared up some of the mystery in which the connection of age and family predisposition was enveloped.   We

---

[1] *A Treatise on Tuberculosis*, p. 385.
[2] 'Duration of Phthisis, and certain Conditions which Influence it,' *Medico-Chir. Trans.*, vol. liv.

have demonstrated clearly that the chief influence which family predisposition exercises is on the aged at which the patient is likely to be attacked.

It hurries the onset of the disease. This conclusion was arrived at by careful investigation of the age at which the one thousand private cases were attacked. This having been ascertained, we pursued a similar inquiry with regard to those affected with family predisposition, and lastly, with reference to those who were free from it, as far as could be ascertained at the time of their being under observation, that is to say, who up to the last date had had no relatives affected with phthisis. The result was as follows :—

	Average age of attack	No. of Cases
Average age of attack in total males	29·47	625
,, ,, ,, females . . .	26·06	375
Average age of attack in males free from Family Predisposition	30·03	355
Average age of attack in females free from Family Predisposition	28·05	181
Average age of attack in males affected with Family Predisposition	27·07	270
Average age of attack in females affected with Family Predisposition	21·51	214

This table shows that the average age of attack in the number affected with family predisposition of both sexes was earlier than the average age of attack in the whole number by $2\frac{1}{2}$ years in the males, and by $4\frac{1}{2}$ years in the females. There is also shown that it was earlier than that of those free from family predisposition by 3 years among the males, and by $6\frac{1}{2}$ among the females.

This is very striking; but, remembering that these statistics were of the upper classes only, and might only apply to certain circumstances connected with them, I determined to pursue the same inquiry among the lower classes, and for this purpose I had recourse to my Brompton note-books. Among 400 out-patients the result was as follows :—

	Average age of attack	No. of Cases
Average age of attack in males free from Family Predisposition	32·01	100
Average age of attack in females free from Family Predisposition	28·47	100
Average age of attack in males affected with Family Predisposition	24·64	100
Average age of attack in females affected with Family Predisposition	24·00	100

The figures confirm the former results as regards the age
of attack being rendered earlier by family predisposition, but
the relative influence on the two sexes differs greatly from
what we found among the richer classes. In the out-patients'
class the age of attack was about the same for males and
females, the males being attacked earlier than among the
rich, the females later. The age of attack in those free from
family predisposition was for both sexes considerably later,
and the influence in the male sex greater than in the female—
a result exactly opposite to that obtained among the upper
classes. This remarkable influence of family predisposition
in hurrying the onset of the disease, does not seem to have
attracted much attention; and the only authority I could
find who seems to have noticed it was M. Briquet,[1] who, in the
smaller number of 95 cases of consumption, arrived at the
conclusion that 'hereditary tuberculosis develops itself in the
form of phthisis at an earlier period of life than the disease
does when acquired. In 89 of the cases with the history of
hereditary transmission, 26 became phthisical before 30 years
of age; while of 56 cases born of perfectly healthy parents,
31 did not become phthisical until after 30.'

M. Briquet's cases were rather too few to afford strong
evidence, and he makes no attempt to separate the sexes, and
trace the influence in each sex; but we have little doubt that
his observations, as far as they went, were correct, and that
the fact 'that family predisposition hurries the onset of
phthisis,' as ascertained by his and our own researches, is one
not to be disputed.

[1] 'Recherches Statistiques sur l'Histoire de la Phthisie' (*Revue Médicale*, 1842).

Dr. Reginald Thompson[1] has by his admirable and laborious investigation of the records of the Brompton Hospital not only confirmed the above conclusion, but carried it a step further, and demonstrated other results of phthisical heredity. He finds that the effect of double heredity in advancing liability is greater than that of either single heredity; in other words, that those affected with double heredity are more susceptible to the disease than those with single, and that this applies to both sexes.   Dr. Thompson has also carefully investigated the influences of the various forms of heredity in each sex; and, as he has gone more deeply into the question than former writers, we propose to quote some of his conclusions.   The chief interest of the inquiry lies in the relative influence of the paternal, maternal, and cross inheritances in the children; and the careful comparison which Dr. Thompson has made of these with the features of acquired phthisis in both sexes renders his conclusions the more trustworthy.   The effect of *paternal inheritance* in *males* is (1) to extend the liability to the disease over all periods of life, increasing the susceptibility between the ages of 10 and 25; (2) to render the disease far more acute and fatal; (3) to diminish the tendency to large hæmoptysis and to hæmorrhage, as a rule, except during the susceptible period.   The influence of the *paternal inheritance* in *females* is to promote the early development of the disease, but to shorten the length of period for liability, limiting it to 50 years of age, and to superadd in some cases a certain amount of resistance to the disease, which is more the characteristic feature of the male than of the female, and is probably a distinct heirloom of the stronger sex.

The influence of the *maternal inheritance* of phthisis in the *sons* is exceedingly strong, and shows itself (1) in the period of liability being extended over the whole period of life; (2) in early susceptibility to the disease; (3) in the more acute and more fatal character of the malady, and (4) in the greater frequency of hæmoptysis.   The *maternal influence* on the *daughters* is displayed by a decided hastening of the age of

[1] *Family Phthisis*, p. 25.

attack, and by the large number of acute cases and deaths occurring between the ages of 15 and 20. Comparing the relative influence of father and mother on the children, it would appear that maternal inheritance is the worst for children of both sexes. It influences the female in an adverse manner to a greater degree than the paternal influence, and as the inheritance from the father to the daughter includes the engrafting on her of some of the resisting power to disease of the stronger sex, so the inheritance from the mother to the son, as involving female characteristics, is far worse than for the daughter.

Double heredity, *i.e.* both parents being affected with consumption, exercises a more unfavourable influence than single heredity, and, as has been before observed, the age of attack is lowered, the disease is rendered more acute and more fatal, also the tendency to hæmorrhage is increased, and all these features are more marked in the case of males than of females.

Let us now consider whether the presence of family predisposition exercises any decided influence over the type of phthisis. Have cases with family predisposition any distinguishing features ?

We must confess that it is hard to trace any feature which cannot also be found in other cases of consumption. In many instances, great transparency of skin, with veins clearly visible, and a delicacy of outline, is noticeable ; in others, marked defect in development, or else distortion, of the thorax ; in many, glandular enlargements come on at an early age. These and other features are to be found, often strongly marked, in hereditary cases ; but they are not invariably present, and, on the other hand, they are to be seen occasionally in non-hereditary cases. Dr. Pollock remarks that in acute cases of phthisis the influence of hereditary predisposition is undoubted, and he says that of one hundred and seventy-nine cases, only thirty-four could positively state that there was no family taint either parental or remote.

Dr. Reginald Thompson's evidence already quoted points to a large number of acute cases, and to a high mortality within the first eighteen months of the disease. As our patients include cases of one year's duration and upwards, it is curious that they do not display this feature, and we can only account for the longer duration by the difference of class, and consequently of care and of protection from fresh exciting causes received, which doubtless had the effect of retarding the rapid progress of the disease.

A characteristic we have noted in many cases of hereditary disease is insidiousness of onset. Often the symptoms of wasting, night-sweats, and even pyrexia precede the cough and expectoration, and the physical signs, at first absent altogether, are more likely to be first found in the supra-scapular regions than in the sub-clavicular, and, when they appear, they indicate rapid changes. Not uncommonly the heredity shows itself in the same lung being attacked in the parents and children, and, as a rule, the sequence of symptoms is similar. I can call to mind several examples of both paternal and maternal inheritance, where the same lung was the first attacked in father or mother, and in several sons and daughters consecutively, and where the disease took the same course in them all. We must bear in mind that one effect of hereditary predisposition hurrying the onset of the disease is, that the children die at an earlier stage than the parent from whom they inherit, and cases do occur where the children have died of the hereditary disease before their parents have shown symptoms of it. In this case the parents have transmitted the disease, and have later on developed it in their own systems.

We subjoin a few well-marked instances of hereditary origin, some of which were attacked early and some later in life. We think they will be found instructive, as showing that the strongest hereditary taint does not prevent the beneficial effect of remedies, if persevered with.

CASE 1.—A lady, aged 34, first consulted Dr. C. J. B. Williams, June 20, 1859, Had lost her father, mother, and ten brothers and sisters from consumption, and had herself been always liable to cough, and had one constantly since December. Three years previously she had hæmoptysis, amounting to a tablespoonful. At the time of her visit the expectoration was streaked with blood. She had lost much flesh and strength, complained of pain in her chest, catamenia irregular and deficient. *Dulness and tubular sounds in the upper part of both sides of chest, most marked on the right, where there was crepitation.*

—Ordered cod-liver oil in a mixture of hydrocyanic and phosphoric acids, with infusion of calumba and orange; counter-irritation with acetum cantharidis, and a linctus containing morphia.

May 21, 1860.—Greatly improved under the above treatment, and has grown stout. Has hardly any cough or expectoration, but lately suffers from oppression of breathing and frequent boils.—Ordered oil in a mixture of chlorate of potash, nitric acid, and glycerine.

July 30, 1861.—Continued to improve till the winter, when had inflammation of the lung; and since the attack the expectoration has been sometimes gritty and sometimes fœtid. Is still stout and strong, but breath short. Has taken oil regularly, combined with strychnia. *Dulness, tubular sounds in upper left back and right front.*

April 19, 1866.—Has always wintered in Cornwall, and out a great deal in the open air; but lately stomach weak, and has not been taking much oil. Has lost much flesh; but cough and expectoration are less than during last winter, when they increased, and there was some hæmoptysis. Suffers from piles. *Dulness, large tubular sounds in upper right chest.*—To take oil, with phosphoric acid and hypophosphite of soda. An electuary of sulphate and bitartrate of potash.

October 1, 1870.—Returned to Cornwall, and improved again so much that for the last two years has taken no oil. Last February took cold, and cough again returned, and has been troublesome since, with frequent sickness. In July went to North Wales, there caught fresh cold, and has been very ill ever since, with sickness, pyrosis, disgust at food, and great loss of flesh and strength. Urine pale and copious; sp. gr. 1010; no albumen. A number of red scaly spots on skin. Cough, and opaque expectoration. *Dulness and tubular sounds in both upper regions. Collapse under left clavicle, with defective breath. Crepitus in lower half of right lung.*

Heard of her death in December, about fourteen and a half years after her first marked symptoms.

CASE 2.—A young gentleman, aged 16, whose mother and sister had died of phthisis, saw Dr. C. J. B. Williams, August 28, 1855. He had had repeated hæmoptysis since Christmas, sometimes amounting to an ounce, and also cough, with some loss of flesh, and lately pain in right shoulder. *Dulness and tubular sounds above right scapula.*—Oil ordered, with sulphuric acid, and infusion of roses, and acetum cantharidis liniment.

September 29, 1856.—Wintered at Hastings, and improved in flesh, strength, and health. Oil has been taken steadily, except for about three weeks, when it was omitted, and he lost flesh. Hæmoptysis has occurred several times, generally after some exertion, and once amounted to ℨvj. *Dulness, tubular and obstructive sounds in upper right chest, with subcrepitus.*

October 9, 1857.—Was ailing in May, but now better and free from cough. In last month has suffered from lowness of spirits, sleeplessness, and confused head. *Chest clearer, but still same tubular sounds, with obscure breathing in upper right chest.*

January 11, 1859.—Sleeplessness gradually improved, and he spent spring at home; but, after three months in the Regent's Park, cough increased, and hæmoptysis came on to the amount of ℨij. *Some dulness, tubular sounds in upper right chest, crepitation sounds above scapula.*

July 11.—Went soon after to Madeira, and rode out a great deal. Lost cough, and ascended the Peak of Teneriffe. Now strong, ruddy, and able to walk six miles.

October 16, 1861.—Last two years have been spent in London. Has been free from cough and hæmoptysis, and his condition has been good, but breath short. Oil has been taken regularly for nearly two years. *Still loud crepitation, left front and right back, with dulness and tubular sounds above right scapula.*

September 7, 1863.—Had been studying at Cambridge for a year, when had hæmoptysis, ℥iij., after bathing; and pain in right chest. After Easter, 1863, nursed father (who died of Bright's disease), and had hæmoptysis to ℥ij. daily, for a week in July; has left off oil and is taking cream, and is much reduced in strength, though not in bulk. *Much dulness, and obstructive sounds upper right back. Tubular sounds both upper sides of chest.*

April 8, 1864.—At St. Leonards; wonderfully improved in flesh, colour, strength, and breath. Walks fourteen miles at a time. Oil in sulphuric acid has been taken all the winter.

October 4, 1864.—Improved at Tunbridge Wells, and was able to walk ten miles briskly: physical signs became drier. He then returned to keep terms at Cambridge. Remained well till on a visit to Liverpool, had hæmoptysis to amount of ℥iv. In February 1865 was sent to Madeira, and gained strength and 7 lbs. in weight, being much out riding; then spent summer in Hants, and returned to Madeira in following winter; remained there till April, and had fever and congestion of the lungs. Since then weaker, with more cough and expectoration, and often chills and heats, and has some expectoration, sometimes opaque, sometimes calcareous. *Tubular and crepitation sounds on both sides posteriorly, most on the right.*

October 15, 1867.—Remained at St. Leonards till February, then went to Mentone till May, and gained 8 lbs.; improving also in strength and appetite. Had been taking oil, ℥vj., once a day. In warm weather improved in weight, strength, and breath, but worse lately. *More crepitation and obstructive sounds in left front and right back.* Nearly thirteen years have elapsed since first attack.

CASE 3.—Miss V., aged 16, who had lost her mother and sister from consumption, was seen by Dr. C. J. B. Williams, March 22, 1859. Had cough and cold in December, which have continued, though to a slight extent, ever since with opaque expectoration, and yesterday some blood. Breath short. *Dulness, large tubular sounds, and gurgle in upper right chest, front and back.* Oil ordered in tonic of phosphoric and hydrocyanic acids, with calumba and orange tinctures, the use of acetum cantharidis liniment, and a morphia linctus.

June 10, 1860. Taken oil, &c., and much better. Last summer had measles; cough was violent for three weeks, but subsided under the use of oil, and has remained moderate since. Wintered at Cannes, and out much. *Still decided dulness, and marked tracheal sounds in upper right chest.* After measles, coarse crepitation was heard over this region.

October 16, 1862.—Well, and free from cough till a month ago, when it was violent for two weeks, and subsided under oil and iodine paint. Now has no cough or expectoration. *Dulness at and above right scapula, with loud tubular sounds.* Back rounded.

1871. Not seen since; but heard of as tolerably well, having passed a winter at Mentone. Twelve years after first symptoms.

CASE 4.—Mr. N., aged 42, consulted Dr. C. J. B. Williams, July 31, 1860. He had lost his father, mother, and a brother from consumption, and had been subject to cough for the last seven years. Two years ago he had hæmoptysis ℥ij. and breath has been shorter ever since. Cough has been constant for the last nine months and hæmoptysis to amount of ℥ss. has occurred once. Has taken oil regularly (℥ss. ter die) for nine months. *Dulness and tubular sounds in upper right chest. Tubular sounds above left scapula.*—Oil was ordered with phosphoric acid and tinctures of calumba and orange; counter-irritation with acetum cantharidis, and a morphia linctus.

April 5, 1867.—Has kept pretty well and stout, but breath has always been short, and he has cough, with opaque expectoration, which have been less in last winter, but patient is hoarse. No oil taken last year.

Still dulness upper right. tubular sounds above scapula. Sub-crepitus and weak breath below.—Ordered oil, with nux vomica, phosphoric acid and quassia.

June 8, 1868.—Tolerably well, but always has slight cough, and some hæmoptysis occurred last spring; oil taken last winter. Sputa opaque and tinged with blood. *Some dulness at and above left clavicle; tubular sounds, but breath weaker above right scapula; crepitation down left front.*

Here the strong family predisposition did not tell till comparatively late in life, and did not prevent the remedies used from having a beneficial effect. The patient was about eight years under observation, and the disease seems to have diminished in right lung, though some increase took place in the left.

CASE 5.—Mr. R., aged 17, whose father died of consumption, saw Dr. C. J. B. Williams, May 7, 1857. For three years he had had swelling and discharge from the cervical glands, and cough off and on for two years; lately has had cough for six weeks, accompanied in last fortnight by pain in the side, and short breath. The scrofulous sores in the neck are still discharging. Oil taken till two months ago. *Dulness, obstructed breath, and much crepitation in front of left chest, mostly in upper portion and above the clavicle. Some dulness above the right scapula.*

Oil ordered in nitric acid and tincture of orange, and a liniment of strong tincture of iodine.

April 13, 1864.—Took oil, and states that he got quite well in six months, and then omitted oil, except an occasional dose. Throughout last winter had cough and expectoration, and three times brought up blood to the amount of ℥ij. Has lost much flesh and strength. *Dulness, cavernous sounds in upper half left chest, with obstruction sounds below.*

What effect has family predisposition on the duration of phthisis? Is the duration curtailed by it? Do patients thus affected die earlier than other consumptive patients? These questions are of considerable importance in the prognosis of the disease; and we have attempted to answer them in the following statistical results, extracted from our tables:—

Average Duration from the Commencement of the Disease in 198 Deaths.

	Yrs. Mths.	No. of Cases
Average duration of cases affected with Family Predisposition	7— 5·8	87
Average duration of cases apparently free from Family Predisposition	7–10·9	111
Average duration of total number of deaths . .	7—8·7	198

Here it is shown that the average duration among the 87 deaths of those affected with family predisposition was 7 years 5·8 months, an average not greatly differing from that of the cases free from this influence, or again from that of the total deaths. A difference of only a few months is noticeable, and does not indicate that family predisposition exercised any decided influence in shortening the duration of the disease.

The evidence of 397 living cases of family predisposition supports the same conclusion; for among these the average duration already reached was 7 years 11 months; that of all the cases free from predisposing taint being 8 years 7¼ months, and of the total living 8 years 2 months.

It would seem, therefore, that family predisposition, without reference to sex, exercises but a slight influence over the duration of phthisis.

Does it influence the duration of the disease in one sex more than another?

We have seen that the age of attack is influenced differently in the sexes, and also that phthisical females live a shorter time than phthisical males; and the following question should now be solved: Is the duration of the disease shorter among phthisical females affected with family predisposition than in others not so affected? This may be seen by a glance at the table below :—

Relative Duration of Disease in the Two Sexes in Family Predisposition.

	No. of Cases	Dead	No. of Cases	Living
		yrs. mo.		yrs. mo.
Males .	54	8 — 2	216	8 — 2·7
Females . .	33	6 — 7·3	181	7 — 6·7
Total . .	87	Total .	397	

The inspection of this table will show that neither sex is influenced as to the actual duration of the disease, and that family predisposition does not curtail it—a conclusion which contrasts with that arrived at with reference to the age of attack; and is entirely at variance with the old views of hereditary phthisis being more rapidly fatal than acquired phthisis.

To this conclusion it might be replied, by those who still believe in the curtailing influence of hereditary taint, 'Perhaps family predisposition, as a whole, does not influence the duration of phthisis, but do not some of its various degrees do so?' We had hoped to answer this point satisfactorily, but, unfortunately, lack of materials prevents our doing it completely. Such information as our cases afford is to be found in the following table :—

	Dead	No.	Living	No	Total
	Average yrs. mo.		Average yrs. mo.		
Grand-parents affected .	[1] 16—7·00	2	[1] 10—6·62	8	10
Father „ .	[1] 7—5·28	7	7—9·25	36	43
Mother „ .	8—3·54	11	7—10·75	56	67
Both parents „ .	—	1	[1] 6—9·11	9	10
Brothers and sisters .	7—6·02	44	8—2·24	180	224
Uncles and aunts . .	[1] 12—6·00	4	7—6·18	44	48

The evidence here furnished is chiefly of a negative kind, and must be viewed as proving that some forms of predisposition do not curtail the duration of the disease, and not as demonstrating what influence other forms may have on it. We see in 11 deaths where the mothers were consumptive the duration was 8¼ years, that 56 living patients similarly situated had lived 7 years 10¾ months. Among these few cases maternal influence had no effect in curtailing the duration, which in the 11 deaths was rather higher than the ordinary average.

In the case of brothers and sisters, where the numbers warrant our speaking more decidedly, what do we find?

[1] Numbers too small to yield fair average.

In 44 deaths the duration was 7 years 6½ months, slightly below the whole average of deaths; while of 180 living cases in this category, the mean duration was 8 years 2¼ months, about the same as the common average. Here again family predisposition seems to have exercised little or no influence. The other numbers are too small to furnish even negative evidence.

Our last point to consider is the *age* which these patients affected with family predisposition live to. Did they die earlier than other consumptive cases? The grounds for determining this question have been already to some extent settled, the age of attack and the duration in these cases having been ascertained, and therefore the following result was easily arrived at:—

		No. of Deaths
Average age reached by males free from Family Predisposition	41·51	65
Average age reached by males affected with Family Predisposition	35·29	51
Average age reached by females free from Family Predisposition	34·92	46
Average age reached by females affected with Family Predisposition	30·71	33

Family predisposition, therefore, though it does not materially curtail the duration of the disease, has considerable influence in shortening the duration of life, such limitation amounting in males to more than six years, and in females to not quite five years. But we must remember to assign this cutting short of the span of life to its proper cause—not to hereditary phthisis being more virulent and rapid in its progress, for that idea has been disproved—but to the fact that those who come of a consumptive stock are liable to be attacked earlier than others whose families are free from taint.

Our conclusions on the subject may be briefly summed up:—

1. Family predisposition prevails among, and exercises a more decided influence on, females than males; and the

former have a greater power of transmission than the latter.

2. Fathers transmit more frequently to sons, and mothers to daughters, than the converse.

3. Paternal inheritance, whilst most unfavourable for the males, is less so for females, as it generally includes an increase of resisting power.

4. Maternal inheritance is unfavourable for both sexes, but most so for the males.

5. Double heredity exercises the greatest influence, and affects sons more strongly than daughters.

6. Family predisposition does not directly shorten the duration of the disease.

7. It precipitates the onset of the disease, and thus shortens the duration of life.

We have entered thus fully into the relation which family predisposition bears to phthisis, because we believe the subject has hitherto been considered as rather obscure, and not because we think the other causes of phthisis less important. But the action of these is better recognised, and therefore needs but a passing mention.

Impure air and *improper food* are well-known general causes of consumption. Among the lower classes, in crowded cities, evidence of their effects is only too common. Of 3,211 men who became inmates of the Brompton Hospital in ten years, 1,812 (more than half) had in-door employments. Among my own out-patients, the numbers of those who plied their trade in close, ill-ventilated rooms, and during long hours, were very great. Clerks, compositors, tailors, shoemakers, among men ; and milliners, dressmakers, among women, were attacked at an early age.

Want of proper food acts by impoverishing the blood and lymph, whether the fault lies in the quality or quantity of food. If healthy pabulum be not supplied for its nutrition it is impossible for the tissue to retain its normal standard ; it starves, and the organ, of which it forms part, fails to fulfil its proper function, and the system becomes the prey of the first attack. Instances of the action of this predisposing cause can be seen in the poorly fed children who attend at the Brompton Hospital, and in whom a very slight exciting cause brings on the consumptive disease. An excellent instance of the effect of starvation in producing phthisis is given by Peter,[1] in a man who produced stricture of the œsophagus in himself by swallowing strong sulphuric acid, and took so little food that he wasted rapidly away and died. After death his lungs were found stuffed with recent tubercle, and in parts excavated and the intestines full of ulcerations; the whole disease having supervened since the injury.

Typhus and typhoid fevers.—Of these, typhus is the least powerful cause, though it exercises some influence; but many cases of consumption are to be traced to an attack of typhoid or pythogenic fever. Murchison says, 'an attack of pythogenic fever is often followed by tubercular deposit in the lungs.' And, again, 'in my experience, acute tuberculosis of the lungs is a far more common complication or sequela of pythogenic fever than of typhus; and it is intelligible why this is the case, when we recollect the more protracted duration of the former malady, and the greater emaciation it entails. Louis records four fatal cases of pythogenic fever, in which the lungs were found studded with recent tubercles. Bartlett also observes that consumption is a common sequela of this fever in America.'[2] Gueneau de Mussy, Villemin, and other French writers question the genuineness of the typhoid fever in these instances, and suggest that the cases were really ones of acute tuberculosis.

It is worthy of notice, that the organs so commonly affected in typhoid fever—viz., the solitary glands, Peyer's patches, and the mesenteric glands—are also frequently the seat of tubercular disease. The wasting, too, which accompanies both these diseases seems to be connected with a disordered state of the lymphatic system, particularly of the lacteals. We know that both typhoid fever and the intestinal forms of consumption largely affect portions of the lymphatic system, but the difficulty of distinguishing between the lesions of the two diseases during life has been solved by the dis-

[1] *Clinique Médicale.*   [2] *Treatise on Continued Fever.*

covery of the bacillus tuberculosis, which can be detected in the stools of intestinal phthisis, and is quite distinct from the micrococcus of typhoid fever.

Scarlatina and measles are very common causes of consumption, chiefly in children. Acting generally on the system, they exhaust the patient, and leave him an easy prey to the first exciting cause that comes. Acting locally on the lungs, by their sequelæ, bronchitis and inflammation, they embarrass these organs with consolidations and epithelial aggregations which prove good feeding- and breeding-grounds for the bacillus.

The *cessation of habitual discharges*, as those from fistulæ in ano and old ulcers, will sometimes give rise to symptoms of consumption, which often diminish on the discharge being re-established. In fistula this is so marked, that many physicians refuse to sanction, and many surgeons to perform, operations for fistula on patients who have shown evidence of consumptive disease of the lungs. How often has a successful operation for fistula been followed by the eruption of miliary tubercle in the lungs!

Miscarriages, bad confinements, and over-lactation are fertile causes of phthisis among the poor, chiefly through the great exhaustion consequent on them.

Mental depression.—Some doubts have been expressed by authors as to this exerting any influence in the causation of consumption. When mental depression arises from any great loss, whether of relatives, friends, or property, it is often followed by irregular habits. Food is not taken regularly, nor in sufficient quantity, and, on the other hand, stimulants are often taken too freely. In these cases, it is doubtful whether we ought to assign as the cause the mental depression, or the irregular living accompanying it. Laennec[1] gives an interesting instance of the effect of mental depression.

'I had under my own eyes,' says he, 'during a period of ten years, a striking example of the effect of the depressing passions in producing phthisis,

---

[1] *Diseases of Chest*, Sir John Forbes's translation, p. 331.

in the case of a religious association of women, of recent foundation, and which
never obtained from the ecclesiastical authorities any other than a provisional
toleration, on account of the extreme severity of its rules. The diet of these
persons was certainly very austere, yet it was by no means beyond what
nature could bear; but the ascetic spirit which regulated their minds was such
as to give rise to consequences no less serious than surprising. Not only was
the attention of these women habitually fixed on the most terrible truths of
religion, but it was the constant practice to try them by every kind of
contrariety and opposition, in order to bring them, as soon as possible, to an
entire renouncement of their own proper will. The consequences of this
discipline were the same in all; after being one or two months in the establishment, the catamenia became suppressed, and in the course of one or two
months thereafter phthisis declared itself! . . . . During the ten years that I
was physician of this association, I witnessed its entire renovation two or
three different times, owing to the successive loss of all its members, with the
exception of a small number, consisting chiefly of the superior, the grate-
keeper, and the sisters who had charge of the garden, kitchen, and infirmary.
It will be observed that these individuals were those who had the most
constant distractions from their religious tasks, and that they also went out
pretty often into the city on business connected with the establishment.'

Damp.—A damp atmosphere may be generated either by
moisture brought to the locality through the prevalence of
certain winds, or by the impermeable nature of the soil underlying it, causing the accumulation of moisture on the surface.
Whether a damp atmosphere generated in the first-mentioned
way gives rise to consumption, there is as yet no decided
proof; but of its origin in the last-mentioned way, viz., from a
damp soil, the investigations of Drs. Buchanan and Bowditch
leave no room to doubt. Dr. Buchanan was appointed by the
Privy Council to investigate the effects on the public health
produced by the improvements lately made in the drainage,
water-supply, &c., of certain towns. He found, with regard
to phthisis mortality, that its diminution or non-diminution
depended on whether the sanitary improvements of the place
had or had not included any considerable drying of the soil.
In fifteen large towns where a diminution had taken place
after the improvements, the death-rates from phthisis had
fallen some 11 to 20 per cent.; in others, 20 to 30 per cent.;
and others again 30 to 49 per cent.; and in many towns this
diminution of deaths from phthisis formed the principal sanitary amendment. 'This,' as Mr. Simon [1] says, ' is extremely

[1] *Report of the Medical Officer of the Privy Council for* 1867.

interesting and significant, when it is remembered that works
of sewerage by which the drying of the soil is effected must
always precede, and do indeed sometimes precede by years,
the accomplishment of other objects—house-drainage, abolition
of cesspools, and so forth—on which the cessation of various
other diseases is dependent.'

These results naturally directed attention to the influence
of the soil in the distribution of consumption, and led Dr.
Buchanan to institute a farther inquiry on this point. By a
careful comparison of the geological formations of the registra-
tion districts of the counties of Surrey, Kent, and Sussex, with
the death-rates from phthisis in these districts and by an
elimination of all probable chances of error, he arrived at the
important conclusion that *wetness of soil is a cause of phthisis
to the population living on it.* This conclusion was supported
by the evidence of the Registrar-General of Scotland with
regard to that country; and by that of Dr. Bowditch of
Boston, U.S., who in 1862 had drawn attention to the inequality of the distribution of phthisis in the State of Massachusetts, and to the connection of this inequality with
differences of moisture of soil. He cited the written statements of medical men resident in 183 towns, which tended to
prove the existence of a law in the development of consumption in Massachusetts, that dampness of the soil of any
township or locality is intimately connected, and probably as
cause and effect, with the prevalence of consumption in that
township or locality ; and he also adduced particular instances
as demonstrating that even some houses may become the foci
of consumption, when others but slightly removed from them,
but on a drier soil, almost wholly escape. The following instance of this law came under my notice, one of the family
being for some years under my care.

The rector of a parish in Essex resided on a clay soil, and
had a large pond immediately in the neighbourhood of his
rectory. He and his wife have always enjoyed good health,
and there was no hereditary disease traceable, either in his own
family or his wife's. Of their twelve children, eight were
born in the rectory, and four in a neighbouring parish : but

G

all spent childhood and youth at their father's house. Six
have died; four of consumption, one of scrofulous disease of
the spine, and one of whooping-cough at the age of 5. Of
the six alive, three are healthy, one is delicate, but I have not
heard from what cause; two have scrofulous disease of the
spine. The three healthy ones have been but little at home
since they have grown up, and one spends much time in
travelling. So that out of twelve children, there are no less
than four cases of consumption, and three of scrofula.

This seems to me a fair instance of phthisis arising from
endemic causes; the social position of the family, who are
rich, precluding many other causes, which we have been discussing, from entering into consideration. We may therefore
conclude that dampness of soil is an undoubted cause of consumption; and in our preventive treatment of the disease we
should aim at either the drainage of the soil, or removal of the
inhabitants to a drier locality.

Local causes act, not so much by predisposing *the system*,
but *the lungs* to tubercular attacks, that is to say, by local
irritation they give rise to alveolar epithelial proliferation,
and to the local pulmonary congestion, which appears to be
the favourite feeding ground for the bacillus tuberculosis.
Such are—

*Trades and occupations giving rise to a dusty or gritty
atmosphere.*—A life pursued in an atmosphere abounding
in small particles of flint, or iron, or coal, or cotton, or flax,
or straw, as is the case with stonemasons, potters, fork-
grinders, needle-grinders, cotton-carders, and chaff-cutters, is
shown by Drs. Greenhow, Peacock, and others to be a short
one; and the cause of death was generally found to be
tubercular phthisis, induced by constant inhalation of the
irritating particles. These have been detected chemically
and microscopically in the lungs. They seem to set up irritation in the larger bronchi, causing thickening; and also in
the lung-tissue, causing induration and consolidation. It was
at first doubted whether these lesions were tuberculous; but
the presence of both grey and yellow tubercle, and the tendency
of the consolidations to soften and form cavities: and above

all, the detection of the bacillus tuberculosis in the lesions, sufficiently demonstrate their consumptive nature. Dr. Greenhow calculated that forty-five thousand deaths occurred from these causes in England and Wales; and he clearly showed that the whole of this mortality was preventible by the introduction of better methods of ventilation and working. Since his reports, Acts of Parliament have been passed which, if properly enforced, ought to totally abolish this cause of phthisis; and it is to be hoped that ere long we may no longer be able to number ' *dusty occupations* ' among our *causes* of consumption.

The rôle that *syphilis* plays in the causation of phthisis has been much debated, and it is undoubtedly the fact that many of the cases now held to be characteristic examples of syphilitic disease of the lungs, as proved by the presence of gummata and specific ulcers, were formerly included, at any rate during life, under the head of phthisis. I have always held that one of the best methods of diagnosing these cases was by treatment. When a case of pulmonary syphilis is submitted to specific treatment by mercurial baths or iodide of potassium, the effect is often magical, cough and expectoration diminish, consolidations of long standing clear up rapidly, and the greater portion of the symptoms vanish. In the case of ordinary phthisis no result follows specific treatment. It is not uncommon, however, for syphilis to be present in cases of undoubted consumption, and the complication is one of the most troublesome we are called upon to deal with. Too often it is mixed up with laryngeal phthisis, and manifests itself in extensive ulceration of the tongue, pharynx, and larynx, differing from the tubercular in the size and shape of the ulcers, and the parts of the larynx affected, and these cases are most intractable. Sometimes syphilis shows itself in the character of the cough—due possibly to enlargement of the bronchial glands—which is remarkably convulsive and brassy, and unattended by expectoration.

The question however is, Does syphilis bear any relation to the causation of phthisis, in the same way as damp, fevers, and other recognised predisposing causes? I am inclined to agree with Lancereaux and Fournier, that it acts as a depressing constitutional cause, and predisposes most powerfully to the bacillar attack. It is not uncommon to see a young man, who has been long a subject of constitutional syphilis, finally become the subject of phthisis, though it must be admitted that this is much more common among the lower classes than the upper, and it therefore might be inferred that want and exposure to weather may have assisted in the predisposition.

Fournier [1] states that the course of syphilitic phthisis is usually a slow one, unless other organs than the lungs become involved, when it becomes rapid. He distinguishes three varieties of the disease : (1) a latent form, where the pulmonary lesions are circumscribed and give rise to few symptoms during life ; (2) another, where lesions suffice to cause some distress of breathing, but not to produce change in the general condition, which seems excellent. The patient thrives and fattens, but complains of slight cough, expectoration, and dyspnœa, and the physical signs indicate a limited induration or excavation ; (3) a third variety combines severe local and general symptoms, and the physical signs show all the features of ordinary advanced consumption. It is not necessary here to lay down rules for diagnosis between syphilitic disease of the lungs, pure and simple, and syphilitic phthisis, but the prognosis and treatment of the two classes of cases are obviously very different.

Alcohol.—In spite of Dr. Walshe and other eminent authorities, I have no hesitation in assigning a large amount of phthisis to the predisposing cause of alcohol. My case books contain numbers of records of publicans, barmen, and barmaids, and especially of commercial travellers, whose habits of excess in this direction are notorious, and the greater part of these have consulted me for the lesions of consumption. It is probable that the number of deaths from consumption among drinkers would be greater, were they not carried off early by the more immediate consequences of their habits, such as

[1] *Gazette Hebdomadaire de Médecine et de Chirurgie*, November and December, 1875.

delirium tremens, gout, kidney and liver disease, and insanity, &c.; but it is the more gradual imbibers who contract phthisis, the change apparently commencing with loss of appetite and inability to take sufficient nourishment, which soon produces wasting, and the loss of all power to resist vicissitudes of weather. A slight catarrh in many of these cases leads to tuberculisation and rapid excavation of the lungs. It is extraordinary how easily in the lungs of topers the tubercular masses break down, and how rapid the course of the disease proves. Beel of New York, Lancereaux, and Hérard and Cornil are all strongly of opinion that the immoderate use of alcohol predisposes to phthisis.

Intermittent fevers.—These have been classed by Boudin, Barth, Hahn, and some of the American physicians as antagonistic to phthisis, and these authors point to a small phthisis mortality in several districts where intermittent fevers are most rife and deadly. It is possible that where the malarial poison is very strong, the class of individuals who would some time or other perish from phthisis fall early victims to malaria, and thus reduce the phthisis mortality. This probably applies to tropical countries, though even here, as in Peru, both forms of disease are found side by side in the same district. My own experience is that an eruption of miliary tubercle often occurs in people who have long resided in tropical countries and have been subjected to malarial fevers, on their return to England. I have lately had under my care an Indian chaplain, who, during a previous visit to England, had an attack of pneumonia, which completely cleared up before he returned to India. After five years of malarial fever, contracted at unhealthy stations, he returned to England, and, at the close of one of his malarial attacks, one of his lungs became affected, and the expectoration contained abundant tubercle bacilli. A cavity formed, and the opposite lung was soon implicated. I had carefully examined him during previous visits to England, and had always failed to detect tubercular disease in his lungs, and was much pleased at the complete resolution of the consolidation, which took place after the attack of pneumonia. The chief interest of the case lay in the patient not being attacked with phthisis until he had been weakened very considerably by intermittent fever. Another case was that of a lady of healthy family history, who, during a residence in India, had several attacks of intermittent fever, and on returning to this country contracted phthisis, of which she eventually died. During the period of her tubercular disease she returned to India, and a fresh malarial attack was followed by rapid advance of the pulmonary mischief.

It may be well, before closing this chapter, to allude to the position occupied by consumption with reference to *infection*, which the important discovery of Koch has lately brought to the front. How far phthisis is contagious is a question which has been discussed for centuries, and has given rise to great variety of opinion in different countries, especially in Europe, the north of which seems to hold its non-contagiousness, and the south, its contagiousness.

It is not unnatural reasoning to urge that a consumptive containing in his breath and sputum the germs of his disease in more or less abundance, according to its stage and activity, must prove a source of infection and danger to his neighbours, and specially to those with whom he lives in intimacy. On the other hand, when we remember that tuberculous sputum has been shown by inoculation experiments to retain its virus intact for months, and consider the amount deposited on the pavements of our towns, and which is dried and wafted as dust, hither and thither, in the atmosphere of our highways and in our houses, and is necessarily liable to be inhaled by a large proportion of the population, we cannot help concluding that though the virus must be ever present, it is not as harmful, as might be expected, to the community in general.

Again, the possibility of infection from inhaling the breath of consumptives, which contains tubercle bacilli, is always present in churches, theatres, and in large gatherings of people, private or public, especially so if, as is usually the case, the ventilation is not properly provided for. Still we are rarely able to distinctly assign to cases of phthisis such an origin.

When we take the evidence of the great London hospitals

for consumption, where large numbers of consumptives are congregated together, including acute and advanced cases, and which consequently ought to be centres of infection, what do we find ? The statistics of the Brompton Hospital, established now forty years, which have been published by Dr. Cotton[1] and Mr. Virtue Edwards and myself,[2] and those of the Victoria Park Hospital, published by Dr. Andrew,[3] directly negative the idea of infection, either to the resident staff or to patients admitted into the wards for other diseases than consumption. The Brompton Hospital staff has been shown to furnish as few examples of phthisis as that of any hospital or public institution not specially devoted to consumption, and the same may be said of the Victoria Park Hospital. Infection in the wards between consumptive and non-consumptive patients is unknown, and whenever the purity of the air has been impaired by defects of ventilation, drainage, or overcrowding, the evil effects have shown themselves in outbreaks of erysipelas and sore throat, and not in tuberculosis. It is obvious, therefore, that phthisis is not contagious in the same sense as scarlatina, small pox, or other acute fevers, but the point to be settled is, whether it is communicable from person to person under any circumstances.

Musgrave Clay managed to collect 111 cases of infection, for the most part due to close attendance on husbands or wives. The Collective Investigation Committee of the British Medical Association, in reply to a circular on the subject addressed to 12,000 members, received 778 negative replies, 39 doubtful, and 261 affirmative, these latter including some imperfect cases, but a few very good instances of infection.[4] But, as Dr. Ransome[5] wisely remarks, 'what are these among the hundreds of thousands of cases in which no sign of contagion has been observed, and how can a disease be regarded

[1] *Lancet*, 1867. [2] *British Medical Journal*, September 30, 1882.
[3] 'Croonian Lectures,' 1884.
[4] It is much to be regretted that this interesting collection of answers should be prefaced by a report which does not do justice to the evidence, and is more remarkable for a bias towards the infection theory than for a calm examination of the accumulated facts.
[5] 'Limits of Infectiousness of Tubercle.'

as ordinarily contagious upon so small a foundation as this '?
Dr. Hermann Weber's [1] cases proved the possibility of phthisis
being communicated from husband to wife, and indicated that
pregnancy considerably increased the danger of infection to
the wife, and the acuteness of the disease thus generated.
According to him, the danger to the husband is comparatively
small. Dr. R. Thompson,[2] in 15,000 consumptives, found
15 cases of wives becoming infected through consumptive
husbands. A most interesting case of infection is furnished
by Reich of a consumptive midwife, who was in the habit of
removing by aspiration the mucous products which obstruct
the primæ viæ of newly-born infants, and if the slightest
asphyxia occurred, of insufflating their lungs with her own
breath. Ten of these children, in whom no family predisposition existed, died of tuberculous meningitis, there being
no epidemic of the kind in the district. The other midwife
in the same neighbourhood, who was healthy, had no cases of
meningitis amongst infants.

My own experience is that for the last twenty years I
have carefully watched for cases of infection in hospital and
private practice, and though I have come across a certain
number of apparent cases, they have never stood the test of
close inquiry, there being always some additional element to
explain the causation of disease. Except a few cases of
husband and wife infection, and a very few instances of sisters
or mothers and daughters sleeping together in an ill-ventilated
bedroom, apparently contracting disease, I cannot record any
infective cases. Another instance [3] recorded is that of a healthy
servant girl, aged 24, who cut her hand with the fragments
of a spittoon containing tubercle bacilli-laden sputum. A tuberculous nodule developed in the wound and the axillary glands
became affected. The subject may be fairly summed up with
the following conclusions :—

1. The evidence of large institutions for the treatment of
consumption, such as the Brompton Hospital, directly nega-

[1] *Clinical Transactions*, vol. viii. p. 141.
[2] *Lancet*, 1880.
[3] *London Medical Record*, August 1886.

tives any idea of consumption being a distinctly infective disease, like a zymotic fever.

2. Phthisis is not, in the ordinary sense of the word, an infectious disease; the opportunities for contagion being most numerous, while the examples of its action are exceedingly rare; nevertheless its communicability is possible under extremely favourable circumstances.

3. In the rare instances of contagion through inhalation, the conditions appear to have been (1) close intimacy with the patient, such as sleeping in the same bed or room; (2) activity of the tubercular process, either in the way of tuberculosis or of excavation; (3) neglect of proper ventilation of the room.

4. In addition to the above, a husband may, though he rarely does so, infect his wife by coition; and this risk is considerably increased in the event of pregnancy.

5. Infection through the milk of a tubercular cow, in cases where the udder is affected, is possible.[1]

6. By the adoption of proper hygienic measures, such as good ventilation, and separation of consumptive from healthy people at night, and by proper care being exercised about the selection of cows for the milk supply, all danger of infection can easily be obviated.

[1] The experiences of Foot, Leube, Kommercil, Lochmann and Chauveau all point to this.

CHAPTER VIII.

CLINICAL HISTORY OF CONSUMPTION.

Early symptoms of Phthisis—Symptoms and physical signs of Tuberculisation Causes of variations in the latter—Apex Pleurisy—Symptoms and
signs of softening—Lung-tissue in Sputum—Micrococcus tetragenus—Click
and croak sounds— Symptoms and signs accompanying cavities—Four
destinies of a cavity—Chronicity--Extension—Pneumothorax—Contraction
--Disintegration—Symptoms of advanced Phthisis—Thromboses of veins—
Œdema—Bedsores—Aphthæ—Ulcers of mouth—Diarrhœa—Modes of
Death.

HAVING now dealt with the pathology and etiology of phthisis,
we can proceed to the symptoms and life history of the
disease; but as these are tolerably familiar, we propose
merely to give a sketch of them as a whole, and later on to
treat at greater length and in fuller detail some of the most
important.

This will be the more advisable, as the number of cases,
which follow, will more than supply any void left in the clinical
outlines.

The course of an ordinary chronic case of phthisis is
this: Cough commences, becoming more and more persistent,
accompanied by mucous expectoration, at this stage rarely
containing tubercle bacilli, followed by loss of colour, strength,
and energy, by emaciation and night-sweats, and occasionally
by loss of hair. The pulse becomes somewhat quickened,
though this is not invariable, and the temperature rises above
the normal in the afternoon and sinks below it in the morning.
M. Peter [1] has noted in a large number of cases a rise in
temperature on the affected side during this stage, but this
feature is not always present; and with regard to the general
temperature of the body, which will be treated more fully
later on (Chapter xii.), though slight pyrexia is often present,
tubercle-formation is quite possible without any rise of tem-

[1] *Op. cit.*

perature, but may be marked by a depression, as Alcock and other writers have shown. Pain in the upper parts of the chest is occasionally present, and the number of respirations is sometimes increased, though this depends on the extent of consolidation which is forming in the lung and the amount of irritation it gives rise to. Some hold that dyspnœa is an early symptom and precedes all others, but the writer has found quite the opposite—that patients do not notice their breath to be short, until their lungs are largely involved. Disturbance of the digestive powers and considerable irritability of the intestinal mucous membrane, with a red streak on the gums, is noticeable in some cases, though chiefly in the acute forms. The tongue becomes white, the bowels torpid, and the urine scanty. The most constant of the above symptoms are the persistent cough, with mucous expectoration, and the progressive emaciation; and in many cases so obscure are the beginnings of the disease that these are the only symptoms discoverable, arising from the presence of the tubercle bacilli in the alveoli. The physical signs [1] of the first stage depend to a great extent (1) on the number and aggregation of the miliary tubercles; (2) on the area of lung thus consolidated, and (3) the irritation which the bacillary changes cause in the lung.

In the greater number of cases, tubercle-formation commences at the apex of one lung, and is detected by the presence of certain physical signs in the supra-scapular, supra-clavicular, or sub-clavicular regions, the signs extending downwards at a later date. The signs vary much in particular cases, but consist at the first in the impairment of the ordinary respiratory murmur by a species of crepitation, differing from the pneumonic crepitation chiefly in its more scattered character, in its being audible at first only with expiration, and in its crumpling nature. It can be heard best after coughing, and often precedes the change in the percussion note by a considerable period of time. Accompanying this, is increased vocal resonance and bronchophony, with more distinct conduction of the cardiac sounds;

[1] For more complete treatment of the physical signs see Chapter ix.

and percussion discovers dulness of varying shades in one of the above-mentioned regions. When a certain definite amount of consolidation has taken place, some impairment of the mobility of one side of the chest may be noticed : this is to be detected under the clavicle, where, if any adhesion of the pleura exists, there may be some flattening. Another significant sign is the dry friction sound, audible generally in the supra-scapular and scapular regions, and indicating limited pleuritis from a nodule of tubercle formed immediately below the pleura. The dulness usually appears first above the scapula, next over the sternal end of the clavicle, and gradually extends downwards, being limited in front generally for a considerable period by the third rib. This limit is specially marked on the right side, and indicates that the upper lobe alone is affected.

When the crepitation and the wheezing—which may be considered as indicative of irritation in the pulmonary tissue, caused by tuberculosis—have subsided, auscultation reveals prolonged expiration, or certain varieties of tubular sounds, which show condensation of the lung-tissue around the neighbouring bronchi, and this is rendered more apparent by the accompanying dulness.

The symptoms which accompany softening of tubercular masses and their subsequent excavation are by no means uniform. Many authors associate this stage with marked signs of pyrexia, with copious night-sweats, and increase of cough and emaciation ; but this is not always the case, for, according to the writer's researches, the process may go on with even sub-normal temperatures, and with gain of weight ; but as fresh formation of tubercle often accompanies the softening process, some of the above symptoms, which have been assigned to softening, may be due to the tuberculisation and pneumonia accompanying it. The symptoms which should be most depended upon for the detection of this process are increase of cough and abundant expectoration of a yellow colour, occasionally streaked with blood. If this latter be carefully collected and boiled, with an equal volume of caustic soda, of the strength of 20 grains to the ounce (Fenwick), and the sediment then placed under a moderate magnifying power of the microscope (400 diameters), delicate filaments of yellow elastic tissue of hook-like shape, or else exhibiting the characteristic form of sections of pulmonary alveoli, may be detected. Bayle states that particles of white matter, resembling rice grains, are found in the expectoration when a cavity is being formed, but the expectoration chiefly consists of pus, and on analysis shows $2\frac{1}{3}$ to $3\frac{1}{2}$ per cent. of albumen, and a large proportion of phosphates. Ponchet found abundance of monads and bacteria in it, and there are doubtless a number of different bacilli found in a completely formed cavity, including the *micrococcus tetragenus*. It is during the process of excavation, too, that the expectoration contains abundant tubercle bacilli, of which the numbers are often so great that in some well-stained sections [1] they may be detected as red masses by the naked eye. The physical signs which these changes give rise to are often more obscure than is usually stated. The percussion sounds greatly vary; sometimes there is an increase of dulness, possibly due to pneumonia, of adjacent lobules. At other times, marked hyper-resonance is present, as if air had taken the place of tubercular masses voided by expectoration. In all these cases much depends upon the situation of the lesion. The formation of a cavity deep in the lung, and far from the chest walls, may take place without being detected, except by the expectoration; whereas the formation of a similar one on the surface gives rise to unequivocal signs. Auscultation reveals—where formerly bronchophony and fine crepitus existed—crepitation of a very coarse character, commencing with a *click* sound, and after awhile developing into a *croak*. When this last note has been reached, loud tubular sounds become audible on coughing, and we soon get the sounds characteristic of a cavity. The great distinguishing features of these moist sounds of softening are their variety, their short duration, and their concentration over one small portion of the lung. It should be remembered that in phthisis, crepitation much more commonly signifies tubercle-formation

---

[1] For methods of staining see Chapter iv.

or pneumonia, than it does softening of already formed tubercular masses.   The formation of a cavity is generally followed
by regular morning expectoration, generally opaque, and
nummular in form, and in the majority of cases, unless
interfered with by treatment, by the usual consumptive
train of symptoms, if these have not already appeared.
These are, night-sweats, slightly elevated afternoon and
evening temperatures, rapid loss of flesh, strength, and
colour.   The drawn look of the face, the pallor of skin, the
hectic spot on the cheek, the pearly white colour of the
sclerotic, the great emaciation, the clubbing of the fingers—
though this sign may be associated with any chronic lung
affection, and specially with chronic empyema—and other
signs which proclaim the confirmed consumptive to the
world, generally belong to this stage, and all more or less
denote blood-infection from the lung products, and sometimes even resemble pyæmic symptoms.

The future history of a cavity may follow one of four
courses.

1. It may remain patent, secreting pus, like a chronic
abscess, but not increasing in size.

2. It may enlarge by caseation and ulceration going on in
its walls, by which process blood-vessels may become exposed.
In this case the expectoration becomes more nummular and
abundant, containing quantities of lung-tissue and remains
of bronchi, and excavation may in time convert the lung
into a mere pleural bag, devoid of lung-tissue, with what
remains of the bronchi opening into it.   The physical signs
attending this increase in size, are amphoric breathing, and
often hyper-resonance on percussion, or cracked-pot sound;
and if the communication through the bronchi be narrow,
coughing will give rise to metallic tinkling.

3. It may open into the pleura and cause pneumothorax
or pyo-pneumothorax; and that this does not occur oftener is
owing to the adhesive pleurisy which so often accompanies
the early consolidations of phthisis, especially if the tubercle
be superficial (*see* Chapter on Pneumothorax).

4. It may contract, and the sides, gradually approaching

each other, form at length a firm, tough cicatrix, causing a
stretching of the surrounding tissue, and often considerable
displacement of the neighbouring organs, as has been described in Chapter vi. This process, which is rarely complete,
is the natural cure of the third stage in phthisis, and is
evidenced in most cases by a flattening of the chest wall,
chiefly in the infra-clavicular space, a disappearance of the
cavernous sounds, and a substitution of deficient or harsh
breathing, and sometimes even of healthy sounds over the
seat of the cavity, owing to adjacent lung-tissue being drawn
across.

Of these destinies of a lung excavation, the two first are
undoubtedly the commonest. Where the cavity remains
quiescent, and no fresh tubercle-formation takes place, the
patient may live on, with only the inconvenience of regular
expectoration and occasional dyspnœa, for years, and preserve
the appearance of actual health. Where a cavity continues to
increase by further ulcerative processes, it is rare for the
opposite lung to escape infection, which may take place either
(1) through inhalation of the cavity secretion, containing
bacilli, into the bronchus of the other lung ; (2) or through the
lymphatic system ; or, (3) as occurs in more distant portions
of the same lung, through the circulation. The infected lung
passes rapidly from consolidation into excavation, the progress
of the disease being rapid, the cough and expectoration increase,
hectic fever becomes more frequent, the patient rapidly wastes
in body and reaches an extreme state of emaciation, the adi-
pose tissue disappears from all parts of the body, the temporal
and malar bones become prominent, the jaws are sharply
defined, the scapulæ, ribs, and sacra all stand out, as if, as is
really the case, they were only covered by skin, and the patient becomes to all appearances a mere skeleton. By an all-
wise arrangement a kind of balance seems to be maintained
between the diminished requirements of the body and the mass
of the blood, for this latter is reduced in bulk in proportion to
the lessened respiratory surface, and the individual thus
gradually dwindles and sinks.

In the last stage of phthisis various symptoms appear

indicative of the disorganisation the blood has undergone, and the manifest lowering of the standard of life. Thromboses form in the veins of the lower extremities, specially in the internal saphena and popliteal veins; œdema of the ankles and feet ensues, and bedsores form on those recumbent parts where pressure is exerted, as, for instance, on the hips, buttocks, and sacrum; aphthæ form on the tongue and fauces, and when removed rapidly spread round the hard palate, buccal surface, and gums.

Ulceration of some part of the mucous membrane of the mouth and pharynx is not uncommon, the part affected being generally the edge of the tongue or the buccal surface in the region of the back molars. Ulceration of the soft palate rarely occurs except in connection with syphilis. Near the end profuse sweats follow the swallowing of all fluids, indicating how completely they partake of the nature of a flux. The breathing becomes quicker, and expectoration more and more difficult.

Diarrhœa, which will be discussed presently, is most prevalent and troublesome at this stage, and often proves fatal before the pulmonary lesions have reached their furthest development.

Death may occur in several ways, either—

1. By apnœa, from the patient's inability to expectorate the accumulating secretion.

2. By thrombosis of the pulmonary artery, inducing lividity and dyspnœa.

3. By pneumothorax, as has been described above.

4. By asthenia or collapse, the heart's action gradually failing, the patient being utterly exhausted, either by the wasting course of the disease, or by the attendant diarrhœa.

Hæmoptysis may cause death, either by collapse from loss of blood, or by suffocation through the blood rapidly filling the air cells.

Some of the principal symptoms of phthisis deserve a fuller description. Such are the physical signs, temperature, diarrhœa, hæmoptysis and albuminuria, and these will now be discussed.

CHAPTER IX.

THERE has been so much said on the pathology and clinical
history of consumptive diseases, that we cannot afford space
for a complete separate description of their signs and symptoms. It must suffice to sketch the most common and
remarkable physical signs which attend the development and
progress of phthisis in its chief varieties.

When a common cough or bronchial cold turns to consumption, there will generally be an increase of the signs of
bronchitis in particular spots, especially in the upper portions
of the lungs. Below a clavicle, or at or above a scapula, a
persistent sonorous or sibilant rhonchus, or still more any
degree of crepitus, is suspicious; and the more so, if these signs
are confined to these parts. In general capillary bronchitis
there is also crepitus; but, then, it is more in the lower than
in the upper regions. All fine crepitus may be taken as a sign

of the parenchyma being either congested or inflamed; and the finer, the sharper, and closer to the ear—the more purely vesicular it is, like the crepitation of pneumonia. But the crepitus of early phthisis is not like this; it is more sub-crepitant, crumpling, or a mere roughening, of inspiratory sound, and often accompanies the expiratory, which the crepitation of pneumonia never does. The natural vesicular breath-sound is impaired, or superseded, by the crepitation, except when it is so slight as only to roughen it. Fine crepitus, with or instead of the breath-sound, signifies some intermitting or vibrating obstruction to the entry of air into the lung-tissue, such as may be produced by swelling and increased secretion of the bronchioles and air cells. Now, wherever these sounds of crepitating obstruction are heard, it may be inferred that some plastic or histotrophic change is going on, whether of good or bad import, is at first uncertain, but the sign at its first appearance demands attention.

Soon other signs follow, indicating the partial consolidation of the lung. The sound on percussion becomes duller, very slightly it may be at first, but still perceptible on careful manipulation, and on comparison of the two sides of the chest. Then may come also the tubular sounds, usually inaudible through the ill-conducting lung texture, but now transmitted through its becoming more solid. These are hardly distinct where and whilst the crepitation prevails; but as this diminishes with increased obstruction, in situations overlying considerable bronchi, below the clavicle, above, within, and at the scapulæ, and in the axillary and middle dorsal regions, the sound of air passing into and out of the tubes is heard, having more or less of a whiffing or sharper blowing quality, which contrasts well with the soft diffused character of vesicular breath-sound. Often too, but not always, the morbid sound differs in an increase in loudness and duration of the expiratory sound, which is hardly audible in natural breath. This is not one and the same thing as tubular breath-sound: for although this commonly includes it, yet expiration is sometimes long and loud, without being tubular. It would take too much space to discuss and explain the whole of this

subject; and it may be stated that, besides the ordinary loud
expiration of tubular breath-sound, transmitted from the large
tubes, expiration may be made audible and prolonged by any
resistance to the escape of air through the small tubes, short
of producing a rhonchus or wheeze (which is a totally different
sound), and such a resistance may be caused by tubercles or
other solids outside these tubes.　So likewise the expiratory
part of tubular breath-sound is increased in intensity by
partial obstructions in the large bronchi, as at the root of the
lungs, from pressure of enlarged bronchial glands; in the
trachea, from goitre or aneurysm; in the larynx, from con-
stricted glottis; and even in the throat by enlarged tonsils.
Exaggerated tubular sounds of this kind may sometimes be
heard through every part of the lungs, where there is no
disease, but then may readily be traced back to their source.
Excluding such extreme cases, tubular sounds near the root
of the lung, especially the right, heard above and within the
scapulæ, are among the earliest and most common signs of
disease in the lungs; and it is rare to find a case with mischief in other parts of the lung of any duration, without this
becoming manifest.　But it may arise from an enlargement
of the bronchial glands, without involving the lung-tissue;
and whilst we recognise its significance, as proving an infection
of part of the lymphatic system, we must not accept it as an
indication of the consolidation reaching into the lung, without
the additional evidence of dulness on percussion, bronchophony, or impaired or crepitating vesicular sound in the part.

Bronchophony, or tubular voice, does not always accompany tubular breath-sound.　It generally requires more consolidation to transmit it, and a greater freedom of the tubes
from constriction and secretion.　It is most heard in the
vicinity of large tubes, like tubular breath; and its combination
with this forms the snuffling, or whispering, bronchophony
so ominous under a clavicle, or above a scapula.　Over smaller
tubes it has often a reedy quality, as in the mammary and
sub-scapular regions.

If phthisical tendency prevails, soon signs of softening and
excavation follow, in the form of increase of crepitation in one

or more spots, looser and coarser, or of more croaking character, generally with diminished breath-sound, and small crepitation around. These spots, soon becoming cavities, form little islands of cavernous voice and breath-sound, first mixed with coarse crepitus or gurgling; afterwards, more croaky and dry, with the characteristic pectoriloquy, and the occasional concomitant, cracked-pot, or chinking percussion. When the cavities become large, the *souffle voilée*, or cavernous puff with the cough, the amphoric resonance or metallic tinkling, which I long ago explained as an echo from the walls of the cavity, give decisive information of the ravages of the consuming disease in the lung.

Thus, in bronchitis passing into phthisis, there is a gradual transition of the signs of the former into those of the latter. In the variety where the tubercle is peribronchial, there is a longer persistence of bronchial rhonchi, sonorous, sibilant, and mucous, giving the disease a wheezy or asthmatic character, until softening ensues and cavities form, which relieve the constrictions.

Acute pneumonia passing into phthisis, from the hepatisation being of plastic or of cheesy nature, is marked by the persistent dulness and by the same loud [1] tubular sounds, and

---

[1] The remarkable loudness of the tubular sounds of a completely hepatised lung has not to my knowledge been satisfactorily explained. The 'consonance' of Skoda is not applicable, inasmuch as it would require a certain relation between the sound of the voice and the size of the tube, as in the case of the reciprocating notes of tubes or chords. But I believe the true explanation to lie in the fact that, whereas the lungs are naturally constructed to destroy the vocal sound by the tubes ending in a spongy texture, which thoroughly damps or chokes all sonorous vibration—no sooner is this spongy tissue made solid, than the tubes become reflecting cavities, capable of reverberating the voice with all the loudness which it has in the trachea, and the vocal vibrations are not only heard, but may be felt by the hand applied over the part. Thus the voice is not only better conducted, as supposed by Laennec, but it is also greatly intensified, by the solidification of the lung. There is yet another acoustic effect developed in the tubes of a consolidated lung, which explains the loudness and almost musical tone of its tubular breath-sound. Naturally the air passes to and fro in the air cells, and although its passage causes the breath-sound, and any accidental rhonchus in the tubes, yet this prevents any longitudinal vibrations in the whole tube. But when the tubes are stopped at their vesicular end by consolidation, the air breathed no longer passes through them, but passing across their open ends, in its way to and from the still

other signs of consolidation giving place to coarse liquid crackling or gurgling, commonly in the central or superior portions of the lung, and the signs of one large or of several small cavities soon follow, to announce the rapid destruction in this form of galloping consumption.

The signs of suppuration of the lung, or abscess ending in phthisis, are those of one or more cavities forming and extending ; and of tubercles forming in other parts—such as crepitus, dulness, and tubular sounds at or near the apex of the opposite lung, which may have been previously sound.

The more common mode in which pneumonia or pleuro-pneumonia terminates in phthisis, is through the chronic consolidation which they leave behind them, instead of dispersing, undergoing caseation through the bacillar action, and partly softening and disintegrating, and partly becoming converted into fibrosis. The site of these tracts of tubercle is marked by the signs of extensive dulness, absence of vesicular breath-sound and motion, and exaggeration of tubular sounds of breath and voice, persisting for months or even years ; the collapse and tight dull sound of the walls of the chest of the contracting portions; the irregular and sometimes cracked-pot or chinking dulness over the parts undergoing caseation and excavation, which also yield their signs of crepitation and gradually increasing cavernous sounds ; whilst in other portions of the lung, the breath-sound may be puerile, or mixed with crackling from emphysematous over-distension, which is also seen in the protrusion of the intercostal or supra-clavicular spaces on coughing.

There is another mode in which tubercular masses may form and inaugurate consumption without any distinct inflammatory attack ; without pain or fever ; with little cough and expectoration ; but generally with shortness of breath and weakness. A peculiar crumpling crepitus invades a considerable portion of one or both lungs, superseding the breath-sound in the part—in some cases gradually followed by dulness

pervious lung, it may cause a hollow whistling sound like that produced by blowing across the open mouth of a panpipe. The same principles are applicable to some sounds heard in cavities in the lungs.

and tubular sounds ; in others becoming mixed with wheezing rhonchi and the clear stroke-sound of emphysema. The primary condition seems to be one of congestion rather than inflammation—hence the absence of active symptoms, and the slow rate at which phthisical processes follow. In fact, they sometimes do not follow ; but that portion of the lung becomes partially emphysematous, and the tendency to further consolidation is thereby restrained. This is a common result of those long-continued congestions at the base of the lungs, resulting from organic diseases of the heart and liver. But, as in predisposed individuals inflammation may create the necessary nidus for the bacillus to exist in, so continued or extreme congestion in the same subjects may produce a similar result. The crumpling crepitus and impaired breath-sound may be caused by congestion alone ; but this does not continue without altering the nutrition further in one way or another : if chiefly around the tubes, producing vesicular emphysema, with its wheezing breath-sounds and clear stroke-sound ; if in the alveoli—dulness, and tubular sounds, added to the crepitus, which eventually passes into cavernous sounds, audible in one or several parts of the lung texture thus invaded.

Spontaneous miliary tubercles, scattered through the lung, without preceding inflammation, are sometimes hardly indicated by physical signs. When numerous, as in acute tuberculosis, they excite more or less general bronchitis ; and the attendant sibilant, sonorous, and mucous rhonchi obscure the special signs of the tubercles, until the increasing obstruction and density of the lung become apparent from interrupted breath in parts, with patches of irregular dulness on percussion. These signs, together with the persistent high temperature—ranging from 100° to 105° F.—and the rapidly increasing weakness, wasting, and oppression, soon declare this frightful form of the disease.

But when the miliary tubercles are few and scattered, they may produce no signs. An increase of numbers, and, still more, their accumulating in a particular spot, will cause a roughness in the breath-sound and a prolongation of the expiratory sound over them. So, likewise, the clustering together of even a few miliary tubercles may, perchance, slightly deaden the sound on percussion in a spot, and transmit more of the voice and heart-sounds than is to be heard in other parts.

The late Dr. Theophilus Thompson, and others, have laid much stress on the *wavy* or *jerking* respiration (*respiration saccadée, entrecoupée*) as an early sign of phthisis; but no one seems to have traced it to its true cause. It is nothing more than the respiratory sound modified or divided by the successive pulsations of the heart. These, on the left side especially, slightly impede the passage of air in part of the lung, and thus give its sound a jerking or interrupted character. The presence of tubercles in the lung increases this effect by transmitting the heart's pressure further, and by narrowing the area of the passing air. Hence, too, this kind of respiration is observed most in females with a narrow chest and a palpitating heart; and in such I have frequently heard the wavy breathing, without any evidence of the existence of disease of the lung at all. With this understanding of the true nature of the sign, we can better estimate its value as indicative of disease in the lung.

The same remark may be applied to the subclavian arterial murmur which was mentioned by Dr. Stokes as a sign of incipient phthisis. It is caused by pressure of the apex of the lung on the artery, and although such pressure is more readily produced when the lung is partially condensed, yet it does occur in some subjects without any disease of the lung.

With the advance of the tubercles to caseation and the infection of new parts of the lung, the various degrees of crepitus, click, and croak become developed, and are the more striking in lymphatic or infected tuberculosis, from not being preceded by the rhonchi or crepitus of inflammation. And the signs which follow—increased dulness, tubular sounds, cavernous croak and gurgle, pectoriloquy, hollow puff or *souffle* with cough, &c.—commonly have a more remarkable character of isolation in this than in the inflammatory form. where the disease is more diffused.

But it must be kept in view that the grey tubercle specially

represents the infective type of disease, and that even where
the original lesion is accompanied by inflammation, the subsequent spread of the disease will be through the lymphatic
or adenoid system, in the form of grey miliary tubercles containing tubercle bacilli. It is this which establishes the identity of consumptive diseases, which not only have all the
degenerative and wasting character, but they all tend sooner
or later to infect the lymphatic system, and break out in the
tubercular form. We have, therefore, to watch for the signs
of these in parts hitherto untouched, especially at the sum-
mits and roots of the lungs, and in the bronchial and other
lymphatic glands. And so long as we find these signs wanting,
we have ground for hope that the disease has not assumed its
most constitutional and destructive form, and is still limited
to the part already invaded.

The consumptive disease may be known to be in an active
or increasing state, when there is more obstruction to the
breath-sound, crepitating or complete ; when the dulness
becomes more marked and extensive ; when tubular sounds
are hollower or louder, or are mixed with a bubbling or moist,
coarse crepitus, and these signs are further enlarged into the
gurgling, churning noises of softening tubercle masses and enlarging cavities. And the increasing size of the cavities may
be judged by the sound of their hollow : when small or
moderate, and communicating with the bronchi, forming
islands full of voice or blowing breath-sound close to the ear
or stethoscope applied to the chest ; when large, giving less
loudness of pectoriloquy, but the more mysterious reverbera-
tions of amphoric blowing or metallic tinkling, which add a
peculiar sepulchral tone to the sound.[1]

It is not necessary here to dwell on the *symptoms* which
mark this last stage of decay. They indicate not only rapid

[1] The physical signs denoting the reinhalation of pulmonary secretions and
the infection of fresh lung centres are generally to be detected in the mammary
and dorso-axillary regions, from these being the points of strongest inspiratory
suction, and from the distribution of the bronchi conducting the sputum in these
directions. Somewhat coarse *râles*, indicating the presence of the sputum in the
smaller bronchi are followed later by patches of fine crepitation, showing the
parenchyma of the lung has become the seat of mischief, and later on areas of
cavernous breathing prove that the secondary cavities have formed.— C. T. W.

degeneration and waste, but often corruption and decomposition, in which septic parasites, vibrios, bacteria, and aphthous fungi, lend their destructive aid. Thus $\phi\theta\iota\sigma\iota s$ passes into $\phi\theta o\rho\acute{a}$. It is not to be forgotten, in connection with the subject of this worst form of decay, the putrefactive, that it sometimes occurs at an earlier stage in the form of gangrene and gangrenous abscess, and the fœtor is strong evidence of its presence. The secretion of dilated bronchi is sometimes also very offensive, from being long retained, in consequence of the mechanical difficulty in expectorating it.

The signs of the cure of phthisis might be expected to be the complete disappearance of those of the disease, but it is rare that the disease and its effects are so completely removed as to leave no trace behind. We can record some cases [1] of incipient disease, chiefly mixed up with inflammation, and of decidedly consumptive character, in which crepitation, dulness, and tubular sounds, have been entirely removed, and the patients have been restored to complete health.

But the commoner degree of what may still be called a cure, is where the general health is recovered, cough, and expectoration, and other symptoms, have ceased; yet the physical signs, whilst showing a cessation of all active disease, still indicate traces of its effects on the lungs and their coverings. Thus, a collapse under a clavicle; a flattening of the upper or lateral walls of the chest; slight variations in the sound on percussion, and in the respiratory movements; a weakness or a mere roughness of the inspiration in the former seat of disease; a remnant of tubular sound, especially above a scapula, in some of its varieties—whiffing, blowing, long expiration, with bronchophony—may permanently remain, evidences of trifling changes left by former diseases, but not materially interfering with function or structure, and therefore productive of no further disorder than perhaps slight shortness of breath, and disposition to cough on exertion and on changes of temperature.

[1] This was written before the treatment of consumption by residence in high altitudes was largely used, which has produced plenty of such instances. —C. T. W.

If the disease has lasted long, and especially if the con-solidations have passed into caseation and softening, more permanent injury is done to the lung textures; and although even these, if limited, may be checked and repaired, there remains more or less injury to the organ, producing various characteristic signs. Some of these have been already noticed in the pathological history of the disease, and will be further exemplified in the abstracts of cases.

In the acute forms of phthisis, the first step towards arrest or cure is by the disease becoming chronic; the high temperature and other febrile symptoms subsiding, the pulse losing its frequency, and some abatement taking place in cough, oppression, pain, and other local symptoms. The physical signs, although more tardily, also show a change; the crepitus becoming less liquid, and more croaking or crumpling, and small degrees of it (sub-crepitation) being heard in parts previously quite obstructed; and the dulness on percussion diminishes, in parts at least, and often is replaced by patches of unusual clearness (emphysema). If excavations exist, their cavernous sounds become more croaking and dry, and albeit often louder, yet more limited in extent; and the crepitus around them and in other parts of the lung diminishes or disappears.

A consideration of the more conservative and reparatory properties of the less decaying new formations will supply a key to a knowledge of some remarkable physical signs developed in cases of arrested phthisis. The development of fibroid, or scar tissue, checks the progress of decay and disintegration; but being itself a shrinking and dwindling material, it causes contraction and puckering of parts of the lung texture, and consequently either collapse of the corresponding chest wall, or the emphysematous distension of neighbouring air cells. Therefore, in chronic phthisis, where the disease is arrested or retarded in its progress, we often see partial sinking or flattening of the walls of the chest, especially below and above the clavicles, whilst near or even at the same spots, a cough or forcible expiration will cause a protrusion of emphysematous lung in the intercostal and supraclavian spaces. In a considerable number of cases, this substitution of emphysema for lung decay eventually converts phthisis into habitual asthma, with its signs of tympanitic stroke-sound, and wheezy dyspnœa and cough, and its symptoms of limited respiration and circulation, and consequent reptile scale of life.

In more partial forms of arrested phthisis, signs of permanent emphysema are common in the vicinity of old cavities or cicatrices ; namely, clear stroke-sound, dry, whiffy breath-sound, and sometimes more or less of a permanent dry crepitus, generally in the middle or lower regions.

Among the cases will be found examples of permanent recovery with signs of a cavity still persistent in the lung. A few also are recorded of pneumothorax, with its unequivocal signs, terminating in complete recovery. Instances of calcareous expectoration are more numerous, and may be referred to here, as affording a kind of physical evidence of the arrest of phthisis by its decaying matter passing into a state of mineral obsolescence. But some of these examples show that even petrified tubercle may excite fresh symptoms ; and it is commonly thrown off in consequence of some new attack.

CHAPTER X.

CLINICAL ASPECTS OF TUBERCLE BACILLUS.

Difficulties in testing sputum—Sources of error—Bacilli scanty in first stage, but
often precede physical signs—Abundant in that of softening and excavation—
Four examples—Their relation to progressive excavation and to extending
tuberculisation—Three cases—Spread of bacilli through re-inhalation of
sputum—Bacilli in pneumothorax—In quiescent cavity—Abundant in
laryngeal phthisis—Their value in diagnosis—Chronic pneumonia, bronchiectasis, asthma, and bronchitis—Emphysema—Cases of obscure pyrexia
—Laryngeal phthisis—Syphilis of larynx—Value of bacillus in prognosis.

THE study of the part played by the tubercle bacillus in
the clinical history of phthisis is far more important than the
study of the parasite in the *post-mortem* lesions, important
though that be; for too often at the autopsy we see, not the
disease itself so much as its results, and it is only by studying it step by step, and investigating during life the various
organs attacked, that we can ever hope to attain a satisfactory knowledge of its true influence on the system, of its line
of attack, and of its extensions throughout the body. During
the last four years, hundreds of examinations of the sputum
of patients in the different stages of phthisis have been
carried out in hospital and private practice, by myself and
my assistants; and an attempt has been made to determine
the relation the bacillus bears to the various clinical phenomena of phthisis.

The great difficulty lies in the fact that, as a rule, we
have had to rely mainly, if not entirely, on examinations
of the sputum, the attempts at testing the blood and lymph
having generally failed, partly, probably, through the bacilli
being only present in small numbers, and partly because
their detection possibly requires a more elaborate system of
examination than the time at our disposal allowed.

It is impossible to judge from the naked-eye appearance of sputum whether or no it contains tubercle bacilli. Often glairy mucous expectoration yields as many bacilli as the typical nummular purulent masses raised from a cavity.

In dealing with sputum, we must be prepared for many failures, unless great care be exercised in its collection. It often happens in a case of phthisis that the greater proportion of the expectoration is bronchial, and not alveolar; and although the smaller bronchi are sometimes the seat of tubercle in early phthisis, they generally escape disease, on account of their being protected by their thicker epithelium. The sputum, if it comes from this source, and especially from the bronchi of the non-tubercular lung, which are affected with catarrh, may be abundant and yet contain no bacilli. Unless, therefore, we can be sure that the sputum comes from the alveoli attacked, we cannot be certain that it will contain tubercle bacilli. The only means of insuring accuracy is to collect the sputum of the whole twenty-four hours, and to test several specimens; but in the early morning expectoration, which is untainted by food and saliva, and is the collected secretion of several hours, we generally succeed in finding bacilli, if they are to be found at all. At least three testings should be made before a conclusion is arrived at. In incipient phthisis, where the physical signs denote very limited consolidations, or clusters of tubercle, and not softening or excavation, bacilli are present, but in small numbers, and some fields of the microscope contain none at all, but provided there is any expectoration at all, a diligent search will generally detect a few.

The annexed cases are fair examples :—

CASE 1.—Beatrice L., aged 15. Admitted January 2, 1886. Family history good.

She had pleurisy on the left side March 1885, and was taken into Great Ormond Street Hospital for six weeks, and left it apparently well, but having lost much flesh and with a persistent cough.

During the six months before admission into Brompton she was an out-patient suffering from cough, night-sweats, dyspnœa and occasional hæmoptysis, amounting on one occasion to half-a-pint. On admission the cough was most troublesome at night, accompanied by pain over the mid-sternum and left side, which was somewhat catching in character. Expectoration scanty, chiefly on waking in the morning.   Catamenia absent for some months, tongue clean, bowels regular.   She is now gaining flesh ; temperature normal, pulse 70.

Crepitation heard on cough on the right side in the first interspace, near the sternum, and, behind, in the interscapular region, extending slightly into the scapular ; no tubular sounds or dulness anywhere.

January 15.—The expectoration contains a few tubercle bacilli.   Ordered o take sulpho-carbolate of sodium in 8-grain doses three times a day.

January 21.—She has gained 6 lb. in the last fortnight.   Cough is less, and the physical signs are the same.   She has been taking 10 grains of sulpho-carbolate of soda three times a day.

January 28.—Sputum still contains a few bacilli.   The sulpho-carbolate caused at first slight purgation, but now is borne well.   Cough less and expectoration has nearly ceased.

February 11.— Patient has gained 3½ lb. more in weight.   Cough same.

Area of crepitation has rather increased posteriorly.

March 18.—Has developed in frame and increased in weight 19 lb. weighing now 8 st. 1½ lb.   Aspect healthy ; cough is very slight and expectoration is generally absent; *crepitation is more scattered, but extends over the same area.*   She has taken sulpho-carbolate of sodium for two months.

February 1887.—Has been re-admitted with increase of symptoms.   *Crepitation now heard at both apices in addition to above signs.*

Here the position of the physical signs might have thrown doubt on the diagnosis of phthisis, which was set at rest by the presence of the bacilli in the sputum.   In addition to the physical signs, the scantiness of the sputum forbade any notion of a cavity.   The gain of weight was enormous, and the fact of its accompanying the sulpho-carbolate of sodium was rather significant.

CASE 2.— Edith C., aged 17, milliner and dressmaker, admitted on December 30, 1882.   No family predisposition.

Has had pain under the right clavicle for three months, which kept her in bed and out of work for one week.   She states that she has no cough, but that some expectoration follows a tickling in the throat.   She has not lost flesh or had night-sweats.   At present her breath is not short.   Tongue clean ; bowels and catamenia regular; temperature and pulse normal ; weight, 8 st. 6½ lb.   Dr. Percy Kidd, who admitted her in my absence, found weak respiration, with prolonged expiration in the first interspace and above the scapula, on the right side.

February 22, 1883. –Has gained 6 lb.   The expectoration is so scanty that it is difficult to obtain any ; it has been examined three times for bacilli; on the first two occasions none were found, but on the third a few were detected.

Tubular sounds audible in the first and second interspaces.

March 15.—Has improved in strength, and a little in weight, being now 8 st. 13½ lb.; bacilli present in the expectoration, but very scanty.   Some fields of the microscope contained from five to ten bacilli, others none at all.

Tubular sounds in first and second interspaces, with occasional crepitation in the first interspace on coughing.   Temperature 98·4° F.

In the foregoing case the signs may be said to have developed while under observation, the crepitation, be it noted, appearing at a later date than the prolonged expiration or tubular sound. The bacilli could only have reached a small number of the alveoli, and detection was consequently very difficult.

The process of *softening* and *excavation* is invariably accompanied by a great increase in the number of the tubercle bacilli in the sputum ; this increase is sometimes noted before the physical signs clearly indicate the change, and generally before lung-tissue can be detected in the sputum. Dr. Hunter Mackenzie,[1] after a series of observations, has come to the conclusion that the most unfavourable cases are where the bacilli are abundant and the lung-tissue scanty. The change from a few bacilli to several hundreds in a field of the microscope is very striking, and the presence of a large number is not necessarily accompanied by either pyrexia or night-sweats, or wasting, though all these may be present. The subjoined four cases illustrate this stage well, the first showing abundance of bacilli only during the excavation process.

CASE 3.—Amelia L., aged 34, married. Admitted November 25, 1885. Her father died of hæmoptysis, and her mother of rapid phthisis at 25. Patient enjoyed good health till two years ago, when cough came on, which has continued ever since, but has decreased lately. For the last eighteen months has been losing flesh, and is now very thin. She had slight hæmoptysis twelve months ago. The cough is troublesome at night; the expectoration scanty and muco-purulent. She has occasional night-sweats, and suffers from pain under the scapulæ and shortness of breath.

Crepitation above the right clavicle. Crepitation of a coarse character throughout the whole of the left lung, most marked in the upper third. No cavernous sounds are to be heard anywhere.

January 7, 1886.—Improved at first and gained 2 lb. in weight. Physical signs the same as on admission. The cough has increased lately, and the expectoration is found to contain large clusters of tubercle bacilli.

January 28.—Cough better. *Physical signs the same.*

February 18.—Cough has been very troublesome lately, and the expectoration has largely increased. She gained in weight up to January 19 from 6 st. 3 lb. to 6 st. 7 lb., but has now lost, her present weight being 6 st. 5¼ lb. She has occasional sharp pains in the left front chest. Has been very bilious lately.

Crepitation on the right side as before: on the left side cracked-pot sound in

[1] *A Practical Treatise on the Sputum,* p. 11.

the first interspace and cavernous gurgle, most audible near the sternum. Cre-pitation below this point, nearly to base. Posteriorly crepitation throughout, but cavernous sounds are audible over an area about the size of half-a-crown in the dorso-axillary region. Bacilli abundant in sputum.

CASE 4.—Mary G., housemaid, aged 20, single. The paternal aunt died of phthisis, and an aunt and uncle are now consumptive. One brother is also reported to be consumptive.

Admitted January 15, 1886.—Has not been well since she came to reside in London, two years ago, and for the last fifteen months has had cough, expectoration, loss of flesh, profuse night-sweats, and pains in the left side, most marked in the axilla. The dyspnœa is great on any extra exertion. During the last three months all these symptoms are greatly increased, and the sputum has been streaked with blood. Catamenia irregular ; bowels constipated ; appetite bad.

Some fine crepitation is audible in the first and second interspaces, on the right side, on the left there is dulness and coarse crepitation, with some flattening to the fourth rib, also crepitation of a finer character over the upper third posteriorly. No cavernous sounds were audible anywhere.

February 5.—Cough and expectoration about the same, but more pain in the left front chest.

Cavernous sounds are now audible in the left interscapular region, and croaking rhonchus in the first and second interspaces. The sputum contains abundant bacilli of great length.

February 18.—She has gained 2½ lb., and states that she feels better. *Cracked-pot sounds can be detected in the first interspace, on the left side, the cavernous sounds in the interscapular region having extended almost to the base.*

March 18.—Has gained 6 lb. in weight and improved in appearance. *Physical signs the same.*

CASE 5.—William S., aged 26, labourer, was admitted February 2, 1883. No family predisposition. Has been subject to winter cough for some years. After a succession of wettings last Easter, cough with expectoration increased, and his breath became shorter. Six weeks ago there was slight hæmoptysis, and he has had night-sweats for three months, and has been losing flesh since Easter, having wasted to the extent of two stone. At present cough troublesome, expectoration muco-purulent and aërated ; breath short, night-sweats severe. Tongue has an irritable look, appetite very poor ; the patient suffers from dyspepsia. Pulse rapid, temperature 100·6° F.

On right side, crepitation to the third rib, with some tubular sounds. On left side, crepitation to the fourth rib : crepitation audible above both scapulæ.

February 12.—Has had diarrhœa for two days ; three motions a day ; is sweating less. Temperature has ranged from 97° to 99° F. in the morning, and from 100° to 103° F. in the afternoon.

February 19.—Diarrhœa ceased ; breath short on exertion ; bacilli in sputum abundant. The evening temperature has ranged from 100° to 102° F.

February 26.—Has complained of pain in the right side for the last few days, and friction is audible over the lower front. The pain has been relieved by strapping ; cough very troublesome ; bowels rather relaxed.

Cavernous sounds audible on the left side, over limited areas in the first, third, and fourth interspaces, where there is some flattening and retraction of the spaces on inspiration. The heart is apparently uncovered by lung. On the right side the crepitation remains the same. Bacilli scanty.

March 5. Cough and expectoration less; appetite improving. Physical signs the same. Bacilli in sputum scanty. The temperature has fallen, ranging from 98° to 99·8° F.

March 19.—Symptoms have become worse; temperature rises to about 101° every night. Cough worse; the patient is unable to lie long on one side. *On the left side cavernous sounds more marked, with less crepitation. On the right side crepitation greatly diminished, with doubtful cavernous sounds under the clavicle.*

April 2.—Cough more troublesome, accompanied by retching; expectoration increased; night temperature ranges from 101° to 102° F.; night-sweats troublesome; on the left side physical signs as above; on the right sid crepitation has extended to the fifth rib; bacilli few and dividing.

April 9.— Cough about the same; expectoration muco-purulent, with patches of pigmentation; temperature at night 100° to 102° F.; bacilli abundant, sixty in the field, either single or in groups of three; they have well-marked beadings, and are apparently in a state of active division.

April 16.—Symptoms about the same; on the right side crepitation is increased, front and back. *Cavernous sounds are audible on the left side to the fourth rib.* Evening temperature 101° F.

May 8.—Patient worse; has wasted greatly; *more crepitation is heard on the right side, front and back; cavernous sounds detected in both lungs.*

The increase of the bacilli seems in this case to have preceded the detection of the excavation by a considerable period a week elapsing between the noting of 'abundant bacilli and the cavernous sounds being heard in the left lung. On the right side the development of the physical signs came before the increase of the bacilli, but, later on, the lung excavation changes were accompanied by great activity of the bacilli. Their numbers increased, they showed well-marked beadings or spores, and appeared ranged in the little faggot bundles which generally indicate, or usher in, fresh tubercular outbreaks. The appearances of division mentioned in this case are when the bacilli show greatly varying lengths.

The following case is an example of excavation taking place in phthisical consolidation following pleurisy, the bacilli being present during the period of lung destruction.

Case 6.—Albert R., a labourer, admitted January 19, 1883, aged 19. No consumption in the family. He had pleurisy at 14, and has since had three attacks, the last being in March 1880. Cough constant since the pleurisy, and worse in winter; it has been increasing lately. Expectoration has diminished, but is thick, greenish yellow, sometimes frothy. He had

I

hæmoptysis to the extent of twelve to fourteen ounces two years ago. Has had occasional night-sweats, but none lately, and has lost flesh since 1880. At present he complains of pain in the right chest, palpitation, shortness of breath, and cough.

Dulness in the lower third of left chest anteriorly, and the upper half posteriorly. Crepitation on cough, audible in the first interspace. Weight 10 st. 7½ lb.

February 5.—Expectoration has been slightly streaked these last three days ; cough less ; bacilli abundant in the sputum.

February 19.—Expectoration again streaked, but contained no bacilli.

March 12.—Gaining weight rapidly and improving ; no bacilli in sputum.

March 20. — Improving ; has gained 9 lb. in weight ; cough slight ; expectoration yellow.

Tubular sound below the clavicle, slight dulness in the first two interspaces, slight crepitation on cough in the second interspace ; cavernous sounds heard posteriorly above the scapula. Sputum contains bacilli.

March 11.—Has improved and gained nearly a stone during his stay in the hospital. Cough moderate ; expectoration slight ; breath short on exertion ; bacilli scanty, about six in the field of the microscope.

When a cavity has formed, and with its formation the cough and expectoration have diminished, the number of bacilli generally diminishes too, and, as a rule, varies in proportion to the amount of expectoration, as is well seen in the last case. When, however, the cavity is increasing in size, and lung tissue is expectorated, the bacilli become very numerous, and often appear in masses resembling those produced by cultivating them in appropriate media. It is rare for these numbers not to be accompanied by pyrexia, hectic, night-sweats, and great wasting. The form of phthisis where excavation proceeds most rapidly is in acute phthisis (scrofulous pneumonia or galloping consumption), and here the bacilli are the most abundant of all. This is quite in accordance with what we should expect, as not only are the epithelial elements, the fuel of destruction, most abundant, but fibrosis -the great obstacle to excavation—is altogether absent. In this form there are generally several centres of active caseation, and these are found to swarm with bacilli. Sometimes, when large cavities have formed, the number of bacilli may decrease with the subsidence of the symptoms, but as a rule bacilli abound from the commencement to the end of this form of phthisis. *Extension* of tubercular disease in the lungs is generally accompanied, and often foreshadowed, by increase of

bacilli in the sputum, and the nature of the extension depends
very much on the channels chosen by the bacilli. In many
cases they advance through contiguous alveoli, generally from
above downwards—as appears to have been the line of extension in case 7.

Case 7. Frank R., aged 26, footman, admitted on March 19, 1883.
His mother died of some disease of the lung, and a brother and sister are
asthmatic. After exposure late one night, a year ago, when out with the
carriage, cough and expectoration came on, and have continued ever since.
Night-sweats followed, and persisted for six months. He lost flesh at first,
but has regained it lately. His breath has been short, on exertion, for some
time. Has never had hæmoptysis. The cough at first was very severe, and
ended in vomiting. The expectoration has been white and frothy, but lately
has become somewhat opaque. At present his appetite is fair ; tongue furred ;
some dyspepsia after food ; bowels confined ; temperature 98·4° F. ; pulse 90 ;
urine normal.
*Slight dulness, with fine crepitation from the claviele to the third rib, and
scattered crepitation audible over the upper half of the left chest posteriorly.*
March 30. —Sputum examined, and found to contain twenty bacilli in the
field, and these apparently multiplying.
April 9.—Cough more troublesome ; expectoration increased ; appetite
bad. The patient states that he does not feel so well.
April 11.—Sixty bacilli in the field, of greatly varying lengths.
April 16.—Cough and expectoration somewhat less ; he complains of
pain in the left chest.
Crepitation somewhat more abundant in the left posterior region.
April 17. —About thirty bacilli in the field, still of greatly varying lengths.
April 23. Physical signs about the same. No cavity can be detected.
Cough and expectoration are moderate, and the patient is gaining weight.
May 7.—The patient is not so well, and coughs more. *Crepitation has
increased, and extends from the claviele to the fifth rib on the left side.* Since
admission the temperature has once risen to 100° F., and the morning records
have as a rule been sub-normal.

The above seems to be a case of advancing tuberculisation
of one lung ; and the interesting feature is that the spread of
the crepitation downwards was preceded by a multiplication
of the bacilli noted on March 30, and by their detection in
greater abundance in the field on April 11, when they were
still found to vary considerably in length, as if there was a
prospect of further multiplication and consequent spread of
disease by contiguity.

Case 8.—Arthur B., draper's clerk, aged 30, admitted on March 19, 1883.
For eight years he has been subject to winter cough, and for four years cough
has been continuous, and during the last two years he has complained of shortness

of breath, which increased so much at Christmas 1882, and was accompanied by such debility, that he had to give up work. Two years ago he had slight hæmoptysis, and night-sweats have occasionally appeared. He has during this time lost more than a stone in weight. At present he complains principally of shortness of breath, especially on exertion: cough severe night and morning; expectoration yellow, but sometimes frothy. *Slight dulness, with crepitation from clavicle to the lower edge of the fourth rib, on the right side. Sonorous rhonchus audible at the base posteriorly.*

March 30.—The sputum contains a few bacilli.

April 2.—Breathing rather easier, but patient weak and languid. Cough less, expectoration scanty. Physical signs on right side the same as before : on left side some wheezing and rhonchus are audible.

April 9.—Cough and expectoration less. Sputum contains a fair number of bacilli, forty in the field. Some single ones, but many in faggot groups of two, three, and four. The bacilli vary in appearance. Some are simple rods without beadings or spores, and others again show the beadings well.

April 16.—Breath still rather short ; expectoration deeply stained with blood during the last few days. *Fine crepitation is audible in the interscapular region, on the left side.*

April 17.—Forty bacilli in the field, still varying in length.

April 23.—Fine crepitation is also audible at the right posterior base.

May 7.—Cough rather worse, expectoration same in quantity, but tinged with blood. Temperature rises at night to 100° F. Physical signs the same.

June 4.—Expectoration has diminished. Temperature high at night. *Physical signs the same on the right side ; on the left there is crepitation at the extreme base anteriorly, and in the scapular and interscapular regions posteriorly. No cavernous sounds can be detected anywhere.* Examination of the sputum shows abundant bacilli, fifty in the field, and with well-marked spores.

In the foregoing case the disease undoubtedly progressed, the physical signs indicating increase of tuberculisation in the same lung, and also extension to the opposite one. No cavity was detected, but with a history of so long standing we cannot be sure of the absence of such. The main feature of the case was progressive tuberculisation, and we may fairly connect with this the large number of bacilli, their increase between March 30 and April 9, and the presence of well-marked bundles which are so characteristic of active tubercular changes.

The following is a most interesting case of the spread of the bacilli through re-inhalation of sputum in very acute disease :—

CASE 9.—Rebecca G., aged 16, confectioner, was admitted December 11, 1885. She had suffered for three months during last winter with pain in the right axillary region, on drawing a deep breath. In June 1885 she noticed that her breath was getting short, and in August cough came on with expectoration, accompanied by night-sweats and emaciation. Cough has continued and the expectoration has been occasionally streaked with blood. The night-sweats have ceased, but the breath is still short on exertion, and she still complains of pain in the right chest. Tongue clean; appetite fair; bowels constipated; expectoration contains a few tubercle bacilli. Dr. Percy Kidd, who had admitted her in my absence, found *on the right side bronchophony and blowing breathing in the first interspace and above the scapula*, also *some crepitation over a rather larger surface than the above. On the left side blowing breath-sound above the scapula.*

December 31.—For the first few days after admission the temperature varied between 99° and 101° F., but for the last three days she has complained of feeling cold in the morning, and there has been a distinct rigor at 10 or 11 A.M., the temperature rising to 100° F. at that time, and continuing high till 8 P.M. *The crepitation has increased under the right clavicle, and can be detected in the right axilla.* The number of bacilli in the sputum have increased.

January 7, 1886.—Two days ago, during a rigor, the morning temperature rose to 104·2° F. The sputum contains a large number of bacilli. *The crepitation under the right clavicle has become coarse, and some crepitation is audible in the interscapular region.*

January 14.—The temperature has continued high since the last report, and has only once fallen to normal. It is now about 103° F., but she has had no rigors. In addition to the above physical signs, *there is crepitation in the left mammary region, below the fourth rib.* The bacilli have somewhat diminished in numbers.

January 28.—The temperature, which has been taken five times a day, still shows a morning rise, the maximum being attained in the afternoon, between 5 and 8. The minimum, which is in the early morning, rarely falls to normal. Cough very troublesome; expectoration rather difficult to bring up and contains abundant bacilli. *The crepitation in the left lung now extends to the base, anteriorly and posteriorly.*

February 11.—Has had slight diarrhœa, and is getting weaker. The breath is shorter.

The crepitation amounts to gurgling in the second space on the right; gurgling sounds are audible in the third interspace on the left side, and in the dorso-axillary region at the same level.

February 18.—Patient complains of great dyspnœa and some pain in the left side. She improved under mistura ætheris c. ammonia, and mistura vini gallici, but was taken home by her friends, and thus lost sight of.

In this case we have attempted to make a careful comparison between the evolution of the disease, as indicated by the symptoms and physical signs, and the numbers of the bacilli detected during frequent examinations of the sputum; and we may fairly come to the conclusion that a certain amount of correspondence existed between them. The rise of temperature was accompanied by an increase in the number of bacilli, and was followed closely by a spread of the crepitation in the right lung.

The large number of tubercle bacilli found on every occasion, except one, which might have been accidental, testified to the rapid increase of these organisms during the fever.

The spread of disease, as evidenced by the physical signs in this case, is very interesting. From January 7 crepitation was detected in the right interscapular region. On the 14th fresh crepitation was heard in the left mammary region, below the fourth rib. This had extended in the course of a week throughout the whole lower lobe of the left lung. By February 11 a cavity had formed in the upper right lung, and another one in the left lung in the mammary region, and a third in the dorso-axillary region in the same lung. Now, when we consider the locality of each of these fresh centres of tubercular disease, the probability of their being due to the re-inhalation of bacilli becomes very great. It is most likely that the left lung was infected by sputum inhaled from the right, descending by the main bronchus into the mammary region, and by the upper posterior horizontal bronchus into the dorso-axillary region, there setting up fresh tuberculisation and ending in the formation of cavities in these two regions, which Dr. Ewart has shown are the most prone to excavation, next to the apex.

It is impossible to explain the lines on which the disease spread by any reference to the circulation, for if it were dependent on the bronchial or pulmonary circulation, the diffusion of tubercle throughout both lungs would have been more equal and general.

Another good instance of the connection between the advance of excavation and the presence of bacilli in abundance is seen in case 10.

CASE 10.—Mary C., aged 35, married; admitted January 20, 1886. Family history free from consumption.

Has been subject to occasional attacks of rheumatism, and in November 1884 had bronchitis, since which time has suffered from increasing cough and expectoration, loss of flesh, and shortness of breath. On two occasions the expectoration has been streaked.

For some months she has had severe night-sweats, and complains of a dull, aching pain in the chest ; the voice has been husky since the commencement of her illness, and at times she has had dysphagia. She has also been subject to palpitation. Her cough is at present most troublesome in the beginning of the night; expectoration is abundant and contains a large number of tubercle bacilli. The evening temperature ranges from 99° to 100° F. The larynx on being examined shows no ulceration, and is very pale.

Expansion of chest very deficient in front. *Cavernous sounds audible in the first interspaces on both sides, and on the right side some crepitation is heard in the second interspace. On the left side crepitation is audible in the third and fourth interspaces; the second interspace being quite free from adventitious sounds. Crepitation above the left scapula.* A loud systolic murmur is audible over the cardiac region, loudest at the apex.

February 4.—The temperature has continued high and the cough troublesome. The crepitation in the left mammary region has become more croaking, especially after cough. Bacilli are abundant.

February 18.—Expectoration has increased considerably. Evening temperature varies from 100·5° to 102·5° F. Cavernous sounds are audible on the right side in the first interspace, and above the clavicle and scapula. Cavernous sounds are heard from the clavicle to the third rib, on the left side, and above the clavicle. Posteriorly they extend downwards over the upper half of the scapula, and are much masked by the presence of mucous rhonchus and *râles.* There is a second spot of cavernous breathing just below the angle of the scapula, behind the vertical axillary line, the sound being only distinct after coughing. The whole chest is hyper-resonant.

Here excavation appears to have proceeded by rapid strides in the left lung. The original apex cavity extended downwards to the level of the third rib and posteriorly, so as to involve a great part of the upper lobe, but a secondary cavity was also formed in the dorso-axillary region, probably through re-inhalation of the secretion from the large primary one. The numbers of the bacilli were very abundant in the sputum all through these changes, and may be very fairly associated with them. Where excavation is proceeding in both lungs at the same time, the number of bacilli is large, but the largest numbers of all are found in the sputum of advanced or dying cases, in which the sputum appears to be little else than masses of bacilli. In cases of pneumothorax, on the other hand, it is not uncommon to find none at all, if the collapse of the lung is tolerably complete. Where, however, it is not so, and some of the cavity-discharge still reaches the main bronchus, and is expectorated, the bacilli are tolerably abundant.

In cases of quiescent cavity the bacilli are never numerous, and often very scanty. This is the more so if the cavity is contracting and the lung undergoing fibrosis. In the sputum of these patients it is rare to find many bacilli, and not uncommon to find none at all, especially if during the process of lung contraction the bronchi have become blocked.   In a case of a young woman with a retracting cavity in the left lung, whose general condition was excellent, I failed at first to detect any tubercle bacilli in the scanty sputum, but on the third or fourth testing I succeeded in detecting a few scattered ones.   In another case, that of a man aged 26, with a small retracting cavity at the apex of the left lung, the bacilli were present, but only in small numbers, though the expectoration was tolerably abundant.   In cases of quiescent cavity of often considerable size, the sputum shows very few bacilli, probably because the walls of these cavities consist almost entirely of fibrosis.

In laryngeal phthisis, we find the number of tubercle bacilli is large, independently of the accompanying lung conditions.   If there be well-marked ulceration of the vocal chords the bacilli are always abundant, even if the disease of the lungs be not specially active, the reason being that, owing to their situation, at no great distance from the mouth, they are not diluted with a large quantity of bronchial secretion.   In the annexed case the amount of pulmonary disease was comparatively small, though both lungs were attacked, but there was well marked laryngeal ulceration.

CASE 11.—Sophia C., married, aged 36, book-sewer; admitted January 1, 1886. Her father died at 64 of haemoptysis : her mother at 56 of apoplexy. Has four brothers—all healthy. She has never been strong, and has had two miscarriages. Was chilled during last pregnancy and got a cough, and after confinement, eighteen months ago, became worse. The child died, when six months old, of phthisis in Great Ormond Street Hospital. After this confinement night-sweats came on, with shortness of breath and some emaciation, and later on her voice became affected. Tongue furred ; bowels costive ; pulse and temperature normal.

Slight dulness and tubular sound, with occasional crepitation in the first interspace, on the right side and above the scapula. The larynx on examination showed thickening and swelling of the inter-arytenoid fold. Both chords were ulcerated. Interior of larynx much reddened and epiglottis swollen.

January 28.- Cough improved since admission, but aphonia complete, and some dysphagia. Some pain in the right side on drawing breath. The sputum contains abundant tubercle bacilli. She was ordered ten grains of sulpho-carbolate of sodium, with tincture of cascarilla, and infusion of orange three times a day, also daily insufflation with iodoform and morphia.

February 11.—Under the insufflation the larynx has improved, and the voice is fairly audible. Cough and expectoration have rather increased. Sputum contains a large number of bacilli. She has gained in weight.

On the right side there are slight tubular sounds in first interspace and above the scapula on the right side; on the left crepitation on cough, with harsh inspiration in first interspace and above the scapula.

March 18, 1886.—Has improved in appearance. Voice nearly normal. Scarcely any cough and expectoration. Physical signs the same. Weight 6 st. 7½ lb., being a gain of 5 lb. since admission.

Diagnosis.—It is probable that the discovery of the bacillus will assist us most in diagnosing phthisis from other lung affections. Where it is most useful, is in separating clearly cases of chronic pneumonia (1) and those of dilatation of the bronchi (2) from phthisis. We often see an instance of chronic pneumonia with extensive consolidation and well-marked crepitation at the apices of one or both lungs. The history is somewhat obscure, and often, especially if hæmoptysis be present, points to phthisis, and the physical signs rather incline us to hold that softening is taking place in an old consolidated lung. The expectoration, if carefully examined and found to contain no tubercle bacilli, at once settles the point. A good instance of this was a coachman, aged forty-six, whom I saw in consultation with Mr. Tweed of Upper Brook Street, in February 1883. He had been a hard drinker, and had had pneumonia three years previously. A week before, a fresh attack had come on, with high temperature and quick pulse. The cough was incessant, the expectoration partly frothy and partly purulent. Crepitation was audible over both lungs, and some dulness at the bases, but cavernous sounds were heard above the right clavicle, which made us both suspect that a cavity had formed in the old consolidated lung, and that the case might have become one of phthisis. Examination of the sputum, which was made more than once, failed to detect any bacilli, and we accordingly came to the conclusion that the case, as it eventually proved, was one of pneumonia, in which previous attacks had given rise to dilatation of the bronchi. It must be borne in mind that had the case been one of phthisis, the bacilli would not only have been present, but very numerous, as they usually are in cavities, where the secretion is abundant. Another instance was a gentleman, aged forty-six, whom I saw in February 1883. He had pleurisy several years ago, and for four or five years had suffered from cough and expectoration, with slight hæmoptysis. The pulse and temperature were normal. The physical signs showed crepitation over the whole right side, with some spots of coarser crepitus in the first interspace and between the fifth and sixth rib. Dulness extended over the lower half of the right front chest. He complained of great dyspnœa on exertion. Under treatment the cough moderated, but the expectoration remained yellow; there was no wasting, night-sweats, or fever. Examination of the sputum showed abundant bacilli, which gradually disappeared as the symptoms diminished.

Another class of cases, which are very difficult to diagnose, are those of dilatation of the bronchi, following on chronic pneumonia or bronchitis. When the dilatation of the tubes occurs in the lower lobes, the physical signs and the convulsive character of the cough, as well as the fœtor of the expectoration, render diagnosis easy, but when the bronchiectases are found scattered over the upper portions of the lungs, as well as in the lower, and when they give rise to cavernous breathing in the first and second interspaces, the diagnosis often becomes exceedingly difficult, and depends chiefly on very careful mapping out of the area of cavernous sounds, such as I have shown elsewhere.[1]   In such cases examination of the sputum removes all doubt.   The following is a good instance :—

CASE 12.—Hannah I., aged 24, servant, was admitted March 3, 1883. Her father died of phthisis; one sister suffers from some chest affection. She had typhoid fever fourteen years ago, followed by scarlet fever, and has been subject to cough ever since; six years ago she had disease of the hip-joint, and became an in-patient of the London Hospital; three years ago she had hæmoptysis, and the cough has been worse ever since; she has had night-sweats for twelve months, and some shortness of breath lately; she has lost flesh for six months; her voice disappeared for the whole of one winter two years ago. At present her cough is incessant, and somewhat hard in character. Expectoration is yellow, 2 to 3 oz. The patient complains of pain in the left chest. Tongue clean; bowels relaxed; pulse 116; respiration 16; temperature 99° F. Catamenia absent seven months; aspect cachectic; weight 7 st. 13½ lb.

[1] *British Medical Journal*, 1881.

On the left side dulness, crepitation to the fifth rib, posteriorly crepitation over the whole side.

March 15.—Cough more troublesome ; expectoration one ounce daily ; her voice has disappeared ; crepitation audible under the right clavicle. No bacilli could be found in the sputum.

March 30.— Her cough is worse ; crepitation less on the left side ; no bacilli in sputum.

April 7.—Cough increasing ; expectoration more abundant, purulent and decidedly fœtid. The patient has herpes labialis. Dr. Kidd examined the larynx and found the vocal chords healthy, but adduction imperfect. The sputum contains no bacilli.

April 19.— Cough somewhat better, but still bad at night, and is remarkably convulsive ; expectoration has a sulphurous odour. On the left side the crepitation has much diminished, but tubular sounds are audible over a small area below the scapula. The sputum has been again examined, with negative result.

The history of this case, especially the facts of the cough and expectoration following hip disease, pointed towards phthisis : the convulsive character of the cough and the hysterical aphonia raised doubts in my mind, which the examination of the sputum strengthened, and which were entirely confirmed by the latter becoming fœtid. It was not a case of phthisis, but one of bronchiectasis in an hysterical subject.

A third group of cases in which the diagnosis will be materially assisted by the bacillus, is where (3) the catarrhal symptoms mask the consumptive ones ; and here, the detection of the consolidation is sometimes difficult, and in the absence of pyrexia, the diagnosis often obscure.

The following are very good instances :—

CASE 13.—William B., a bricklayer, aged 46, was admitted February 5, 1883. His mother and two brothers and sisters are asthmatic. History : The patient has had cough for two winters, which has become persistent since last March, and is accompanied by expectoration. He has had shortness of breath for two or three years, which has lately increased. For twelve months he has lost flesh to the extent of a stone. He has never had hæmoptysis or night-sweats. Cough troublesome ; breath short on exertion ; tongue slightly furred ; temperature and pulse normal ; weight 8 st. 2¾ lb.

Chest hyper-resonant ; wheezing and sonorous rhonchus audible over the whole lungs ; slight dulness in the first intercostal space on the left side. Expectoration frothy, with some yellow streaks, and contains abundant bacilli.

February 26.—Cough more troublesome ; expectoration abundant ; some crepitation in the left side, in the fourth interspace, on coughing ; wheezing and rhonchus audible everywhere ; bacilli abundant and dividing.

March 5.—Has suffered much from cough and wheezing, but now the former is easier and the expectoration much less; bacilli few in number.

April 2.—The breathing has improved during the last few days. The expectoration is very abundant, yellow, and comes up easily. Pulse 104; temperature 99·2° F.; respirations 22. Bacilli very few in number.

Sonorous rhonchus diminished over the entire lungs, and now chiefly confined to the left side. Doubtful cavernous sounds audible above the left clavicle on coughing.

April 9.—Cough less; breathing better; expectoration diminished; appetite good; *distinct cavernous sounds above the left clavicle;* bacilli few, four or five in the field, with an occasional group of three.

This case is interesting with reference to its diagnosis, as well as regards the progress of the cavity-formation. There was a strong family history of asthma, and dyspnœa appears to have accompanied, if it did not precede, cough, in the history. Had it not been for the wasting and slight dulness in the first interspace on the left side, we should hardly have expected phthisis, especially in a man of forty-six; and while the abundant expectoration made us suspect excavation, the examination of it settled all doubt, though it was not till the wheezing and rhonchus had entirely cleared up, that we were able to hear the cavernous sounds. Possibly, had we examined for lung tissue, this might have been made out earlier. The reduction in the number of bacilli, when excavation was complete, was very marked.

CASE 14.—Hugh P., aged 41, railway porter; admitted January 4, 1886. Several of his brothers suffered from bronchial asthma. He had whooping cough when a child, and during the last five years has been subject to nocturnal dyspnœa, the attacks sometimes lasting an hour or two, but not as a rule accompanied with cough. The short breathing lately has not been confined to the night, and is increased by exertion. Since August 1885 he has had occasional pains in the left breast, and for the last two months has had a severe cough, with copious expectoration. He has not lost flesh; his temperature is 98° F.; respirations 28.

The whole chest is hyper-resonant, there being no cardiac or hepatic dulness, sibilus and sonorous rhonchus are audible everywhere. Crepitation of a scattered kind is heard in the left front chest, from the clavicle to the fifth rib.

January 11.—The breathing is easier, the chest measurements show a circumference of 17½ inches for each side, below the nipple. The sibilus and sonorous rhonchus have almost disappeared. *Crepitation has diminished in the left front, but is coarser in the first and second interspaces. Cavernous sounds are audible above the clavicle.*

January 18.—The sputum contains abundant bacilli. Physical signs as above.

Three months after admission this patient had gained one stone in weight, and his cough had diminished, but the signs of a cavity were still distinct, and the sputum contained bacilli.

Here the presence of emphysema and the history of asthma pointed to the probability of pulmonary congestion, but the presence of the tubercle bacilli in the sputum settled all doubts, and the detection of cavernous sounds at a later date confirmed the diagnosis.

We must never forget that emphysema is the ending of many cases of chronic phthisis, and that in the *post-mortem* room we often find in the middle of emphysematous lungs old puckered cicatrices of cavities, and, more often, cavities still secreting. In the latter case the examination of the sputum during life will give us a clue to the real nature of the case, and I have often succeeded in thus detecting phthisis beneath the mask of emphysema. (4) Another group of cases in which the bacillus test aids diagnosis, is that of obscure pyrexia. Sometimes pyrexia accompanies doubtful pulmonary signs, and acute cerebral symptoms supervene. Some of these patients develop tubercular meningitis, which is rapidly fatal; here the sputum, if there be any, generally shows tubercle bacilli. But we may have other patients with similar head symptoms, with pyrexia, with cough, and expectoration, and a history of having been weakened by losses of blood, as after a confinement. I have had two such cases under my care. Repeated examinations have failed to detect tubercle bacilli, and the patients have completely recovered, showing that the pyrexia and brain symptoms were probably due to disturbances in the brain and spinal cord from diminution of blood supply, and not from tubercular or other lesions.

Another type of obscure pyrexia is that which accompanies tuberculosis in the old. It is not uncommon for patients of advanced age to die of tuberculosis, of which the principal manifestations are fever and wasting. There is generally cough and expectoration, but these, being common in the aged, are mistaken for those of chronic bronchitis, and attention is generally directed to symptoms proceeding from the state of the liver or intestines, and thus the true lesions come to be overlooked. The examination of the sputum in these, as a rule, shows abundant tubercle bacilli and explains the pyrexia. In a man aged fifty, recently under my care at the Brompton Hospital, who had suffered for years from the effects of lead poisoning, and principally complained of colic and dyspepsia, the temperature was high, and showed the curve of acute phthisis. Physical examination showed some old pleurisy and an enlarged liver. Still he had cough, expectoration, and was wasting, and the sputum on examination showed abundant tubercle bacilli. About six weeks after admission signs of disease were detected in the upper lobe of the left lung, and a cavity quickly formed.

(5) Cases of laryngeal phthisis at an early stage present occasional difficulties of diagnosis, partly from the mucous membrane of the larynx being swollen, and not presenting distinctive ulceration, and partly because, respiration being a painful act, the patient does not breathe sufficiently deeply to give auscultation a fair chance. Here again, the examination of the sputum helps us considerably. A gentleman, aged thirty-five, with a decidedly syphilitic history, came to me in February 1883, with loss of voice, which he stated was only an occasional symptom, wi h cough and abundant expectoration, loss of flesh, and night-sweats. Examination of the chest showed obscure breathing in almost every part, with fine crepitation under the right clavicle. The larynx presented a red and somewhat tumid mucous membrane, and at a later date the crepitation above mentioned passed away. The sputum was examined and found to contain abundant bacilli. About a month later crepitation was detected at the left apex, and a cavity rapidly formed, the larynx now exhibiting the characteristic features of tubercular ulceration. He died some months after of advanced laryngeal phthisis. In this case it is true that the general evidence pointed towards phthisis, but the paucity of the physical signs and the history of the patient made me strongly suspect syphilitic disease of the lungs and larynx, which the detection of the bacilli and the after-history of the case negatived.

(6) Not only in cases of doubtful syphilitic laryngeal disease, but also in the diagnosis of syphilitic disease of the lung, is the bacillus most useful. In patients with a marked syphilitic history and physical signs indicating consolidation or congestion of the lower lobes of one or both lungs, we are apt to conclude somewhat hastily that the cases are specific. In several of these, tubercle bacilli have been detected in the sputum, thus indicating the true nature of the disease. Doubtless, as this excellent test is more largely applied, other and no less important points of diagnosis will be achieved, but enough has been said to show how valuable it is not only in its positive aspect of determining what *is* phthisis, but also in its negative, in showing what is *not*. We cannot too strongly impress on our readers the importance in cases, where the slightest suspicion exists, and where there is expectoration, and especially if the expectoration be profuse, of examining for tubercle bacilli.

(7) The last classes of cases worth notice are those of pyo-pneumothorax, where it is desirable to ascertain whether the disease is due to a tubercular cavity opening into the pleura, or to an empyema bursting into a bronchus. In the former case tubercle bacilli are found in the sputum and the pleuritic fluid. In the latter they are not present.

Prognosis.—The value of the bacillus tuberculosis in the prognosis of phthisis can be to a certain extent deduced from what has been already stated. Wherever it abounds in the sputum it will be well to give a guarded opinion, and should the large number be accompanied by pyrexia and increase of symptoms of irritation, and above all, by an increase or extension of the physical signs, the prognosis is, for awhile at any rate, unfavourable. A rapid increase in the number of the bacilli indicates either excavation or extension of tubercular disease, or both, and conversely their diminution points to a tendency to quiescence. The appearance of bundles or masses of the bacilli is also by no means favourable.

The disappearance of the bacillus from the sputum is of course a highly favourable sign. At the same time, it must be confessed that the numbers of the bacilli in the sputum are too variable to allow of their being of much use for accurate prognosis, and we are inclined to agree with Dr. Hunter Mackenzie,[1] 'that the bacillary test is extremely valuable, as giving an accurate indication of the existence of pulmonary tubercular disease, but it sheds little light on the course which the disease will pursue,' and with M. Germain Sée,[2] that the multiplicity of the parasites does not in any way indicate the gravity of the lesion.

[1] *Op. cit.*   [2] *Bacillary phthisis.*

CHAPTER XI.

HÆMOPTYSIS AND THE HÆMORRHAGIC VARIETY OF CONSUMPTION.

Hæmoptysis—Its significance—Views of Louis, Laennec, Andral, Watson, Austin Flint—Niemeyer's Explanation of large Hæmoptysis—Rarity of Bronchial Hæmorrhage—Comparison of Bronchial Hæmorrhage with Epistaxis inappropriate—Differences in the Bronchial and Nasal Tracts—Niemeyer's Views of the relation of Hæmoptysis to Phthisis discussed—Origin of Phthisis from Hæmoptysis improbable—Author's experience—Various causes of Pulmonary Hæmorrhage—Sources of Hæmoptysis in Phthisis—Fatty Degeneration of Vessels—Ulceration—Aneurysms of Pulmonary Artery—Their Varieties and Pathology—Influence of Age and Sex in Hæmoptysis—Influence of Stage—Illustrative Cases—Hæmorrhagic Phthisis—Its characteristic Symptoms and exciting Causes—Examples—Effects of Hæmoptysis on the Lungs—Insufflation and Gravitation of effused blood—Blood the carrier of Tubercle bacilli—Eruption of Tuberculosis—Pyæmia—General and local Pneumonia—Blood residues—Illustrative Cases—Influence of Hæmoptysis on Duration.

So many cases of phthisis are accompanied by hæmoptysis in some part or another of their course, most commonly in the early stages, that spitting of blood, or 'bursting a blood-vessel,' as it is popularly called, has long been considered by the public, and to some extent by the profession, as an indication of consumption. The connection of hæmoptysis and phthisis, though simple in the stages of softening and excavation, is by no means always so in the early stages of the disease, and especially in cases where large hæmoptysis takes place, and but slight, if any, physical signs are detected at the time of its occurrence. Here the existence of consumptive disease is often denied; and when at a later date it develops itself more clearly, its cause is sought for by some in the blood effused into the bronchi during hæmoptysis, which is considered to have given rise to inflammation and destruction of the lung substance. We propose in this chapter to examine the views held by various writers on the relation of hæmoptysis to phthisis, and to state the conclusions which our own experience has led us to adopt on this subject.

K

What then do authorities say as to the significance of hæmoptysis? Louis states, that excluding cases of amenorrhœa and mechanical injuries to the chest, he did not find a single instance of hæmoptysis among twelve hundred cases, unconnected with tuberculous disease of the lung. Laennec holds much the same opinion, and Andral states, that of persons who have had hæmoptysis, one-fifth have not tubercles in the lungs; but he does not state whether any cardiac or other lesion existed to account for the hæmorrhage. Sir Thomas Watson[1] says, 'If a person spits blood, who has received no injury to the chest, in whom the uterine functions are healthy and right, and who has no disease of the heart, the odds that there are tubercles in the lungs of that person are fearfully high.' Austin Flint[2] says: 'Hæmoptysis, the hæmorrhage limited to the bronchial mucous membrane, and not dependent on disease of the heart or an injury of the chest, is always presumptive evidence of existing pulmonary disease.'

On the other hand, Niemeyer,[3] after stating that bronchial hæmorrhage is the 'most frequent cause of hæmoptysis,' explains that it 'proceeds from rupture of the capillaries, caused either by over-distension, or else by a morbid delicacy of the walls, a result of perverted nutrition.' He remarks very justly, that 'trifling capillary hæmorrhage, such as occurs in bronchial catarrh, violent irritation of air passages, and in the circulatory disorders attending organic disease of the heart, proceeds from the first of these causes: but that in most hæmorrhages in which large quantities of blood are poured into bronchi, to be ejected by hæmoptysis, they are due to the latter condition.' It is much to be regretted that Niemeyer uses such vague terms as 'morbid delicacy' of walls of vessels: but as far as we understand him, in regarding the vascular walls as the seat of disease, and their fragility, as the cause of large hæmoptysis, we agree with him. Why he should assign such hæmorrhage to the bronchial trunks and capillaries, we are at a loss to understand, as he gives no fact to support his statement, and we know that simple bronchial hæmorrhage

[1] *Practice of Physic*, vol. ii. p. 200.   [2] 'On Phthisis.'
[3] *Text-Book of Practical Medicine*, vol. i. p. 141.

has rarely been demonstrated by *post-mortem* examination—the instances being aneurysm of the bronchial artery, noted once by Dr. Church,[1] and a case of hæmorrhagic diathesis, mentioned by Dr. R. Thompson.[2] The appearance of the bronchi after death from hæmoptysis, containing large clots and fluid blood, with the mucous membrane stained, and even softened by it, has sometimes given rise to the idea that the bronchial membrane is the source of the hæmorrhage; but, as Dr. Reginald Thompson[3] points out, this condition of the bronchi is equally present after death from the bursting of an aortic aneurysm, or the lesion of a pulmonary vessel, or pulmonary aneurysm, or again, in fatal hæmorrhage from the stomach, it being simply due to regurgitation of blood into the lungs at the moment of death, whatever be the source of the blood. Niemeyer's comparison of bronchial hæmorrhage with the large hæmorrhage from the nasal mucous membrane, which occurs in profuse epistaxis, does not hold good, as may be shown by structural differences in the two tracts. The Schneiderian membrane in parts, as on the septum nasi and over the spongy bones, is very thick, partly through the presence of glands, but chiefly, as Todd and Bowman[4] say, 'from the presence of ample and capacious sub-mucous plexuses of both arteries and veins, of which the latter are by far the more large and tortuous. These serve to explain the tendency of hæmorrhage in case of general or local plethora.' The bronchial mucous membrane, though undoubtedly vascular, cannot be said to present in its structure any explanation of copious hæmorrhage like that of the nasal tract.

Niemeyer[5] sums up his views on the relation of hæmoptysis to phthisis in the following paragraphs:—

' 1. Bronchial hæmorrhage occurs oftener than is generally believed in persons who are not consumptive at the time of the bleeding, and who never become so.

' 2. Copious bronchial hæmorrhage frequently precedes consumption, there being, however, no relation of cause and effect between the hæmorrhage and pulmonary disease. Here

---

[1] *Pathological Transactions.*   [2] *Pulmonary Hæmorrhage*, p. 36.
[3] *Op. cit.*   [4] *Physiological Anatomy*, vol. ii. p. 3.   [5] *Op. cit.* p. 14

both events spring from the same source—from a common predisposition, on the part of the patient, both to consumption and bleeding.

‘3. Bronchial bleeding may precede the development of consumption as its cause ; the hæmorrhage leading to chronic inflammation and destruction of the lung.

‘4. Hæmorrhage from the bronchi occurs in the course of established consumption more frequently than it precedes it. It sometimes, although rarely, appears when the disease is yet latent.

‘5. When bronchial hæmorrhage takes place during the course of consumption, it may accelerate the fatal issue of the disease, by causing chronic destructive inflammation.’

With regard to the first class, Niemeyer states that hæmoptysis occurs, though rarely, in young persons in blooming health and of vigorous constitution, and that there is absolutely no explanation of the disorder, which is often followed by such sad results. Here we question the facts, both of the hæmorrhage being really bronchial, for the reasons we have given above, and of the health of these persons being really sound. After severe injuries to the chest, and occasionally after operations, people have been known to expectorate blood, and even in large quantity, and to have perfectly recovered. Again, cases of large hæmoptysis have been recorded in topers or drunkards, where bleeding from other organs has also been noted, such hæmorrhage being due, as in hæmophilia, to which we will allude later on, to the depraved state of the blood and not to the condition of the vessels. But such individuals can hardly be included in the category of ‘blooming health’ and a vigorous constitution. Excluding the above, in most cases of pulmonary hæmorrhage, even when the individuals appear so well, we have generally detected the physical signs of disease of the lung, present over very limited areas of the lung, and often rather obscure. Austin Flint[1] appears to have been similarly successful, for he says : ‘In cases where, at the time of the bleeding, nought can be detected, and it is generally difficult to thoroughly explore the

[1] *Op. cit.*

chest at this period, a week or two later very decided signs can often be detected.' Niemeyer admits that exceptional instances occur in which tubercles and inflammatory processes form in the lungs in a manner so latent that no tokens of disease are manifested by the individual affected, until he is suddenly attacked by a fit of hæmorrhage; but he denies that this is the case in the great majority of instances, where the first attack of hæmoptysis has not been preceded by cough, dyspnœa, or other sign of pulmonary disorder. How he distinguishes between those two classes, or on what grounds he supposed that they are separate classes, does not appear; but he continues, ' that bronchial hæmorrhage is by no means so rare an event where there is no grave disease of the lungs, is shown, moreover, by the tolerably numerous cases in which persons, after suffering one or more attacks of pneumorrhagia, regain their health completely, and indeed often live to an advanced age, and after death present no discoverable traces of extinct tuberculosis in their lungs.' Unfortunately, none of these 'tolerably numerous cases' are given to support this statement, which can hardly therefore be considered to be supported by satisfactory evidence.

Our own experience is exactly the reverse; and we have generally been able to detect signs of disease during life in the lungs of those patients who have had extensive hæmoptysis, unconnected with heart disease, injury to the chest, disordered state of blood in the form of hæmophilia or alcoholism, or vicarious menstruation.

Niemeyer also finds ' a strong tendency to profuse capillary hæmorrhage from the bronchi in young persons between fifteen and twenty-five, whose parents have died of consumption, and who have suffered in infancy from rickets or scrofula, have often bled at the nose, and grown rapidly tall.' He is tempted to refer the remarkable frequency to a deficiency of vital material which has been immoderately expended in the maladies of childhood or in the process of growth, and therefore does not leave sufficient to maintain the normal nutrition of the capillary walls. He remarks that this does not explain why the seat of the hæmorrhage should be, first in the nose,

and secondly in the bronchi, &c. Now we would ask our
readers, could a case for probable consumptive origin of
hæmoptysis be more clearly made out than it is here by
Niemeyer himself? The family history, predisposing influences and diseases and structural features are complete. The
exciting cause of inhaled bacilli only is wanting to determine
the outbreak of disease in the lung, and in the case of the
scrofulous not even this is required, and those patients will
run the course of phthisis. Again, Niemeyer seems to forget
that there is a common tendency in childhood to coryza and
epistaxis which may continue in youth, if disease in the vessels
of the lung do not cause a diversion. The argument which has
been often used, that because people who have had large hæmoptysis recover without the usual development of the symptoms
of phthisis there is no tubercle, is no argument at all, as it
is quite possible for small masses of tubercle to form in the
neighbourhood of large vessels, erosion to take place, and
after the hæmorrhage, the disease to become quiescent and
arrested. We are quite ready to confirm the accuracy of
Niemeyer's fifth paragraph, and can cite many cases illustrating the destructive effects on the lung of large hæmorrhage
during the progress of phthisis.

With regard to paragraph three, where the doctrine of
development of phthisis from hæmoptysis is laid down, we
are entirely at issue with Niemeyer, who says, ' Unbiassed and
careful observation of patients, who, without warning, and
often in the midst of exuberant health, have been attacked by
pneumorrhagia or hæmoptysis, and who, without rallying,
have perished in a few months of a phthisis florida, or " galloping consumption," has taught me that such patients scarcely
ever succumb to a pulmonary tuberculosis in its stricter sense,
but that they usually die of a form of consumption as yet but
little thought of, and of which bronchial hæmorrhage is the
immediate cause.'

We must confess we have never in the course of practice,
or in consulting the extensive records at our disposal, met
with a case of phthisis florida arising from simple hæmoptysis,
without any signs or symptoms of phthisis: though many

instances of large hæmoptysis, accompanying very limited
induration, have come under our notice, and some (*see*
case 25) ended in death, with pyæmic symptoms. We have
also known men without the symptoms of lung disease bring
up quantities of blood and recover, without permanent cough.
These patients were generally middle-aged, and often had arcus
senilis, from which we are inclined to think that fatty degeneration may have been present in some of the pulmonary
vessels, as well as in the cornea, and thus have caused their
brittleness and rupture. As recovery is the rule in these
cases, an opportunity seldom occurs of investigating the state
of the vessels after death. If the hæmoptysis leave no congestion or injury to the lung, we have generally found it give
relief to previously existing oppression or cough; but in cases
where the physical signs of consolidation and breaking down,
and not merely of bronchial plugging, exist, there follow
increased cough and fever, and the lung passes rapidly through
its career of destruction. Again, looking at Niemeyer's view
from the aspect of probability, if persons have been really in
exuberant health, without any previous disease of the lungs
or of their vessels, why should they, without warning, suddenly
spit up quantities of blood and go into rapid consumption,
when really healthy persons may be subjected to all sorts of
violent exertion without such results, or even if they do spit
blood, do so without further symptoms?

Having thus discussed Niemeyer's views, we will now state
our own conclusions on hæmoptysis, founded on the experience
of many thousand cases. Setting aside mere streaks or
tinges in the expectoration such as are seen in pneumonia and
bronchitis, blood-spitting may arise from (1) alterations in
the blood composition, as in scurvy or purpura, or again
markedly in hæmophilia; also (2) from congestion of the
lungs through cold or alcoholism, when very large quantities
of blood are often brought up. In alcoholism the organs are
all gorged with blood and friable, and more than a pint is
often expectorated, and in congestion of the lung from cold I
have known the hæmoptysis to exceed that quantity, and no
pulmonary lesion to be found after death. (3) Cancer of the

lungs and hydatids of the lung may cause fatal hæmoptysis, two instances of the former being recorded by the late Dr. Theophilus Thompson,[1] and one of the latter by Dr. Percy Kidd,[2] where the separation of an hydatid membrane from its cavity in the lung led to fatal hæmorrhage. A common cause (4) is disease of the heart and great vessels, especially aneurysm of the aorta. Another (5) is strain on the heart from overexertion, giving rise to pulmonary congestion. We may add, too, (6) embolism and thrombosis of the pulmonary vessels. Lastly, (7) diseases of menstruation, and (8) injuries to the thorax and lungs.

Now, if we except the above causes and also bronchiectasis in which large blood-spitting may take place, we may lay down as a law that hæmoptysis exceeding one ounce in amount is due to changes in the pulmonary blood-vessels connected with phthisis. What are the changes in the blood-vessels in phthisis?

First. Fatty degeneration was demonstrated, by Dr. Radcliffe Hall, to exist in the small blood-vessels. In a series of careful observations ' On the varieties and metamorphoses of tubercle,' well illustrated with plates, to which the reader is referred,[3] Dr. Hall records having detected fatty degeneration of the blood-vessels in phthisis in four instances, ' near to, but not mixed up with, tubercle in one ; within grey tubercles on the enclosed wall of an air-vesicle in one; in the wall of a large cavity in two.' Alluding to the case where the vessel was connected with grey tubercle, he states that it was studded with dim granules and oil dots of various sizes, and that the patient had suffered from copious and repeated hæmoptysis two years before his death. Dr. Hall remarks, ' Considering how many cases of phthisis spring into noticeable activity coincidently with an attack of hæmoptysis, the patient so commonly declaring that his chest was strong, and his health good, until suddenly he broke a blood-vessel, this fact of fatty degeneration of blood-vessels, where no inflammation exists, at the earliest stage, and connected with the highest form of

[1] *Pathological Transactions*, vols. vii. and viii.
[2] *Medico-Chirurgical Proceedings*, New Series, vol. i. No. 8.
[3] *British and Foreign Medico-Chirurgical Review*, April-October, 1855.

tubercle, while as yet it occasions neither pain nor organic irritation of the lung—is valuable and explanatory.'

Second.   Ulceration and erosion of the pulmonary vessels take place during the tubercular changes.   This is no doubt the cause of the early hæmoptysis of phthisis, and is probably due to the infiltration of the walls of the blood-vessels with tubercle, and to ulcerative changes taking place in them subsequently.

The observations of Weigert, Mugge and Percy Kidd show the presence of bacilli in considerable masses in the walls of the small pulmonary veins in acute miliary tuberculosis, and Koch and Kidd have also noted them in the small arteries. Here they are seen in large numbers in the infiltrated and thickened wall, in connection with commencing caseation, and while in one part the calibre of the vessel appears encroached on by the infiltration of the wall (*see* plate IV. fig. 2) in another the bacilli are seen passing through the wall, and in Koch's [1] case they had extended into the interior, and thus might have mingled with the blood corpuscles and serum in the still patent vessel.   The significance of Ribbert's researches in connection with hæmoptysis has been already dwelt on.   While these observations demonstrate, as has been noted above, the spread of tuberculosis in the system, they also explain the so-called initial hæmoptysis of phthisis, for it is manifest that its presence or absence and its amount in any case, would depend on whether the primary tubercle, caused by the entry of the bacilli into the alveoli, lay in close proximity to vessels, or at some distance from them, and whether their vascular walls were perforated by the bacilli, or were pressed on by the infiltration and thus occluded : or again, closed by thrombi formed in their diminished lumen. The implication of a large vessel would account for the large hæmoptysis which often accompanies early phthisis, in which the physical signs are sometimes difficult of detection.

In connection with this, I may notice that tubercle bacilli were **first** recognised in expectorated blood in 1883, by my

---

[1] Watson Cheyne, *Practitioner*, April 1883, p. 292.

clinical assistant, Dr. Perez, and myself, at the Brompton Hospital in two patients of Dr. Tatham's.

Third. Aneurysm of the branches of the pulmonary artery is perhaps the best ascertained cause of fatal hæmoptysis in phthisis. The changes which the vessels undergo to arrive at this condition have of late years received considerable attention from pathologists, and great light has been thrown on the various steps of the process. Rokitansky remarked on the frequent patency of branches of the pulmonary artery lining the walls of cavities, and gave an excellent account of pulmonary aneurysm in his 'Pathological Anatomy.' We believe that Dr. Peacock[1] and Mr. Fearn, of Derby,[2] were the first to record instances of pulmonary aneurysm in this country.

Of late years numerous cases have been described by Drs. Cotton,[3] Quain,[4] Moxon,[4] Douglas Powell,[4] West,[4] Percy Kidd,[4] and others. Dr. Rasmüssen,[5] of Copenhagen, has treated very fully of the modes of formation and bursting in these vascular expansions, as exemplified by nine cases, carefully investigated by himself. Dr. Douglas Powell[6] gives a table of fifteen cases of fatal hæmoptysis occurring at the Brompton Hospital between 1865 and 1870, in all of which cavities of various sizes existed; in thirteen the ruptured vessel was detected, in one there was rupture without dilatation of the vessel, and in twelve, ruptured aneurysms were found. Dr. S. West[7] furnishes twenty-five fatal cases from the Victoria Park Hospital for Diseases of the Chest, in sixteen of which the source of the hæmorrhage was found, and proved to be an ulcerated vessel in six, and pulmonary aneurysm in ten. In all, cavities were present, and for the most part of old standing, as evidenced by their fibroid walls. Dr. Percy Kidd,[8] in an account of thirty-five cases of fatal hæmoptysis from the Brompton Hospital, occurring during the years 1882 to 1885, traced the source of bleeding in all

[1] *London and Edinburgh Monthly Journal,* 1843, p. 383.
[2] *Lancet,* 1841.   [3] *Medical Times and Gazette,* January 13, 1866.
[4] *Path. Trans.,* vols. xvii. xviii. xxii. xxxv.   [5] *Dobell's Reports* for 1869.
[6] *Path. Trans.,* vol. xxii.   [7] *Medico-Chirurgical Transactions,* vol. lxviii.
[8] *Medico-Chirurgical Proceedings,* vol. i. No. 8. New Series.

but two, and in one of those cases an unruptured aneurysm
was found. In thirty, ruptured pulmonary aneurysms were
discovered. From the accounts of the above-mentioned
observers, and from several cases we have witnessed ourselves,
the course of events appears to run thus : When the processes
of softening and excavation are going on in the lungs, branches
of the pulmonary artery are laid bare, and are to be occasionally seen, as yet undilated, in the walls of cavities in phthisical
patients who die from other causes than pulmonary hæmorrhage. This state of the vessels, if there is much obstruction
to circulation in the lungs, does not last long; either a rupture
takes place, as occurred in one of Dr. Powell's cases, or the
want of support causes the vascular wall to dilate on the
side towards the cavity, and a protuberance varying in size
and shape according to that of the unsupported portion of
vessel, takes place. The slightest form of aneurysm is that
denominated *ectasia*, by Dr. Rasmüssen, where the vessel
touches the cavity for only a limited extent, and a small
oblong dilatation is formed, due partly to the expansion of the
bore of the vessel, and partly to the wall growing thicker.
According to Dr. Kidd the thickening involves all three coats
of the artery, but in the early stage the intima is mainly
affected, the muscular coat is ruptured, and a thrombus
separates the intima and the greatly expanded and attenuated
adventitia. In more advanced aneurysms there is less
marked thickening, but a cellular infiltration of all the arterial
coats. The bursting takes place on the boundaries between
the vessel and the wall of the cavity, and the point of the lid
always lies in the direction of the blood current. The
aperture is generally triangular, but sometimes a longitudinal
slit. A very good example of ectasias was recorded by
Dr. Quain,[1] where in one cavity two small branches of the
pulmonary artery exhibited several varicose dilatations, one
of which had burst, causing fatal hæmoptysis. The other
form of aneurysm is sacculated, or sometimes fusiform, and
varies in. size from a pea to a small orange, being sometimes
large enough to nearly fill a small cavity ; a larger amount of

the vessel is exposed than in the ectasias, and the pouch contains sometimes clots of blood, sometimes laminated fibrin,
the perforation, according to Dr. Rasmüssen, being found
on the most protruding portion of the dilatation. Several
aneurysms or ectasias are sometimes found in the same lung,
though their presence is generally confined to cavities. The
number varies greatly, sometimes two or three, sometimes
six, and in a remarkable case of Dr. Kidd's twenty-two were
found in the same lung. Sometimes they are found in cases
where there has been no marked history of hæmoptysis, and
the acuteness or chronicity of the disease does not always
affect their formation, for though most commonly accompanying chronic cavities, they are found occasionally in acute ones
devoid of lining membrane and with soft caseous walls. They
are more common in small cavities than large ones, and are
rather more frequent in the left lung than the right. The
aneurysms are found chiefly in the lower two-thirds of the
lung, and very commonly near the base, and the fact of their
greater frequency in this part of the lung is assigned by
Dr. Reginald Thompson to the more active movements of
respiration in this portion, which hinder thrombosis and
favour ectasia of vessels. This is still more the case when,
as often happens in advanced phthisis, the diaphragm and
lung are adherent. The cause of the formation of these
aneurysms does not always lie in any general diseased condition of the pulmonary vessels, for these, except at the points
of dilatation, where there is endarteritis, have been hitherto
found to be healthy, but it owes its existence principally to
the want of support of the wall of the vessel turned towards
the cavity, and partly to heightened intervascular pressure,
from the tendency of blood to fill the vacuum caused by the
contraction of the lung, especially around the cavities, and
also from the obliteration of so many branches of the pulmonary artery. A considerable share may be assigned to the
more active movements of respiration in the lower lobes.
Hæmoptysis from a ruptured aneurysm is not rarely fatal,
although the first attack is not necessarily so; for in eight of
the Brompton Hospital cases, there had been copious hæmo-

ptysis on more than one occasion, previously to the attack
which terminated life. I remember one well-marked case of
limited cavity, where each attack of hæmoptysis was accompanied by gurgling sounds over the region of the excavation,
and on the subsidence of the hæmorrhage, the cavernous
sounds always became dry. The attacks were frequent and
profuse, amounting to pints of blood at a time ; but the patient
recovered strength in the intervals, and although the hæmo-
ptyses extended over a period of ten years they eventually
became less frequent and almost ceased, the patient dying of
albuminuria and dropsy. I have no doubt that here the cause
of the hæmorrhage was a ruptured pulmonary aneurysm, which
was closed by the formation of a clot, the recurrence of the
hæmorrhage being due to its disintegration or to the bursting
of another pulmonary aneurysm in the same cavity. The
gradual contraction of the cavity through fibrosis of the lung
obliterated the vessel, and the patient died of a common
sequence of phthisical disease. Aneurysm of the pulmonary
artery may also take place in the wall of a bronchus and
burst into it, as has been shown by Drs. Powell and Reginald
Thompson.[1]

Having thus pointed out the pathology of large hæmoptysis
in phthisis, we may state that the expectoration of small
quantities of blood is probably due to congestion or inflammation of one or other of the two systems of vessels existing
in the lung, similar to what occurs in heart disease or in
pneumonia. For such slight oozings of blood it is not
necessary that there should be any rupture of vessels, as it
has been ascertained that the red corpuscles can, like the
white corpuscles, migrate through the vascular walls and
pass into the surrounding tissues, but it is extremely improbable that any considerable hæmorrhage can take place in
this way.

It will be useful now to consider certain other points connected with hæmoptysis, such as

(1) The influence that age and sex exercise on its occurrence.

[1] *Op. cit.*

(2) The stage and form of the disease in which it prevails most.

(3) What effect it exerts locally and generally on the course of the disease—what complications it gives rise to.

(4) Whether or not it curtails the duration of life.

The influence of *sex*, according to the first Brompton Report,[1] is very trifling, the females showing excess over the males of only 3 per cent. In 1,000 cases tabulated from private practice, the majority of whom have been under observation for many years, we arrived at a different result. The total number of cases of hæmoptysis in both sexes was 569, or about 57 per cent.; but among the males the percentage was 63, and among the females 47, giving an excess to the males of 16 per cent.

Again, in 283 cases where the amount of blood expectorated at any one time exceeded an ounce, the liability of males was still more marked. Large hæmorrhage was found to occur in 34·76 per cent. of the males, and in only 17·67 per cent. of the females, this greater liability of males to spit blood in large quantities being amply confirmed by Dr. Pollock and the second Brompton Report.

Age cannot be said to exercise a very marked influence, except that, according to the first Brompton Report, in the case of females, hæmoptysis occurred more frequently under 35 years of age than between 35 and 70, and in the case of males more frequently between 25 and 45.

The influence of the *stage* of the disease is of much greater account; and here the first Report demonstrated, from observations on 696 cases, that hæmoptysis is more frequent (as 3 to 1) before softening than after that process has taken place. Our own observations amply confirm this; and with a view of carrying these investigations farther, we have classified our 283 cases of profuse hæmoptysis into stages, annexing the number of deaths from this cause to each stage. It must be borne in mind that presence or absence of excavation in these cases rests solely on the evidence of physical signs, and it is quite possible that in so-called first-stage cases minute

[1] p. 28.

cavities may have existed without detection. Dr. Reginald Thompson's statistics,[1] on the other hand, give the period of softening as that where haemoptysis is most common :

The following are our results : —

Stage.	Cases.	Deaths.	Percentage of Deaths.
1st	187	26	13·95
2nd	65	16	24·61
3rd	31	21	67·74
	283	63	

The figures demonstrate most convincingly the significance of large haemoptysis in the third stage of phthisis, and its comparatively slighter importance in the first stage ; but a few clinical cases to illustrate these points may not prove unacceptable. The first three are instances of haemoptysis in the third stage.

CASE 15.—Sarah B——, a single woman, aged 26, was admitted into the Brompton Hospital, under the care of Dr. Quain, Nov. 21, 1870. She had lost a brother and sister from consumption, but gave no history of illness previous to two months before admission, when she was attacked with cough, streaky expectoration, followed by great wasting, feverishness, and night-sweats. When seen, her cough was constant, with muco-purulent expectoration, and she had some dyspnœa. Tongue furred; bowels active; appetite good; catamenia regular till last period, but scanty, now absent about two weeks. Pulse 120.

On examination of the chest, *cavernous gurgle was audible over the right side of the chest, front and back, and some crepitation over the left side.*

A few days later I saw this patient, and, in Dr. Quain's absence, had charge of the case. She became worse, losing flesh, in spite of taking oil and nourishment, and about January 12, was attacked with haemoptysis, which was checked by styptics, but recurred again on the 16th, and she brought up large quantities, *i.e.* more than a pint at a time, for several days, the haemorrhage being only slightly influenced by treatment. She died suddenly on the 21st, after bringing up a large quantity of blood. The temperature during the last few days of her life, was, according to Dr. Giffard, the clinical assistant's observations, as follows :

	Morn.	Even.
January 17,	99·3° F.	102° F.
„ 18,	99°	101·2°
„ 19,	99°	100°
„ 20,	99°	

On *post-mortem* examination by Dr. R. Douglas Powell,[2] the lungs appeared highly emphysematous. The right was firmly adherent at its base, and the

[1] *Pulmonary Haemorrhage*, p. 113.
[2] I am indebted for some of the notes of the *post-mortem* examination to Dr. Powell.

pleura thickened and gelatinous. The base of the lung was consolidated, and contained a cavity, projecting into which was the pulmonary aneurysm which gave rise to the fatal hæmorrhage. The aneurysm was situated on a vessel which originally crossed the cavity, and at the time of death projected into it about an inch, having, at this point, broken across in the progress of the disease. The arterial channel was preserved to the end of the truncated vessel, where it was dilated into a bulbous extremity. One side of this channel was formed by the brittle degenerated wall of the vessel, and presented a longitudinal rod-shaped deposit of lymph, the opposite side being made up of a considerable thickness of laminated coagula, filling up the sac of an aneurysmal dilatation. Another cavity was found in the upper portion of the lung containing clots of blood, and an aneurysmal dilatation of a small pulmonary vessel which had burst.

In this case the rapid breaking down of the lung tissue had led to the dilatation of the vessel from loss of support and degeneration; and it would appear that the same process took place in two cavities at nearly the same time.

CASE 16.—A young gentleman, aged 15, was seen by Dr. C. J. B. Williams, in consultation with Mr. W. Jones, on May 5, 1853. Had suffered from cough for six months, and during the last month from pain in the left shoulder, and considerable loss of flesh and strength. *Dulness, and crepitation, fine and coarse, in upper left chest.*—Ordered oil, acid tonic, and counter-irritation.

November 28. - Vastly improved, and has gained 14 lbs. Walks eight miles, and for the last three months has been shooting in the country. *Dulness and tubular sounds in upper left chest as low as third rib.*—Ordered oil in sulphuric acid tonic.

May 17, 1854.—Wintered well, with only slight cough and expectoration and short breath. Has been riding a good deal, and a week ago was caught in the rain, and rode home in wet clothes. Has since had pain in the left side, with night-sweats and diarrhœa. *Increase of dulness and moist coarse crepitation in upper left chest.*—Ordered a blister, and mercurial and chalk powder, with effervescing saline; afterwards to return to the oil, which had been discontinued for six months.

August 9.—Has been taking oil in the country, and is improved; but has cough and yellowish expectoration, occasionally tinged with blood. *Liquid cavernulous sounds below left clavicle.*

The patient continued better; but in September, after having been out shooting for five hours, was attacked with hæmoptysis to the amount of two pints. This recurred again and again, and he died in a week.

On examination after death there were found only a few miliary tubercles at the apex of the left lung, and a cavity of the size of a walnut, lined with a thick membrane, deficient at one point, whence the hæmorrhage had proceeded. The rest of the lung was congested. The right lung was healthy. But for his imprudent exertion, this patient would probably have avoided this fatal hæmoptysis, and might have completely recovered.

CASE 17.—James W., aged 28, potman, admitted into Brompton Hospital under Dr. Theodore Williams, October 30, 1873. Has had cough since Christmas, 1872, increasing in last three months, followed by slight hæmoptysis two or three months ago, and expectoration has often been streaked. Has lost flesh for six months. Night-sweats came on five months ago, but have ceased during the last six weeks under treatment. At the beginning of the illness he had shortness of breath and some rigors, which disappeared and have lately returned, accompanied by discharge from the right ear. At present complains of pain in the lower left front chest and under the clavicle; troublesome cough; thick dark brown sputum and nocturnal rigors. *Crepitation lower right chest posteriorly. Dulness over lower half of right chest anteriorly. Cavernous sounds audible over upper third, and also at base.* Four days later hæmoptysis came on, accompanied by pyrexia and quickened pulse and respiration, continuing daily in quantities varying from ʒiij. to ʒiij., accompanied by a pulse of 112 to 160, respirations from 44 to 60, and a temperature ranging between 98·4° F. and 103·2° F. Cavernous sounds were heard over the whole left lung, and on November 15 (*i.e.* eleven days after the beginning of the hæmoptysis) he expectorated one pint of blood and expired with symptoms of asphyxia.

Autopsy.—Post-mortem, 60 hours. Both pleuræ adherent. On left side adhesions, old at apex, but evidently recent at base. The right lung had a cavity the size of a walnut at apex, and the upper lobe and the posterior portion of the middle were consolidated with catarrhal pneumonia and contained a few patches of miliary tubercle surrounded by red hepatisation. The left lung contained several cavities, a thin walled one, the size of a hen's egg, at the

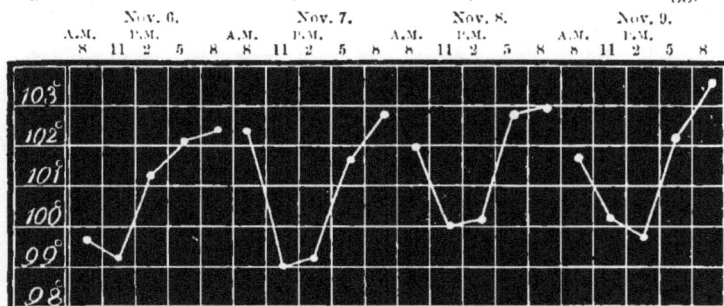

apex, another, half the size, at the base, and several others scattered through the organ, which was infiltrated with catarrhal pneumonia. On the inner side, in a small cavity near the apex, were two small aneurysms of the pulmonary artery, the size of a hemp-seed; two were found in another cavity, and again, two in a third, which they exactly fitted. One aneurysm was nearly as large as a walnut, the other communicated with it and was ruptured, containing a clot. Blood-clots were found in all the cavities and in the bronchial tubes.

This case was one of acute phthisis (scrofulous pneumonia), passing into rapid excavation with only a small amount of miliary tuberculisation. During the process of excavation a large number of blood-vessels were exposed in which aneurysms formed in the manner described above, one of which burst and caused death by asphyxia through the effused blood. From the number of aneurysms in separate cavities, the case was interesting, but its chief interest lay in the thermal chart of the last days, which is annexed. This presents the *typus inversus* which sometimes accompanies fatal hæmoptysis and which gives high morning and evening and lower mid-day temperatures, the exact reverse of the ordinary chart of phthisis.

The following is a good example of hæmoptysis accompanying the first stage, in which the patient remained well for many years, though eventually excavation of the lung took place:—

CASE 18.—A gentleman, aged 36, of strong, large frame, consulted Dr. C. J. B. Williams first on June 19, 1850. Had been in good health, with no cough; but after much exertion in May he expectorated half-a-pint of blood, and yesterday about the same quantity. Had been bled and leeched on the right chest with diminution of the dulness; signs of congestion were found in the right lung. Pulse quiet; no cough; bowels confined.—Ordered gallic and nitric acids, and digitalis; an aperient of sulphate of magnesia with sulphuric acid.

June 24.—Hæmorrhage returned slightly next day, but not since. *Dulness and tubular sounds in the upper right chest front and back.*—A blister to the scapula; nitric and hydrocyanic acid mixture.

July 20.—No return of hæmorrhage; cough slight. Has taken cod-liver oil with acid mixture a fortnight. *Less dulness, but still tubular expiration in the upper right chest, front and back.*

September 26.—At Norwood. Taking oil, &c., regularly. Much improved in flesh and strength. Cough and expectoration slight. *Dulness and tubular expiration less marked.*

June 21, 1851.—Wintered in Madeira. Well, but frequent colds from damp weather. Continued oil and acids till the last three months. *Breath harsh and loud below right clavicle; little dulness and less tubular sounds in right back.*

September 25, 1852.—Has continued well.

November 24, 1853.—Passed last winter with only one attack of cold and cough, when the sputa were discoloured. Has been pretty well since; but has had an occasional feeling of oppression in the right lung, and three weeks ago, after much exertion of voice, brought up four ounces of blood, and some on the following days, when it was checked by gallic acid. Still much cough and wheezing. Urine scanty, red, and turbid. *General wheezing rhonchi in right lung; little dulness or tubular sounds.*—Ordered iodide of potassium, carbonate of potash, squills, stramonium, and liquorice, and a croton-oil liniment.

Under this treatment he soon got the better of this illness, and continued in good health, except occasional attacks of the same kind; more wheezy and asthmatic than phthisical. He continued active in business, and often gave lectures and spoke at public meetings, although warned against doing so.

February 13, 1863.—Three months before had coughed up four ounces of blood, and about the same quantity three days ago. Breath had been shorter since, with more expectoration. Feels the cold more, but voice and muscular strength good. Urine often thick. *Dulness and tubular sound at and above right scapula; moist crepitus and mucous rhonchus above and below.*—To take oil, with sulphuric acid, calumba, &c.; saline at night.

June 18.—Several times hæmoptysis this year; now about half-an-ounce daily for five days. Physical signs the same.

December 27, 1864.—Hæmoptysis recurred every two or three weeks till August, when he went to Bournemouth; none since. Has regained some flesh and strength. *More extensive dulness, obstruction, and coarse crepitus over greater part of right lung; cavernous sounds in scapular region; dulness and loud bronchophony now also at and above left scapula.* These signs have been increasing during the last twelve months, with more constant cough, copious opaque flocculent expectoration, pulse 100, and loss of flesh and strength.

A few months later the cavernous sound in the right lung became amphoric, with signs of increase of disease in the left lung also. In the autumn of 1865 he went to Madeira, and died on the voyage back, early in 1866.

From 1853 to 1863 the disease was so much arrested that he resumed active habits, and ceased to regard himself as an invalid. The disease in consequence recommenced its activity, and ran its destructive course, but the total duration of the case amounted to sixteen years.

While hæmoptysis may accompany all stages of consumption, and may vary considerably in amount, it may also be entirely absent throughout the whole course of the disease. In the acute forms, and particularly in acute tuberculosis, hæmoptysis rarely occurs, and the relief in the more chronic varieties of the malady often derived from a copious hæmorrhage has led some physicians to think that the great congestion noticeable in acute cases might be relieved by local or general blood-letting. It does not appear that the hereditary element exercises any influence on its frequency, the number of cases of hæmoptysis not greatly differing in hereditary and non-hereditary phthisis.

Whilst the most rapid cases of phthisis are nearly free from hæmoptysis, there is a class of patients already alluded to in whom the amount of the disease is small, and large and repeated hæmorrhage the principal feature. We described a few of these cases some years ago under the term of the hæmorrhagic variety of consumption,[1] and it is no small satisfaction to us that so careful an observer as Dr. Peacock also recognised in them a separate class, which he calls ' the hæmoptysical variety,'[2] and he gives an excellent description

[1] *Lancet*, June 1868.   [2] *St. Thomas's Hospital Reports*, 1870.

of its chief features.   As far as we, ourselves, have observed, the symptoms are as follows :—The patient may have had a slight cough for some time, or shown signs of failing health, but very often this is not the case, and he may be in apparently fair health when he is suddenly attacked with hæmoptysis, generally of a profuse character, lasting often many days, and causing much reduction in flesh and strength.   Cough and expectoration follow, and a careful examination of the chest detects only slight physical signs, and these in the supra- or inter-scapular regions, or else beneath the clavicle.   The patient gradually improves, and will sometimes recover his strength and lose his cough before the hæmoptysis recurs, which it is pretty sure to do sooner or later.   The period before its recurrence may vary from days to years, and with it, to a great extent, the prospect of ultimate recovery, as the fre- quently recurring attacks considerably weaken the patient and soon usher in the train of symptoms usually attendant on active consumptive disease.   The cough becomes persistent, the expectoration muco-purulent, when   not   sanguinolent, wasting and night-sweats appear, and the physical signs which first evidenced slight consolidation, and later on, the same in a greater degree, now show softening and excavation.   This is, however, rare, for the changes in the lung are generally limited to increase in the amount of consolidation, and what appeared at first rather obscure and was even denied, now be- comes unmistakable.   The blood brought up is generally florid, but occasionally dark in colour, and partially coagu- lated ; the amount varies greatly, but it is usually large, and in one case under observation amounted to seven quarts at a time.   But the quantity expectorated is no measure of the danger resulting to the patient, as in this form of the disease a large amount of blood may be brought up without fatal or even pernicious results.

Our 1,000 tabulated cases furnish 72 instances of this form of phthisis, of whom 60 were males and 12 females, a proportion of 5 to 1 of the former to the latter, which forms a contrast to the proportionate numbers of the two sexes in all forms of consumption, which was as 2 to 1.   Thus we see that males are far more liable than females to the hæmorrhagic form of phthisis, and this for a reason that will presently appear. The patients were attacked rather later in life than the generality of consumptives; the average age of attack for the males being 30 and for the females 27. Family predisposition was not usually present, appearing in only 25 instances, and perhaps this may account for the later age of attack. In 42 cases the hæmoptysis was preceded for a shorter or longer time by cough, but in the rest hæmorrhage was the first symptom. In 45 no exciting cause was recorded. In 27 an explanation was to be found either (1) in the patient having been subjected to some great bodily exertion, as preaching, lecturing, acting on the stage, rowing, climbing, or running; or (2) in his having been inhaling an atmosphere either mechanically or chemically irritating to the lungs, as that of a laboratory or workshop; or (3) in his having been exposed to decidedly depressing conditions, as chills from getting wet through, great mental worry, fasting, too close application to a sedentary occupation, severe attacks of certain lowering diseases, as dysentery, measles, and syphilis, the latter being, in one case, followed by mercurialisation.

We see, therefore, that in a large number of these cases the attack of hæmoptysis did not occur, without, as Dr. Peacock expresses it, some more or less decided exciting cause, and it is probable that had great attention been paid to this point in the history, an exciting cause would have been traced in many more cases. Of the 72 instances of hæmorrhagic phthisis, we subjoin a few examples:—

CASE 19.—A clergyman, aged 34, first consulted Dr. C. J. B. Williams on July 17, 1848. After hard duty fourteen months ago he had hæmoptysis to the amount of two ounces, and in smaller quantity a week or two afterwards. Was much reduced by treatment, and wintered in Madeira with great benefit, and was well enough to ascend the Peak of Teneriffe without difficulty about six weeks ago; but has had cold and cough ever since. *Dulness and tubular expiration in upper right chest. Bronchophony and mucous rhonchus below right clavicle.*—Ordered iodide of potassium and carbonate of potash, and a croton-oil liniment.

In August he had hæmoptysis in church at Brighton to the amount of an ounce and a half, which ceased under acetate of lead with a daily aperient draught. After this he took cod-liver oil, and got much better.

July 18, 1849. – Well till last few weeks, when he had temporary cough.

July 9, 1852.—Was well and doing double duty till a year ago, when he caught severe cold, and had hæmoptysis to the extent of three ounces; for which he was leeched, and much lowered. Has lost much flesh and strength, has night-sweats, and expectoration has become yellow. *Dulness and loud tubular sounds in the left mammary region.*—Ordered oil in tonic of nitric and hydrocyanic acids, with tinctures of hop and orange.

July 22.—Had hæmoptysis (four ounces), yielding to treatment as before.

In 1854 he went to Madeira, and passed a second winter with benefit. Married in 1856.

March 1868.—Heard that he was quite well, with a family of six children.

May 1871.—Continues well. Since last visit he has had two or three attacks of hæmoptysis, and some years ago severe bronchitis and congestion of the lungs. At present he is in excellent health, doing parish duty, which he has not omitted, for more than two months at a time, for the last twelve years. Upwards of twenty-four years have elapsed since this patient's first symptoms.

CASE 20.—A clergyman, aged 42, first consulted Dr. C. J. B. Williams on August 15, 1850. Nine months ago, after over-exertion, he lost his voice, and had cough, accompanied by some wasting and failure of strength, but continued to take three services every Sunday all the winter. After more exertion a month ago, for four days his morning expectoration was streaked with blood. *Slight dulness ; moist crepitation, with slight tubular sounds below left clavicle, more marked above ; slight dulness and tubular voice in right interscapular region.* —Ordered oil in a tonic of nitric and hydrocyanic acids, with the tinctures of hop and orange, and a liniment of acetum cantharidis.

September 16.—Much improved, under the oil, in flesh, strength, and breath, and able to walk fast and up-hill. Came from Hastings to-day, and after much talking, on lying down at night, expectorated blood to the amount of two ounces, with irritating cough. Is in too much tremor from fear to bear examination ; pulse 90.—Ordered gallic acid, with syrup of poppy, followed by an aperient draught of sulphate of magnesia and sulphuric acid.

September 21.—No return of hæmorrhage ; *dulness, with crepitus above left scapula.*

September 30.—Hæmoptysis to a pint and a half in last two days, and cough still troublesome.—Ordered a blister and a morphia linctus, together with the gallic acid.

April 17, 1851.—Has wintered at Hastings, taking oil regularly, and gradually recovered from a state of great weakness ; moderate cough and only slight expectoration. Has so much improved that now he can walk for five or six hours daily. Physical signs diminished.

May 4, 1854.—Has taken oil regularly, and been quite well all the winter, without cough or expectoration ; but has lately had neuralgia of the eye and occasional flushes of blood to the head ; urine thick, with lithates ; *very little dulness or tubular sounds, but breath obscure above left scapula, and whiffy in left scapular region ; tubular voice above right scapula.*—Ordered oil in nitric acid and tincture of orange, and an occasional effervescing saline.

January 21, 1858.—Has done duty well since. Only occasional cough, but always some mucous expectoration. Is strong and active, although of late has been much worried by a lawsuit. Signs diminished.

May 18, 1860.—Active and well till a month ago, when cough returned, with opaque expectoration, which has since improved. *Still some dulness and tubular breath in both scapular regions, chiefly in left, and also in left front chest.*

September 26, 1866.—Wonderfully well, and generally does duty. Cough and expectoration very slight. Weighs twelve stone—heavier than ever.

This patient was alive and well in January 1871, more than twenty years after his first visit, and more than twenty-one since his first symptoms ; and for the last thirteen years has done duty regularly.

CASE 21.—A gentleman, aged 32, consulted Dr. C. J. B. Williams, August 9, 1854. Two years ago he had hæmoptysis to the amount of half-a-pint, and afterwards every three months he continued to bring up a smaller amount, till June, when he states that in five days he brought up 14 pints ! He has had a bad cough for the last month, and has lost much flesh. Bowels costive. *Dulness, deficient breath, and tubular sound in upper right back. Obstructed breath and crepitation below. Tubular sounds above left scapula.*

Oil was ordered in a mixture of sulphuric, gallic, and hydrocyanic acids, with tinctures of hop and orange. Also, for a time, a morning aperient draught of sulphate of magnesia and sulphuric acid. Counter-irritation with a cantharides liniment.

November 9.—Has gained much flesh and strength, and only once had hæmoptysis, amounting to ʒiij. *Tubular sounds above right scapula, breath still obscure below, and mucous croak on deep breath.*

December 1867.—Heard that he got quite well, except that his breath has continued rather short. Only lately has had more cough and expectoration.

CASE 22.—An officer in the army, aged 20, was first seen by Dr. C. J. B. Williams March 12, 1860, and gave the following history :—He had been subject to cough in winter, but otherwise enjoyed good health till the previous September, when he had a chancre, followed in December by sore throat, and an eruption on the legs. Four weeks ago cough came on, and a week later he had hæmoptysis to the amount of three drachms. Feels much weakness. *Slight dulness ; tubular sounds in upper part of right chest.*—Oil was ordered in a tonic of nitric acid, iodide of potassium and tincture of orange, and a night draught of iodide of potassium and bicarbonate of potash.

June 8.—Has been at Pau since. The cough is better ; but he has had occasional hæmoptysis to the amount of a drachm ; and he still has syphilitic eruptions and pains. Sub-maxillary glands swollen.

November 12.—Has been living in Ireland, and was better till he caught cold in the camp, and fresh cough came on, with increased expectoration, and occasional hæmoptysis to the amount of several ounces. *Dulness and tubular breath and voice above right scapula. Crepitation, and obstructive sounds in the upper third of left chest.*—To continue oil in acid tonic, and winter at Torquay.

September 18, 1861.—Wintered well at Torquay till April, when he brought up 4 ounces of blood, and continued spitting some blood for ten days. He remained well till eighteen days, when hæmoptysis recurred to the amount of 4 ounces, and he discovered he had recent syphilitic symptoms. *Dulness and deficient breathing ; and some croaky sounds in upper left chest.*

July 1862.—Wintered at·Madeira, and remained well, and married there ; but after some exertion he brought up an ounce of blood, and some slight amount since.   There is still *dulness, with rough breath in upper left chest, harsh behind.   Dulness increased above right scapula.*

October 2, 1863.—Passed winter travelling in Italy, Sicily, Spain, and Egypt, with little cough, and only had hæmorrhage to amount of half-an-ounce, Took oil pretty regularly.   In the summer went to Turkey and Malta, and enjoyed the heat.   Still *dulness above right scapula, and croak on deep breath.*

December 28, 1864.—Was last winter at Bonchurch, and able to walk twelve miles a day, and lost cough.   Lately has been in Edinburgh, and tolerably well, except occasional colds and pain in the side.   Oil omitted from May till November.   *Weak breath, and some friction sounds in left back.*

June 2, 1868.—Remained well, living in Hampshire till last summer. When in Paris, spit half-an-ounce of blood, and in November brought up half-a-pint.   Has had cough ever since.   He lived low and lost flesh; but has regained lately with better fare.   Throat sore and tonsils enlarged.   *Tubular sounds at and within left scapula, and below left clavicle.*—To resume the oil, and take every night 10 grains of iodide of potassium.

May 22, 1869.—Has returned to the oil, and passed last winter in Hampshire, hunting, with slight cough, and gained 4 lb.   Still *deficient and tubular breath in both upper lungs, mostly in left, where slight crepitus.*

January 6, 1870.—Has been pretty well, except occasional colds and cough ; and once had hæmoptysis, a drachm, which relieved cough, and again later half-an-ounce.   Has suffered chiefly lately from enlarged tonsils.   *Weak breath and whiffy sounds audible in upper left lung.*

June 1870.—In the spring following had an attack of diphtheria, followed by pneumonia of right lung ; but recovered well with tonics and oil, and is now stronger than before the attack.   *Only tubular sounds and harsh breathing audible at left apex.*

The lung affection in this case probably originated from syphilis, traces of which were perceptible in several parts of its course.

CASE 23.—Thomas M , aged 43, admitted into the Brompton Hospital under the care of Dr. Theodore Williams, August 5, 1871.   Had suffered from rheumatic gout formerly.   He had cough and expectoration with wasting for nine months, and five months ago had hæmoptysis to the amount of one and a half pints, the bleeding returning in smaller quantities every six weeks. Complained of pain in the side and dyspnœa on exertion.   Pulse 88 ; respirations 20.   Temperature 98·8 F.   *Crepitation in the first and second intercostal spaces on the right side.*   He remained in the hospital three months, and had two attacks of hæmoptysis amounting to half-a-pint at a time, but left it considerably improved in every respect.   He remained fairly well for three years.   Was readmitted in April 1875, after an attack of hæmoptysis, which had recurred every day for two months, and varying in amount from 1 ounce to 1 pint. His cough had increased after the hæmorrhage had ceased.   *Crepitation extends over the whole right front chest, cooing sounds are heard over the posterior surface. Some crepitation is audible in the left mammary region.   Some flattening of the upper right chest.*

This case demonstrates how these patients in time drift into the ordinary type of phthisis.

Let us now consider the effects of hæmoptysis, both general and local, on the course of phthisis, and the complications it gives rise to. We know that a very profuse attack may cause immediate death by exhaustion, as in some of the cases already given; or by suffocation as in Case 17 ; or it may prove fatal in a short period through the supervention of pyæmia, as in the very remarkable instance we give below (Case 25). These are, however, exceptionally severe cases of hæmoptysis of rare occurrence, and probably not greatly to be influenced by treatment ; but it will be useful to notice the effects on the system generally, and on the lung locally, of moderate hæmorrhage. To many patients its occurrence seems beneficial rather than otherwise, for the congestion is thus relieved, and the system not materially weakened by the loss of blood. In extensive hæmoptysis, blood, passing from the ruptured vessel into the bronchi and alveoli, is either effused into them by the force of gravity, or is insufflated, i.e. reaches them through the spasmodic movements of inspiration. In the former case, the lower lobes of the lungs are generally infiltrated : in the latter, the blood is found in the upper, as well as the lower, lobes and following the line of the strongest inspirations. The effusion of blood into the smallest bronchi and air cells does not necessarily cause any general irritation ; for after hæmoptysis the presence of blood in the air vessels may often be detected by physical signs ; but these disappear in a few hours or days, and no evil results follow. On the other hand, in some instances its presence irritates the lung considerably, giving rise to catarrhal pneumonia and even to fresh tuberculisation. A rise takes place in the temperature ; and sometimes while the hæmoptysis is going on, the pulse becomes more frequent, the breathing more rapid, the cough more troublesome, and the expectoration, after the bleeding has ceased, becomes rusty and tenacious, though often of a dark tint. Crepitation is heard in the lower lobes of one or both lungs, intermingled with a good deal of sibilant rhonchus, and in time dulness and bronchophony make their appearance.

Fresh lung consolidation may arise in this way, and

when it does so, absorption, as a rule, takes place more or less
gradually. But this is not always so. It sometimes happens
that the fresh consolidation gives rise to symptoms of local
and general irritation, and in time undergoes the retrograde
changes of softening and excavation. Or it may be the start-
ing-point of a fresh outbreak of the tuberculosis in portions of
the lung hitherto unaffected, or again, the effused blood may
undergo certain changes, be partly absorbed, and then give rise
to pyæmia.

The pathology of pulmonary hæmorrhage and its local
effects on the lung tissue have been most ably treated by Dr.
Reginald Thompson,[1] who has specially drawn attention to
the presence after death, in cases where there has been a
history of profuse hæmoptysis, of old blood residues. These
take the form of fibrinous nodules of various sizes (*see* chapter on Pathology), which are the remains of the clots of the
effused blood. They vary in colour, the recent being of a
light red colour, and the older like mottled ivory, with old
blood pigment which collects around the bronchioles in small
black masses of fibrin, which are very characteristic and
have a tendency to separate by traction from the adjoining
pulmonary tissue.

Sometimes softening takes place in the centre of these
masses, which then have a yellowish appearance, somewhat
resembling honey, or else are converted into a mortar-like
substance, and occasionally they open and discharge into a
bronchus, a well-marked fibrinous capsule remaining behind.

These nodules are found under the microscope to consist of
blood corpuscles, filling the alveoli and infundibula, and coagulated fibrin. According to Dr. R Thompson the special *habitats*
of these nodules are the summit and middle part of the upper
lobe, the middle axillary region, between the third and fifth
ribs, close to the pleura : the anterior and inferior border and
the middle of the base corresponding to the summit of the
diaphragm.

Why the effusion of blood into the air passages should in
some cases cause no disturbance, and in others give rise to

[1] *Pulmonary Hæmorrhage*, 1879.

serious lesions, has always been difficult to explain.   The experiments of Perl and Lipmann,[1] and those of Sommerbrodt[2]
show that healthy blood when injected into the lungs of
guinea pigs produced some infiltration of the alveoli, which at
the end of a few weeks had entirely disappeared, and this was
also the case when the blood was injected in a coagulated
state.   There is evidence, too, from cases of injury to, or
operations on, the chest, in which large amounts of blood
have descended into the bronchi, and have been absorbed
without setting up secondary changes.

It is reasonable, therefore, we should look for something
beyond the mere presence of blood in the bronchi and alveoli
to account for the local irritation.

The pulmonary congestion often accompanying large
hæmoptysis, and of which it is the relief, has been assigned
as a cause, but it is much more probable that the effect of the
blood effusion on the lung tissue depends on the nature of the
blood itself.   If it be healthy, pure blood, proceeding from a
mechanically ruptured vessel, we may expect that complete
absorption will take place, but if—as is frequent in phthisis
more or less advanced—it is mixed with secretion from a
cavity, or ulcerating surface, it becomes a carrier of tubercle
bacilli or other organisms, and is thus liable to set up inflammatory and infective processes in whatever parts of the lung
it reaches.   In this way the spread of lung tuberculisation
after extensive hæmorrhage takes place.

These changes are accompanied, and often heralded, by
well-marked pyrexia, which is always of ill omen in connection
with hæmoptysis.   Sometimes, when the amount is small and
the effused blood only reaches a limited portion of the lung,
the pyrexia is moderate and ceases after the first irritation has
subsided, but where the bleeding is profuse and a large proportion descends into the lungs, the pulse becomes rapid and
feeble, the breathing quickened, there is orthopnœa, the
tongue becomes brown, the temperature rises high and presents a very fitful chart (as seen in Case 26), delirium and
great mental excitement follow, and the patient dies with

[1] Ziemssen, vol. v. p. 306.   [2] *Ibid.*

symptoms of pyæmia. The phenomena in these cases we may explain partly by the great loss of blood causing a great irritability of the nervous system, partly by the sudden reduction of the respiratory space through the rapid effusion of blood, and partly by the inflammatory changes set up by the spread of the infective processes.

To sum up, the results of hæmoptysis on the lung are the following. (1) If sufficiently profuse to invade the greater proportion of the lung surface, death by suffocation results. (2) If the blood be healthy and invade only a limited number of alveoli, it may be absorbed without causing consolidation or leaving blood residues. (3) It may slowly disappear, leaving fibrous nodules behind. (4) It may, through being mingled with secretion, cause irritation to the alveolar wall and set up catarrhal pneumonia. (5) It may be the carrier of tubercle bacilli, and lead to fresh tuberculosis. (6) It may, by setting up infective processes, local and general, cause death by pyæmia.

The first case given below illustrates well some of the local effects of hæmoptysis, pneumonia being set up and resolving. In the second case, the general effects were more marked, the patient dying of pyæmia. The third case is a good instance of the most terrible effect of profuse hæmoptysis, the infective processes being most virulent.

CASE 24.—A French valet, aged 34, was seen by Dr. Theodore Williams January 28, 1871. He stated that six years ago, whilst waiting on a gentleman, he had brought up some ounces of blood, and that since that time he had had two attacks, one two years ago, in which only a slight amount was expectorated, and the other three months ago, when he brought up a pint. He had been subject to cough off and on, and had lost some flesh since the second attack, two years ago, but was never laid up, or discontinued his work, till two days ago, when he was again attacked with hæmoptysis, and brought up a pint; and to-day about an ounce. Is now expectorating bloody mucus. Pulse 68; temperature 97·8° F.; respiration normal. Has gained flesh in the last three months through taking cod-liver oil. *Deficiency of expansion and tubular sounds in upper part of right front chest. Some bronchophony above the scapula.*

The patient continued to spit up blood for a week, the amount at a time varying from a teaspoonful to upwards of a pint; various styptics were used, and generally checked the flow of blood, but, if omitted, the hæmorrhage returned. There was no rise of temperature or pulse, but the breathing became slightly more hurried, and the cough worse. On February 6 the physical signs were *dulness, tubular sounds in upper right chest, some crepitation audible in mammary region. Harsh tubular sounds scattered over both lungs.*

He improved on tonics and cod-liver oil for ten days, but on February 16 got excited about the war taking place in his country, and copious hæmorrhage came on. Bowels costive. Cough very troublesome. Temperature 99·8° F.; pulse 80. Ergot of rye and a blister were ordered.

February 17.—Blister rose well, and cough better. No more hæmoptysis has occurred. Bowels still costive; pulse 90; temperature 101·2° F. *Crepitation audible in lower right lung.*

February 19.—Worse. Has wandered much in his mind. Aspect blanched; breathing embarrassed; bowels open. *Crepitation audible in lower left lung as well as in right.* Pulse 108; temperature 101·6° F.; respirations 32. A blister was applied to the left side.

February 20.—Blister rose well, and gave relief to breathing and cough; but, after severe fit of coughing, slight hæmoptysis came on, and now cough and other symptoms are worse. Pulse 108; temperature 101·2° F.; respirations 36. —Ordered oil again.

February 22.—Slight hæmoptysis, and cough has increased. Pulse 100; temperature 101·1° F.; respirations 28.

March 3.—Patient was very restless, cough very troublesome, and breath very short till a few days ago, when he expectorated largely of thick muco-purulent matter and some frothy mucus, and since then his aspect is much brighter. *Crepitation in lower left chest. Rhonchus and crepitation (scattered) over whole right side.*—Quinine, in two-grain doses, twice a day, was ordered to be combined with the oil, and patient took it for two days, and then refused to take any more.

March 6.—Worse, much paler; tongue, which has been generally slightly furred, is now glassy and red. Pulse 120; temperature 100° F.; respirations 32.— Ordered sulphurous acid with tincture of orange, to be taken with the oil.

March 9.—Aspect much brighter; tongue cleaner and moister; appetite improved; and patient gets up for some hours in the day. Cough only at night and in the morning, and expectoration frothy. *Crepitation in right lung has diminished.* Pulse 100; temperature 99·2° F.; respirations 28.

March 28.—Has steadily improved in the last three weeks, and is gaining flesh and strength. Oil and sulphurous acid have been taken the whole time. Breath remains short. Some *dulness above the right scapula, with increased vocal fremitus, and slight crepitation in the upper front; some crepitation at the posterior base of both lungs. Harsh tubular sound with crepitation scattered throughout the left lung.* Pulse 96; temperature 98·1° F.; respirations 28.— Ordered oil, with diluted nitro-hydrochloric acid.

April 21.—Now walks about, though finds breath rather short. Cough only in fits. Has taken oil regularly. Pulse 80; temperature 98·4° F.; respirations 24. *In right chest slight dulness and crepitation to third rib. Some scattered crepitus behind.* In left chest crepitation has quite disappeared, but harsh tubular sounds are audible in parts.

In this case hæmoptysis occurred several times without giving rise to any inflammatory symptoms; but at length, after an attack, pneumonia was set up, which happily ended in resolution.

The following is a remarkable instance of hæmoptysis, ending in fatal suppuration.

CASE 25. – A nobleman, aged 48, of a gouty family, first consulted Dr. C. J. B. Williams July 18, 1855. He had suffered from gout more or less for years, and in order to escape the attacks he had been to the West Indies, and two winters ago to Italy. In the following spring he passed blood by the bowels, and in the summer had hæmoptysis to extent of an ounce, which continued to a less degree in spite of cupping and leeching, but was stopped by gallic acid. Passed the last winter well, and free from cough, in Egypt, until chilled by a ride to Damascus ; and in May, while in Paris, brought up an ounce and a half of blood. Cough returned and increased on reaching London in June. *Slight dulness, slightly louder breath, with some tubular sound in upper left back. More tubular voice and obscure breath in upper right.*—Ordered oil in tonic of nitric and hydrocyanic acids, with tinctures of hop and orange, and counter-irritation.

August 2.—Takes the oil well, and with benefit. *Dulness and tubular sounds, mostly in upper back, with slight crepitation.*

October 5.—Two months ago, after feeling much better and attending quarter sessions, had a cold ; and a few days later slight hæmoptysis, and a sensation of weight in the sternum. Gallic acid and acetate of lead were given On the 2nd and 3rd of October, without effort, he spouted up altogether two quarts of blood, and became very faint. Yesterday and to-day (October 5) he brought up only a few ounces. A draught of infusion of roses and sulphuric acid, containing half-a-grain of morphia, was given every four hours. Bowels not open for three days. Patient almost free from cough. Pulse 90. *Breath clear and puerile in left front ; much obstruction, with obscure crepitation in right front. Too weak to bear examination in the back.*—Ordered an aperient draught of sulphate of magnesia and sulphuric acid, and a mixture of gallic and sulphuric acids, &c. The patient continued in an extremely weak state, with frequent pulse and respiration, hot skin, and low fever; expectorating bloody pus in increasing quantities. A few days later symptoms of pyæmia came on ; tongue became dry and brown ; pulse very rapid and running. Slight delirium ensued, and the patient died within a week.

CASE 26.—Mr. W. H., aged 23, a banker's clerk, whose paternal uncle had died of phthisis. He consulted Dr. Theodore Williams November 22, 1879, having suffered from cough and expectoration for five weeks. Lately the expectoration had been tinged, and night-sweats had come on, with some wasting. Mr. H. assigns much of his illness to the fact of his living in the country and travelling backwards daily to London. Height 6 ft. ; weight 10 st. 8. lb. *Crepitation under the left clavicle and above the left scapula, and some crepitation and loud tubular sounds in the interscapular region.* He was ordered to remove to London and live nearer his work ; to take cod-liver oil with dilute sulphuric acid and strychnia, and a morphia linctus, and to paint the left side of the chest with acetum cantharidis.

December 6.—Cough less frequent, but fits of coughing very violent. Expectoration more abundant.

February 9, 1880.—Has had hæmoptysis Oss. and gained 5 lb. in weight. *Crepitation over the upper half of the chest anteriorly and posteriorly.* Temperature 98° F. ; pulse 84.

February 26.—Has gained 1½ lb., weighing now 11 st. ¼ lb. The cough has been very troublesome in the morning. Expectoration scanty. Breath is short on exertion. Night-sweats are present. There is well-marked shrinking of the left side, and at the ensiform level it measures 14½ inches against 16 inches on the right side. *Crepitation diminished, but cavernous sounds audible in scapular region.*

March 24.—Saw him in consultation with Mr. Martin, of St. John's Wood, and found he had improved till yesterday, when, after being out late the night before, he had had hæmoptysis Oj., and was still bringing up clots. Hæmorrhage was checked by ergot. *Crepitation increased in left lung.* Pulse 80; temperature (afternoon) 98° F. Respirations normal.

April 1.—Heard that the hæmorrhage stopped, but recurred a few days later in large quantities and was accompanied by rise of temperature to 102° F. and of pulse to 120, the respirations varying from 28 to 55. About a fortnight later I saw him and found him apparently suffering from pyæmia, with rapid pulse and respiration, profuse sweats, and some delirium; the dyspnœa was the most marked symptom, and was worst at night. He had orthopnœa, and spent most of his time in a reclining chair, which he dreaded to quit, even for an instant, for fear of suffocation. *Coarse crepitation heard all over both lungs, with well-marked gurgle in the upper left front.* He became highly delirious, and talked a good deal about committing suicide, but sank on May

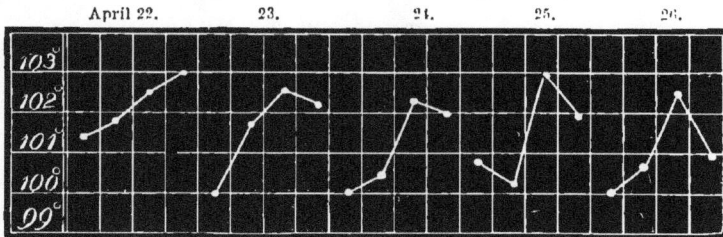

10, the dyspnœa being present to the end. No autopsy was permitted. The annexed chart of his temperature a few days before death shows the fitful character of the pyrexia of this class of cases.

In this case the hæmorrhage was probably due to the rupture of a pulmonary aneurysm laid bare in the cavity which formed in the left lung, but the blood penetrated by gravity and insufflation into a large portion of both lungs, where it probably set up, not only pneumonia, but general pulmonary tuberculosis. These lesions, in addition to the excitement to the nervous and circulatory systems arising from the rapid loss of blood, caused the general septic condition of the system and resulted in a fatal issue.

What influence does the occurrence of hæmoptysis exercise on the duration of phthisis? Does it tend to curtail it? Reliable information on this important point will be acceptable to both physician and patient, affording fair grounds of prognosis to the one, and holding out comforting hopes to the

other.   Our 1,000 tabulated cases include 63 fatal ones,
where the patients had hæmoptysis of an ounce and upwards
on one or more occasions.   Among these the average duration
was 7 years 6 months; an average only differing by a
few months from that of the total deaths.   Again, in 200
living cases of similarly extensive hæmoptysis, the average
was 8 years 3·23 months—about the same as that of the
living cases generally.   These facts certainly do not indicate
that hæmoptysis exercises any curtailing influence over
phthisis viewed as a whole.   When, however, we classify the
deaths according to the stages the patients were in when the
hæmorrhage took place, we perceive the significance which
attaches to the state of lung at the time of its occurrence:—

Stage.	No. of Deaths.	Average Duration.	
		Yrs.	Mths.
1	26	9	2·11
2	16	7	4·62
3	21	7	1·42

We here see that hæmoptysis occurring in the second or
third stages is more likely to curtail the duration of the disease
than in the first; and in the first it is comparatively un-
important.

CHAPTER XII.

THE TEMPERATURE OF CONSUMPTION.

Due to two agencies—Pyrexial—Collapse—Mode of investigation—Temperature of first stage (active)—Pyrexial and sub-normal—First stage (passive) non-pyrexial—Temperature of second stage not necessarily pyrexial—Characteristics of third stage (active)—Chart shows great extremes—Suppuration and tubercle formation—Afternoon pyrexia, morning collapse—Temperature of chronic third stage—Comparison with normal temperatures—Relation of temperature to weight—Tuberculisation and local rise of temperature—Peter's and author's observations.[1]

THE temperature has always—and justly—been considered a very important symptom in phthisis, and its relation to the various pathological states has, from time to time, been much discussed by numerous observers, among whom may be mentioned Jochman, Lebert,[2] Wunderlich,[3] Sydney Ringer,[4] Wilson Fox,[5] and Peter.[6]

Its value, both in the diagnosis and prognosis of the disease, is doubtless great, though Dr. Alcock's and the author's researches have somewhat diminished its importance in the former department.

We will first consider the relations (1) of tuberculisation to temperature, then those (2) of softening and excavation to the same symptom, and, later on, (3) the influence on the temperature of some of the complications of phthisis.

To understand the temperature chart of phthisis we must bear in mind that it is due to two principal agencies.

1. Excessive action of the processes by which the heat of

---

[1] This chapter is for the most part an abstract of a paper contributed by the author to vol. lviii. of *Medico-Chirurgical Transactions.*
[2] *Deutsch. Archiv Klin. Med.,* November 1872.
[3] *Temperature in Disease.*   [4] *Temperature in Phthisis.*
[5] *Medico-Chir. Trans.,* vol. lvi.   [6] *Clinique Médicale.*

the body is maintained, the processes of oxidation and disintegration, combined with a weakening of the inhibitory action
of the nervous system in these phenomena, such as is present
in all inflammations and fevers.

2. A collapse of the constitutional powers, which characterises the natural course of consumption. These two influences act on the patient, and according as one or the other
preponderates, is shaped the course of the temperature. When
they are equally balanced, a chart hardly differing from the
normal, results. When the collapse prevails, sub-normal temperatures appear, and when inflammatory processes are in
the ascendant, pyrexia shows itself. Thus it will be seen
that pyrexia is not a necessary accompaniment of phthisis,
and certainly not of all its steps, and that Dr. Nathaniel
Alcock's opinion[1] that ' the earliest and most appreciable sign
preceding the development of tubercle is an inability to
maintain the temperature up to the normal standard,' is quite
as true, as the view that the formation of tubercle is always
accompanied by rise of temperature. It has been proved that
tubercle may form and extend, and that softening and excavation may take place without any rise of temperature whatever,
the explanation being that in these cases the influence of
collapse predominated over that of pyrexia. When, however,
pyrexia exists at all in phthisis, it assumes a certain definite
course, as characteristic of the disease, as the temperature
curve of typhoid fever is of that malady, but which escaped
the notice of observers, previous to the author, on account of
the mode in which their records were taken ; these not being
sufficiently frequent to include the whole series of thermic
phenomena. Instead of observations being made morning
and night in the usual routine method, the temperatures were
noted by the author and his assistants at least five times a
day, and in many cases eight or twelve times. Also, in a
number of typical cases, the observations were taken every
hour for a cycle of twenty-four hours, this being in some
cases checked by a second cycle taken a few days later on.

[1] *Nature and Varieties of Destructive Lung Disease included under the
head of Consumption.*

Many of the records were carried on for months, and though for the special points of investigation in this chapter about one hundred and four cases and four thousand two hundred and twenty-six observations only have been tabulated, we are in possession of several hundreds of cases and of thousands of records which amply confirm the conclusions based on the smaller number.[1]

In these 104 patients both lungs were attacked in 66 (almost two-thirds), one lung in 38 (about one-third), the right alone in 17, and the left alone in 21. Of the 31 in the first stage, 16 had both lungs affected, and 15 a single one; of the 10 second-stage cases, 5 had both lungs involved, and 5 one lung only; and of the 63 in the third stage, 45 had both lungs attacked, of which 10 had double cavities.

We will now proceed to consider each class of cases and the temperature prevalent in them, and first let us take Class I., *first stage, active.*

The physical signs generally noticed here were (1) those indicating the commencement of disease in the lung hitherto free, or (2) the extension of disease in a lung, part of which was already consolidated ; and great care has been taken to exclude all cases where anything besides consolidation or scattered tubercle was suspected. The crepitation was generally of a fine description, and, as regards locality, often scattered over the upper lobes. That it did not pass into the coarse crepitation of a croaking character typical of softening, was confirmed by subsequent observation in each instance.

The following instance of the *development of disease* and its accompanying thermal phenomena will help to illustrate the features of this class :—

Case 27.—W. G. W., aged 14, was admitted into Wallace Ward April 29, 1873.

History.—Father died of consumption. Sister suffers from chest disease.

---

[1] I must here acknowledge the very valuable assistance I received from the gentlemen who were my clinical assistants at the time of the inquiry, and sacrificed time and comfort to its completion. They were Drs. Crocker and Murrell and Messrs. Bartlett, Parry, McKinlay, Williams, Hartley, Peacey, Boys, Bernays, and Kelly.

A week after Christmas, was attacked with giddiness and pain in right chest, and night-sweats; three days after he complained of cough with scanty expectoration, which has continued since; has not lost much flesh; night-sweats have persisted since Christmas; complains of cough and dragging sensations in the right shoulder; bowels costive; tongue furred; appetite fair; skin rather anæmic; pulse 136; temperature (8 P.M.), 102° F.; weight (May 8), 6 st. 12½ lb. He was carefully examined on May 5, and no physical signs could be detected. The temperature was taken five times daily, and showed the usual pyrexia of first-stage phthisis, i.e. lower or normal records at 8 A.M., slightly raised at 11 A.M., and high after 2 P.M.

On May 8 *dulness was detected over the left clavicle and crepitation below it.*

Date	8 A.M.	11 A.M.	2 P.M.	5 P.M.	8 P.M.	Remarks
May 5	98·8	100·	100·6	101·	100·2	No physical signs
„ 6	98·4	98·2	100·6	101·	101·4	
„ 7	98·3	99·	100·4	100·2	100·	
„ 8	99·4	99·8	101·	102·	101·3	Dulness detected over the left clavicle, and crepitation below it.
„ 9	98·	99·4	100·	100·5	101·3	
„ 10	98·2	99·	99·6	101·	100·4	
Mean	98·5	99·2	100·3	100·9	100·8	

June 26. *Dulness of the whole left side; crepitation to third rib; crepitation above right scapula.*

July 17. –*Crepitation to fourth rib on left side; coarse under clavicle;* weight, 7 st. 1 lb., having never fallen since admission.

This case was a well-marked instance of the development of tubercle, and, as the rise in temperature preceded the appearance of physical signs, was confirmatory of Dr. Ringer's observation that we can often diagnose tubercle by the temperature before we can do so by physical signs.

The effect of *extension of disease* in the same lung is exemplified by Case 28, where the crepitation which had been confined to the apex of the lung, extended to the whole of the lung during the period when the temperatures were being taken, and at a later date to both lungs. It will be observed that though there was a rise in the afternoon, yet, with all the active disease going on, the temperature did not rise high, and only twice reached 100° F.

CASE 28.—Richard D., aged 25, admitted into the Brompton Hospital December 23, 1872.

History.—Mother and brother died of consumption; cough with expectoration two and a half years, succeeded six months later by an attack of slight pleurisy, followed by night-sweats and wasting. Night-sweats disappeared, but have returned within the last six months and are now profuse. Has wasted more or less for six months. Hæmoptysis to the amount of ʒj. eighteen months ago, and the expectoration has been streaked on several occasions.

At present.—Cough troublesome; tongue clean; appetite fair; bowels regular; pulse 80; respirations 24.

At the end of the first week of January the physical signs were *crepitation to fourth rib on the right side.* The subjoined temperatures were then taken:—

Date	8 A.M.	11 A.M.	2 P.M.	5 P.M.	8 P.M.
January 8 . . .	98·	98·2	99·2	99·3	100·2
„ 9 . . .	97·	98·	98·8	98·4	99·6
„ 10 . . .	97·2	98·8	99·5	99·4	99·3
„ 11 . . .	97·1	98·6	99·4	99·6	99·4
„ 12 . . .	97·	97·8	98·6	99·2	99·4
„ 13 . . .	96·8	98·1	99·1	99·4	99·6
„ 14 . . .	96·8	98·2	99·	99·4	100·2
Mean . . .	97·1	98·2	99·1	99·2	99·5

January 30, 1873.—*Crepitation extends over the whole right front, and over the right interscapular region.*

March 6.—*Crepitation has diminished in the upper portion of the right lung, but is now audible, scattered throughout both lungs, front and back.*

The above cases demonstrate the effect of tubercle-formation in causing a definite rise in the temperature, which may cease when this process is complete. It is, however, by no means necessary that any rise should take place, and in several of my cases, where active disease has been going on in one or both lungs, no rise has taken place, but rather a low range of temperature has prevailed. Of this the subjoined is an instance.

CASE 29.—Janet S., aged 46, admitted into the Brompton Hospital December 2, 1873.

History.—Subject for the last four years to cough in winter, which for two years has been continuous, and accompanied by loss of flesh. Night-sweats for one year; slight hæmoptysis four years ago, and again, two years ago, she brought up a pint of blood.

At present.—Cough troublesome, sputum yellow and frothy; night-sweats profuse; pains under both clavicles, most under right; tongue clean; appetite good; bowels regular.

On January 1, 1874, the physical signs were—Right : *Dulness; crepitation upper third, front and back.* Left : *Crepitation audible over whole front and over interscapular region.* The following temperatures were taken some days later :—

Date	8 A.M.	11 A.M.	2 P.M.	5 P.M.	8 P.M.
January 14 . .	98·	98·2	98·8	98·6	99·
„ 15 .	97·	98·	98·3	98·	98·4
„ 16 .	98·2	98·	97·8	98·	98·2
„ 17 .	97·6	98·	98·6	98·6	99·
„ 18 .	97·9	97·	98·4	98·	98·8
„ 19 .	98·	98·2	97·6	98·4	98·4
„ 20 . .	97·6	98·	98·6	98·4	98·2
Mean .	97·7	97·9	98·	98·3	98·6

January 5.—Hæmoptysis to the amount of ʒiss.; no increase of cough followed.

January 19.—Profuse night-sweats.

January 30.—Physical signs the same.

Upwards of 200 observations on the class of *first stage* (*active*) were made on twenty-six patients, at the hours 8, 10, 11 A.M., 2, 5, 8, and 11 P.M., and the results are embodied in Table I., which records, under the various temperatures, the number of times each was noted, as well as the relative percentages. The last five columns express the *maximum, mean,* and *minimum* temperatures, taken at each hour, as well as the *number of cases,* and number of observations made on them.

Attention is specially drawn to the fifth and twelfth columns, as these contain the totals of observations of the pyrexial (100° and over) and of the sub-normal (under 98°) types, with their respective percentages, and a glance at these will give the reader of this and the succeeding tables at once the gist of the contents.

TABLE I.—*Analysis of Temperatures.*

FIRST STAGE (ACTIVE).

Hours	103°	102°	101°	100°	100° and over	99°	98°	97°	96°	95°	94°	Under 98°	Maximum	Mean	Minimum	No. of observations	No. of cases
8 A.M.	—	—	—	2	2	10	76	70	19	2	1	92	100·	97·8	94·2	180	26
Percentages	—	—	—	—	1·1	5·56	42·23	—	—	—	—	51·12	—	—	—	—	—
10 A.M.	—	—	1	6	7	2	3	8	1	—	—	9	101·5	98·6	96·	21	3
Percentages	—	—	—	—	—	—	—	—	—	—	—	—	—	—	—	—	—
11 A.M.	—	—	1	9	10	43	82	21	4	1	—	26	101·1	98 5	95·	161	23
Percentages	—	—	—	—	6·21	26·70	50·93	—	—	—	—	16·14	—	—	—	—	—
2 P.M.	1	3	12	41	57	56	55	15	2	—	—	17	103·	99 3	96·	185	26
Percentages	—	—	—	22·16	30·81	30·27	29 72	—	—	—	—	9·18	—	—	—	—	—
5 P.M.	2	8	15	44	69	52	53	9	2	—	—	11	103·4	99·4	96·	185	26
Percentages	—	—	—	23·78	35·67	28 11	28·64	—	—	—	—	5·94	—	—	—	—	—
8 P.M.	—	9	18	56	83	58	44	5	—	—	—	5	102 5	99·6	97·5	190	26
Percentages	—	—	—	29·47	43 68	30·52	22·10	—	—	—	—	2·63	—	—	—	—	—
11 P.M.	—	—	1	1	2	6	7	1	3	—	2	6	101·	98·3	94·	21	3
Percentages	—	—	—	—	9·52	28 57	33 34	—	—	—	—	28·57	—	—	—	—	—

The observations taken at 8 A.M. exhibit remarkably low temperatures ; 51 per cent. are below 98°, 42 per cent. between 98° and 99°, while only 5½ per cent. rise to between 99° and 100°, and 1 per cent. reaches 100° ; and as this last temperature (100°) was only recorded in one case on two occasions, it may fairly be considered as accidental.

The mean temperature is 97·8°, while the minimum falls as low as 94·2°.

The notes at 10 A.M. show a slight rise from the sub-normal state, which is more marked at 11 o'clock, when we

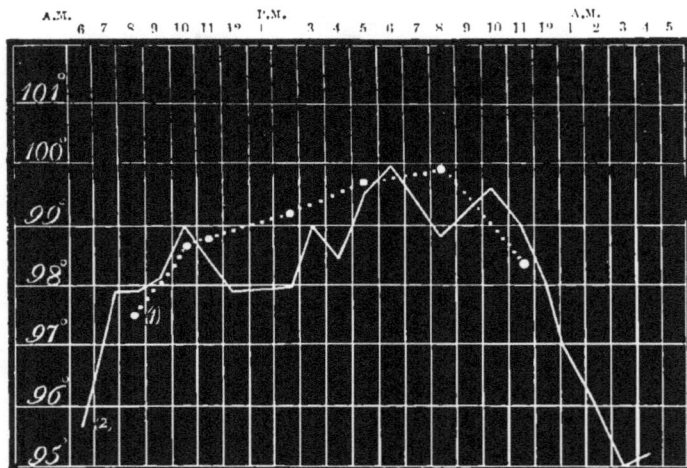

FIG. 1.- Phthisis, Stage I., active. No. 1 (dotted line).—Average of twenty-six patients during seven periods of the day from 8 A.M. to 12 P.M. No. 2 (continuous line). Twenty-four hours' cycle of one case.

find only 16 per cent. below 98°, 50 per cent. between 98° and 99°, and 6 per cent. above 100° ; the mean temperature being 98·5°, the minimum 95°. The 2 P.M. and 5 P.M. observations show a steady continuance of this rise, the sub-normal temperatures diminishing and the pyrexial increasing, until at 8 P.M. the climax is reached, only 2½ per cent. being sub-normal, and more than 43 per cent. exceeding 100° ; the maximum being 102·5°, the mean 99·6°, and the minimum 97·5°.

The 11 P.M. notes show a rapid fall, 28½ per cent. being

sub-normal, 33 per cent. normal, and only 9½ per cent. above 100°. The dotted line (No. 1, Fig. 1) expresses roughly the day-temperature course of Class I. as far as can be ascertained from observations taken seven times *per diem* in these twenty-six cases; but in order to display the night phenomena and to give a fuller insight into the day variations a second continuous line (No. 2) has been annexed, which shows twenty-four hourly observations on one of the same patients, and this may be compared with the dotted one.

Though this case was hardly a typical one, the general course may be said to confirm the dotted line, especially in respect of the rise, which, it would appear, dates from the morning hours. It shows a slight exacerbation at 10 A.M. and a consequent fall, and then a nearly continuous though slight rise from 2 P.M. till 10 P.M., when a rapid fall commences, which reaches its lowest ebb between 2 and 3 A.M. Temperatures of 94° and 95° are then not unfrequently recorded.

The temperature course of first stage (active) seems to be characterised by :—

(1) A wide range of temperature extending from 94° F. to 103·4° F.

(2) Afternoon pyrexia.

(3) Sub-normal temperatures in the early morning indicating the collapse influence.

The next class, Class II., is that of *the first stage, quiescent*, of which a chart will be seen in Fig. 2, No. 1.

This is the ordinary 'chronic first stage,'—where tubercle has formed and become quiescent,—of authors, of which an instance is subjoined, though its common occurrence in practice renders it familiar to everyone.

CASE 30. Julia E., aged 21, admitted into Hall Ward under Dr. Quain, April 14, 1874. Maternal aunt died of phthisis. Cough with expectoration and wasting six months; worse for four months; hæmoptysis, ʒj., four months ago, and to slighter amount once since. Night-sweats occurred at the same period, but have since ceased. Appetite bad; tongue furred; catamenia scanty and irregular for nine months; bowels confined; pulse quiet.

May 13, 1874.—*Dulness and tubular sounds over upper third of right lung; no night-sweats; expectoration slight.*

The subjoined observations were then taken : –

Date	8 A.M.	11 A.M.	2 P.M.	5 P.M.	8 P.M.	11 P.M.	Pulse
May 14 . . .	—	98·2	98·4	—	98·	—	64
„ 15 . . .	97·6	98·2	98·4	97·	97·1	—	64
„ 16 . . .	98·4	97·6	97·2	97·2	98·6	—	64
„ 17 . . .	97·4	97·	96·2	97·6	97·4	97·2	60
„ 18 . . .	98·4	97·4	97·8	98·2	97·6	—	72
„ 19 . . .	97·4	97·8	98·2	97·	97·4	—	60
„ 20 . . .	97·2	97·2	97·	98·6	97·	—	56
„ 21 . . .	97·4	97·	99·6	101·2	99·4	—	68
Mean . . .	97·7	97·5	97·8	99·5	97·8	—	—

May 22.—Physical signs the same.

Here the temperatures seem mostly sub-normal, showing a strong preponderance of the collapse influence.

I have made a series of hourly observations, in one case for

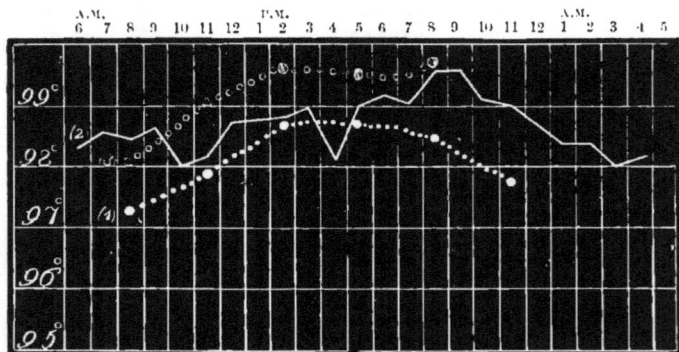

Fig. 2.—Stage I. (quiescent), and Stage II.   No. 1 (dotted line).—Average of five cases of Stage I. (quiescent).   No. 2 (continuous line).—Twenty-four-hours' cycle of one case.   No. 3 (open dotted line).—Average of ten Stage II. cases.

twenty-four hours, to gain some insight into the night temperatures, and the result is seen in the continuous line (No. 2).

The patient was a male, aged 36, with well-marked physical signs of consolidation over the upper third of the left lung, and no advance in the signs had been detected during the two months he had been under observation. The temperatures are chiefly normal in character, with the slight tendency to afternoon rise which seems to be often present. The continuous line in this woodcut (No. 2) gives the cycle in this instance.

Table II. is based on between thirty and forty observations made on five patients of this class at each of the following hours, viz.—8 and 11 A.M.; 2, 5, and 8 P.M.; a few were made at 11 P.M. which confirm the general course (*see* page 18).

The 8 A.M. observations indicate sub-normal temperatures, the mean being 97·3°, the minimum 95·2°, and the maximum 98·5°; 63 per cent. fell below 98°, and the remainder are all below 99°. At 11 o'clock the temperatures are still low, the average being 97·9°, the minimum 95·2°. There is a slight increase, however, in the percentages under the heads of 98° and 99°.

The further notes show a steady rise at 2 P.M., which reaches its maximum at 5, an average of 98·7° being then attained, for though at 2 P.M. 8½ per cent. exceeded 100°, the average was only 98·6°. The curve of the five patients is shown by the dotted line No. 1 in Fig. 2.

At first sight we might conclude that the temperature course in this class does not differ from the normal, but a comparison with the normal chart (Fig. 3), and with that of *first stage active* (woodcut 4) will show that this class partakes of the features of the latter class, the afternoon rise and the early morning fall, though the phenomena are exhibited within smaller limits.

The next class is *second stage* (Fig. 2, No. 3), and under it are included only cases where no cavity has hitherto been detected, but where the course of the physical signs clearly demonstrates one to be in process of formation. The signs principally relied on have been the increasing coarseness of the crepitation, the hoarse and croaking rhonchus, confined to a portion of one lung and combined with tubular sounds, and often high-pitched percussion note. In most cases the large increase of expectoration and the detection of elastic tissue in it have confirmed the physical signs. In many of the cavity cases, which we shall presently consider, softening was proceeding in the opposite lung, or in another part of the same lung, but here, as a suppurating surface already existed in the form of a cavity, which might considerably influence

TABLE II.—*Analysis of Temperatures.*

FIRST STAGE (QUIESCENT).

Hours	103°	102°	101°	100°	100° and over	99°	98°	97°	96°	95°	94°	Under 98°	Maximum	Mean	Minimum	No. of cases	No. of observations
8 A.M.	—	—	—	—	—	—	12	13	5	3	—	21	98·5	97·3	95·2	5	33
Percentages	—	—	—	—	—	—	36·36	—	—	—	—	63·63	—	—	—	—	—
11 A.M.	—	—	—	—	—	6	16	8	2	3	—	13	99·6	97·9	95·2	5	35
Percentages	—	—	—	—	—	17·14	45·71	—	—	—	—	37·14	—	—	—	—	—
2 P.M.	—	—	—	—	3	11	12	8	2	—	—	10	100·4	98·6	96·2	5	36
Percentages	—	—	—	—	8·33	30·55	33·33	—	—	—	—	27·77	—	—	—	—	—
5 P.M.	—	—	1	1	2	11	10	9	2	—	—	11	101·2	98·7	96·4	5	34
Percentages	—	—	—	—	5·88	33·35	29·41	—	—	—	—	32·35	—	—	—	—	—
8 P.M.	—	—	—	2	2	13	8	14	—	—	—	14	100·4	98·4	97·	5	37
Percentages	—	—	—	—	5·40	35·16	21·62	—	—	—	—	38·	—	—	—	—	—

the temperature, it was deemed advisable to separate these
cases from other cases of softening and to class them under
the heading 'Active third stage.'

It must be understood that we do not recognise any very
hard and fast line of demarcation between the second and
third stages of phthisis, and that we have only classed these
cases as Phthisis II. to study the process of softening in its
thermal aspect, which we will now do.

The observations were made on ten patients, at the same
hours as in the preceding class, between seventy and eighty
records being taken at each period, and of these, two examples,
where the physical signs can be compared with the temperatures, are annexed. In the first one (Case 31) the formation
of a cavity was proved by physical signs and the presence of
yellow elastic tissue in the expectoration, the excavation being
accompanied by a rise of temperature. In the second (Case
32) excavation was proved by physical signs alone, and gave
rise to no pyrexia.

CASE 31.—Elizabeth B., aged 28, single, admitted into the Hospital for
Consumption January 21, 1873. Mother died of phthisis. Subject to winter
cough for six years, worse last winter, and continuous since. Dyspnœa for
seven months. Slight hæmoptysis and night-sweats last May, and has been
losing flesh ever since. Catamenia ceased in May, but reappeared in October,
and are now regular. Cough troublesome, with expectoration; pulse 80.
Crepitation is audible over the upper third of the left front chest.

February 17, 1873.—Cough worse, with muco-purulent expectoration;
respirations 32; *croaking crepitation audible in the first interspace on the left
side; coarse crepitation from the second to the fourth rib; croaking crepitation
in the upper half of the left back.* The subjoined temperatures were then
taken :—

Date				8 A.M.	11 A.M.	2 P.M.	5 P.M.	8 P.M.
February 18	.	.	.	98·	99·	99·2	100·	101·4
,, 19	.	.	.	98·6	99·	99·	100·	101·
,, 20	.	.	.	99·6	99·6	99·	100·2	100·
,, 21	.	.	.	98·2	99·	98·6	99·8	100·8
,, 22	.	.	.	97·8	100·4	99·4	98·6	98·8
,, 23	.	.	.	98·	99·2	99·	100·2	100·2
,, 24	.	.	.	98·	98·	100·6	98·8	99·
Mean	.	.	.	98·3	99·2	99·3	99·7	100·2

February 20.—Yellow elastic tissue detected in the sputum, which is abundant, amounting to half-a-pint daily.

March 5.— *Well-marked cavernous gurgle in upper left chest.*

April 15.—*Dry cavernous sound to the second rib; some croaking respiration above the scapula;* cough easier; expectoration small in amount. Has lost 6½ lb. during her stay in the hospital.

CASE 32.—Edwin N., aged 23, carpenter, admitted into the Hospital for Consumption January 29, 1873.

History.—Continuous cough for one year. Slight hæmoptysis last May, accompanied by pain in the left chest, and has been wasting ever since. Night-sweats on and off since May, with dyspnœa on exertion. Tongue clean; appetite bad; bowels regular; pulse 100; respirations 24; cough troublesome.

On admission the physical signs were *dulness and crepitation in the first interspace on the left side.*

On March 18 the disease had extended, and *crepitation was audible over the upper half of the left chest posteriorly and over the whole left front, and was of a croaking character in the upper half.*

The temperature was then taken five times a day, and the subjoined is the chart:—

Date	8 A.M.	11 A.M.	2 P.M.	5 P.M.	8 P.M.
March 20	98·8	99·4	99 6	98·4	97·6
„ 21	98·	99·	97·	99·	98·
„ 22	98·	99·	98·8	98·6	98·2
„ 23	95·6	98·	98·	98·2	98·
„ 24	96·	98·6	99·	99·4	100·
„ 25	97·	99·	98·2	99·	99·
„ 26	98·	98·2	98·4	98·2	99·2
„ 27	97·4	98·	98·6	—	—
Mean	97·4	98·6	98·4	98·7	98·5

March 27.—*Cavernous gurgle detected in upper left front.* Patches of urticaria have appeared on the patient's body.

The statistics of the second stage as set forth in Table III., on the next page, show that at 8 A.M. the temperatures were somewhat higher than in the first stage, 37½ per cent. being below 98°, 42½ per cent. between 98° and 99°, and 5 per cent. exceeding 100°; the mean being 98·1°, and the minimum 95·6°.

At 11 o'clock a rise has taken place, the temperatures exceeding 100° being 27½ per cent., and those below 98° 10½ per cent.; the mean 99·1°, the minimum 96·8°.

A steady rise appears to take place till 8 P.M., the percentages above 100°, at 2, 5, and 8 P.M., being respectively 40, 42¾, and 52, and the means 99·6°, 99·5°, and 99·7°; the highest temperature recorded, 103·6°, being noted at 8 o'clock.

TABLE III.—*Analysis of Temperatures.*

SECOND STAGE.

Hours	103°	102°	101°	100°	100° and over	99°	98°	97°	96°	95°	91°	Under 98°	Maximum	Mean	Minimum	No. of cases	No. of observations
8 A.M.	—	—	1	3	4	9	35	23	5	1	—	29	101·4	98·1	95·6	10	77
Percentages					5·2	11.7	42·85					37·66					
11 A.M.	—	2	8	11	21	23	26	6	2	—	—	8	102·4	99·1	96·8	10	76
Percentages					27·63	30·26	34·21					10·52					
2 P.M.	—	4	15	12	31	26	15	4	1	—	—	5	102·5	99·6	96·4	10	77
Percentages					40·26	33·76	18·					6·5					
5 P.M.	—	1	10	16	27	13	17	6	—	—	—	6	102·	99·5	97·2	10	63
Percentages					42·85	20·63	27·					9·52					
8 P.M.	2	3	10	24	39	15	18	2	1	—	—	3	103·6	99·7	96·2	10	75
Percentages					52·	20·	24·					4·					

The chart of this class has been depicted with open dots (No. 3 in Fig. 2), and seems to indicate a continuous ascent during the day.

The observations, however, resemble those of the first stage in the low morning temperatures, but differ from them in exhibiting a more continuous rise from 8 A.M. to 8 P.M.

The *third stage active class* (Fig. 3) has been subjected to more thermometrical observations than any other, because the clinical and pathological phenomena which distinguish it are more distinctly phthisical than those of other stages.

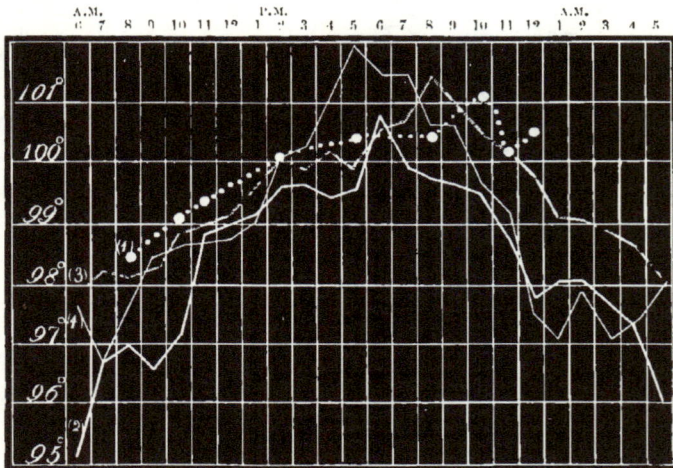

Fig. 3.—'Stage III., active.' No. 1 (dotted line). Averages of forty-three patients during nine periods of the day from 8 A.M. to 12 P.M. No. 2 (thick continuous line).— Twenty-four-hours' cycle of one of the forty-three cases. No. 3 (toothed line). Average of twenty-four-hours' cycle of seven similar cases. No. 4 (thin continuous line) Single case of temperature taken in the mouth.

The rhonchus and sibilus which often usher in tuberculosis may be confused with asthma or bronchitis, and even examination of the expectoration may fail to detect tubercle bacilli until the rapid increase of the consolidations forces itself into notice, but the physical signs and clinical symptoms of an active third stage case are unmistakable. The large amount of purulent expectoration, the hectic aspect, the wasting, the night-sweats, combined with the existence of cavernous sounds in the upper portions of the lungs, are referable to nothing else than a phthisical cavity discharging matter from its surface, or extending itself by further excavation, this being rendered indisputable by the presence of lung tissue as well as tubercle bacilli in the sputum.

The position of the signs, the character of the expectoration, and other clinical phenomena, at once separate cases of bronchial dilatation from this category.

While, however, it is easy to decide on the existence of a cavity, and to assign many of the symptoms to suppuration going on in the lung, a difficulty arises when, in addition to this, there is evidence of fresh tubercle forming either in the same or in the opposite lung. The question then arises, to which are we to assign the thermometrical phenomena—to the abscess symptoms, if I may so call them, or to the tubercle symptoms ?

It is hard to say, though probably the purulent processes exercise the greater influence, but a comparison between the *first stage active* and *third stage active* may throw some light on the subject.

I wish, however, to be clearly understood that in some of these forty-three cases the discharging cavity was accompanied by an increasing amount of tuberculisation.

The thermometry of the ' active third stage ' shows greater extremes than that of any other, and includes both the highest and the lowest temperatures, the actual maximum being 104·6°, the minimum 91·6°, a range of 13 degrees! The existence of these extremes, the marked pyrexia, and the subsequent collapse, show a close approximation to the thermal course, as far as we know it, of suppuration and of pyæmia.

Let us briefly summarise the results of the third stage active, as shown in Table IV. (*see* page 178).

The cases were 43, and observations were taken 13 times a day, viz., at 8, 10, 11 A.M. ; 2, 3, 5, 6, 7, 8, 9, 10, 11, and 12 P.M., between 300 and 400 being taken at 8, 11, 2, 5, and 8. At 8 A.M., the majority of the temperatures are normal and sub-normal; 36 per cent. were between 98° and 99°, and 28½ per cent. below 98°, the minimum being 93·6°,

TABLE IV.—*Third Stage (Active).*

Hours	104°	103°	102°	101°	100°	100° and over	99°	98°	97°	96°	95°	94°	93°	Under 98°	Maximum	Mean	Minimum	No. of observations	No. of cases
8 A.M.	—	—	6	10	35	51	69	122	67	16	9	3	2	97	102·8	98·5	93·6	339	42
Percentages	—	—	—	—	10·32	15·04	20·35	36·	19·76	—	—	—	—	28·61	—	—	—	—	—
10 A.M.	—	—	2	7	13	22	26	11	10	7	—	—	—	17	102·7	99·2	96·5	76	7
Percentages	—	—	—	—	—	28·94	34·21	14·47	—	—	—	—	—	22·36	—	—	—	—	—
11 A.M.	1	4	12	45	69	131	77	92	26	9	5	2	—	42	104·	99·4	94·	342	38
Percentages	—	—	—	13·15	20·17	38·30	22·50	26·90	—	—	—	—	—	12·28	—	—	—	—	—
2 P.M.	—	14	33	78	79	204	84	42	12	3	—	1	—	58	103·8	100·2	94·8	346	39
Percentages	—	—	—	22·54	—	58·96	24·27	12·13	—	—	—	—	—	16·76	—	—	—	—	—
3 P.M.	1	5	8	7	9	30	4	1	—	1	1	1	—	3	104·6	100·6	94·	38	3
Percentages	—	—	—	—	—	79·	—	—	—	—	—	—	—	7·9	—	—	—	—	—
5 P.M.	1	10	43	79	98	231	86	32	6	1	1	—	—	8	104·1	100·4	95·6	357	40
Percentages	—	—	12·04	22·12	—	64·70	24·09	8·96	—	—	—	—	—	2·24	—	—	—	—	—
6 P.M.	—	—	—	8	6	14	2	—	—	1	—	—	—	—	101·3	100·9	98·4	16	2
Percentages	—	—	—	—	—	—	—	—	—	—	—	—	—	—	—	—	—	—	—
7 P.M.	1	4	11	9	11	36	5	—	—	—	—	—	—	—	103·9	101·5	—	41	2
Percentages	—	—	—	—	—	87·82	12·19	—	—	—	—	—	—	—	—	—	—	—	—
8 P.M.	—	20	47	85	110	262	70	29	9	4	—	—	—	13	103·6	100·5	96·	374	39
Percentages	—	—	—	22·72	29·11	70·05	18·71	7·78	—	—	—	—	—	3·47	—	—	—	—	—
10 P.M.	—	2	16	33	15	66	3	1	—	1	—	—	—	2	102·5	101·2	96·2	71	6
Percentages	—	—	22·53	46·47	21·12	92·95	4·22	—	—	—	—	—	—	—	—	—	—	—	—
11 P.M.	—	—	2	5	—	7	12	—	—	1	—	—	—	—	101·4	100·1	96·8	—	—
Percentages	—	—	—	—	—	—	—	—	—	—	—	—	—	—	—	—	—	—	—
12 P.M.	—	2	11	10	16	39	10	3	—	—	—	—	—	—	102·8	100·6	98·4	52	5
Percentages	—	—	—	—	—	75·	19·23	5·76	—	—	—	—	—	—	—	—	—	—	—

the maximum 102°, and the mean 99·5°; 20 per cent. were between 99° and 100°, and only 15 per cent. above 100°. In the few cases in which 100° was reached or exceeded, one was accompanied by acute ulceration of the larynx, another by rapid tuberculosis of the opposite lung. Even here a low morning temperature occurred, which was detected at 5 o'clock in the morning, but did not last until 8, the rise commencing early.

A third exceptional case was one where large hæmoptysis induced catarrhal pneumonia, and subsequently death. Thus are explained to some extent those isolated high temperatures among a great majority of normal or sub-normal ones.

The rise, however, in this stage begins early, and at 10 o'clock 29 per cent. exceed 100°, 34 per cent. exceed 99°; while those under 98° are 22 per cent.

At 11, 38 per cent. are above 100°, 1 reaching 104°; 13 per cent. exceed 101°, while only 12 per cent. remain below 98°, the mean being 99·4°. The rise continues through the 2 P.M. records. Nearly 58 per cent. exceed 100°, 22½ per cent. 101°. The observations at 3 show a persistent rise; at 5 the percentage above 100° is 64, i.e. two-thirds of the whole number, the maximum being 104·1°, the mean 100·4°; the minimum 95°. Only 2¼ per cent. fall below 98°.

The high temperatures are continued at 6 and 7, and at 8 P.M. we find 70 per cent. exceed 100°, and 22¾ per cent. exceed 101°, while only 3½ per cent. fall below 98°, the mean being 100·5°.

The notes at 10 P.M. show the highest temperature, but at the same time a tendency towards equalisation appears, 93 per cent. exceed 100°, the majority ranging between 99° and 102°, only one observation falling below 99°. The mean is 101·2°.

The 11 and 12 P.M. notes indicate a subsidence, and a still more marked avoidance of extreme pyrexia or collapse. The mean at 12 is 100·6°.

The average maximum of the 43 patients was reached at 10 o'clock P.M.; but this is not invariably the case, for it

N 2

may be attained at any hour between 5 and 10 P.M., and in
isolated observations it has been reached even earlier, as will
be seen in some of my twenty-four-hour cases.

This is of no great consequence, as the temperature having
once risen, often remains for several hours pyrexial in character,
but the point of importance is to ascertain the commencement
of the rise and of the fall.

Fig. 3 (page 176) shows pretty fully the temperature-
course of this important class of cases.

The dotted line (No. 1) gives the averages of the 43
patients, during the nine periods of the day already mentioned, viz., from 8 A.M. to 12 P.M.   The thick continuous
line (No. 2) is a twenty-four-hours' cycle of one of these
cases.   The toothed line (No. 3) is an average of 7 similar
cases taken hourly for twenty-four hours, and, like all averages,
is not quite so striking as a single case.

A glance at these curves must convince the most sceptical
that the pyrexia follows a definite order and law. The general
form of the line representing the mean of a large number of
daily observations, tallies with similar periods in a twenty-four-
hours' round of one of these cases ; and if further proof were
wanting of the close correspondence, it is found in the average
of 7 other twenty-four-hour cases.

The rise and fall are well marked in all, though the
maximum is not always reached at the same time.   The
dotted curve (No. 1) reaches it at 10 P.M. ; the toothed line
(No. 3), or the average of 7 cases at 8 P.M. ; and the continuous line (No. 2), at 6 P.M.   When once reached, a fall,
for the most part unbroken, takes place, extending through
the night into the early hours of the morning, reaching 95°
and even lower temperatures, and the effects of this we see
in our 8 A.M. observations, which are often normal or sub-
normal.

The sub-normal temperatures being very striking, I thought
they might be possibly due to the axilla not being properly
closed, and accordingly tried the mouth in one twenty-four-
hours' case and depicted it by the thin line (No. 4).   The
general features of this curve are the same, the gradual ascent

from the normal fall to the maximum of 101·9° (reached a little earlier than in the axilla cases), the rapid descent with a few slight interruptions to the minimum 96·4° reached at 7 A.M.; and then the recovery, as the day advances.

The general characteristics of 'third stage active' temperatures may be summed up :—1st. Afternoon and evening pyrexia ; 2nd. Rapid fall during night and early morning ; 3rd. Recovery in the later morning hours, and consequent normal temperatures.

A careful comparison of this class with the charts of empyema and chronic abscess show a remarkable similarity and render it in the highest degree probable that the special features of this class are chiefly due, not to the tuberculisation, but to the suppuration present.

The class of *third stage quiescent* or chronic cavity (Fig. 4) is one of the commonest met with in practice, and has for its distinguishing thermometrical feature a remarkably low temperature course. The range of sub-normal temperatures is very striking, and will be exhibited in the following case (Case 33), where cavities existed in both lungs, but no extension of disease took place, and some weight was gained. The temperatures ranged from 95·8° to 98·4°, and the mean temperatures are all below 98°, the 8 A.M. mean being 96·6°.

CASE 33.—C. J— (Phthisis, III., with sub-normal temperatures).

	8 A.M.	11 A.M.	2 P.M.	5 P.M.	8 P.M.	Physical signs
March 18 .	96·4	97·2	97·8	97·6	98·	Cavernous sounds in
„ 19 .	96·	97·8	97·2	97·	98·4	both lungs.
„ 20 .	96·2	96·6	97·4	98·4	97·2	
„ 21 .	97·4	97·4	97·8	97·8	98·2	
„ 22 .	96·8	98·2	96·8	98·2	97·6	
„ 23 .	97·	97·2	97·6	98·4	97·6	
„ 24 .	96·8	97·	95·8	98·2	98·	
Average .	96·6	97·3	97·2	97·9	97·6	

The observations on the 'third stage quiescent' class were made on twenty patients, five and six times a day. About 150 are recorded at each period. Table V. (on p. 183) shows an analysis of them.

Of the 8 A.M. temperatures only $6\frac{3}{4}$ per cent. reach 99°; $32\frac{1}{4}$ per cent. reach 98°; while 60 per cent. fall below it; the minimum being 95°, and the mean 97·7°.

A few observations taken at 10 A.M. show no marked change.

The 11 A.M. notes indicate a slight rise; 23 per cent. reached 99°, 49 per cent. 98°; 40 per cent. fell below it; the mean being 98·2°.

The 2 P.M. notes are slightly higher, and in a few exceptional cases exceed 100°; the mean is 98·4°.

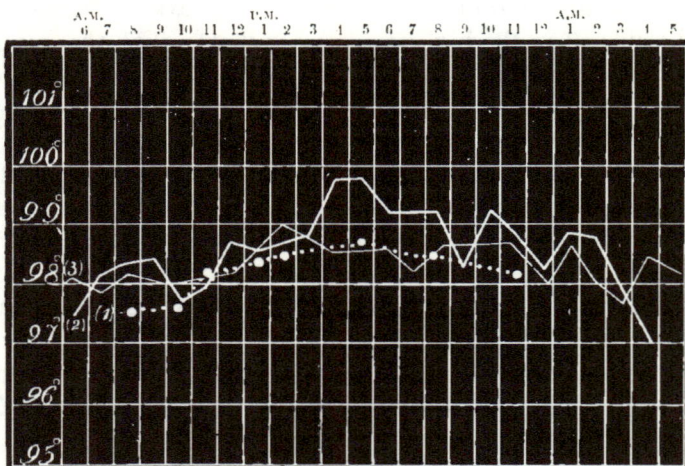

FIG. 4.—'Stage III., quiescent' and 'normal.' No. 1 (dotted line). Average of twenty cases taken eight times a day. No. 2 (thick continuous line).—Twenty-four-hours' cycle of one case. No. 3 (thin continuous line).—Twenty-four-hours' cycle of normal temperature.

At 5 P.M. the maximum is reached; 35 per cent. exceed 99 , and 16 per cent. fall below 98°, the mean being 98·6°.

After this there is a slight subsidence, and at 11 P.M. the mean is 98·3°. The dotted line in Fig. 4 (No. 1) represents the mean of twenty cases and shows the tendency of this class towards a sub-normal character; the mean temperature of these twenty patients hardly reaching the normal 98·6°.

The continuous line (No. 2) is that of the twenty-four-

TABLE V.
Third Stage (Quiescent).

Hours	102°	101°	100°	100° and over	99°	98°	97°	96°	95°	94°	Under 98°	Maximum	Mean	Minimum	No. of observations	No. of cases
8 A.M.		—	—	—	10	48	69	19	2	—	90	99·2	97·7	95·1	148	20
Percentages	—	—	—	—	6·75	32·43	47·	—	—	—	60·81	—	—	—	—	—
11 A.M.	—	—	1	1	33	70	29	10	—	—	39	100·	98·2	96·	143	18
Percentages	—	—	—	—	23·07	48·95	—	—	—	—	27·02	—	—	—	—	—
2 P.M.	1	—	2	3	46	62	33	11	1	—	45	102·	98·4	95·8	156	20
Percentages	—	—	—	—	29·48	39·74	21·15	—	—	—	28·84	—	—	—	—	—
5 P.M.	—	—	10	—	53	63	23	2	—	—	25	100·2	98·6	96·2	151	20
Percentages	—	—	6·62	—	35·10	41·06	15·23	—	—	—	16·55	—	—	—	—	—
8 P.M.	—	1	12	13	47	57	32	6	—	—	38	101·8	98·5	96·	155	20
Percentages	—	—	7·74	8·38	30·32	36·77	—	—	—	—	24·50	—	—	—	—	—
11 P.M.	—	1	1	2	14	19	4	2	1	—	7	100·	98·3	94·	43	6
Percentages	—	—	—	—	32·55	44·18	—	—	—	—	16·25	—	—	—	—	—

hours' cycle of a case in this category, which exhibits in addition to the above features a slight afternoon rise, somewhat resembling that of the active 3rd stage, though far less marked. As usually in phthisis, the morning fall is present.

This class resembles in its thermic features the 'cold abscess' of French writers. Though these patients discharge quantities of pus, they appear not to present the ordinary symptoms of suppuration, but their aspect at once bespeaks their temperature. Pallor often gives place to slight lividity, accompanied by coldness of hands and feet. It is probable that in these cases, much of the respiratory area being blocked by tubercle or utterly destroyed by emphysema, the residue hardly suffices for the proper oxygenation of the blood, and the patient consequently exists—he can be hardly said to live—at the level of batrachians.

Having thus reviewed the thermal features of these cases, as represented by the statistics, illustrative cases, tables and charts, two main facts stand out regarding consumption generally :

1st. The post-meridian character of the pyrexia, when pyrexia exists at all.

2ndly. The remarkable fall at night and the sub-normal temperatures of the early morning.

It is obviously of great importance to ascertain the relations which these bear to a normal standard, especially whether or no the fall is peculiar to the disease ; and for this purpose I have consulted various authorities in order to arrive at that difficult problem : viz., the diurnal thermic course in a healthy man.[1]

---

[1] My colleague Dr. Pollock made a number of observations on cases of phthisis about the same time that mine were being carried on, and his conclusions, as stated in his lectures at the Brompton Hospital (*Med. Times and Gazette*, July 25, 1874) were that 'in ordinary quiescent phthisis (stage not specified) the minimum temperature, 98° or 97°, is in the early morning (3 to 7 A.M.), and the higher temperature occurs from about 3 to 7 in the afternoon.'

Dr. Jürgensen, of Kiel, has made a number of hourly observations on the day and night temperatures of healthy men, extending over several days. These give a chart slightly differing from the annexed one (Fig. 4, No. 3), which I prefer to retain, because it is as well in comparing standards to aim at as nearly similar conditions as possible, and the fact of Dr. Kelly being

Dr. Wm. Ogle's[1] careful experiments give some clue, but from ranges of hours and not exact periods being given, they are hardly suitable for a chart. I have availed myself of Dr. Parkes' standard,[2] agreeing in the main with Dr. Ogle's, and used for testing the effects of exercise and alcohol on the body. The observations were carried out on a healthy young soldier, aged 30, of powerful muscular frame, and were taken every two hours from 8 A.M. to 8 P.M.; the figures representing the average of six days. Dr. Parkes' observations extend no further than the day, and being at a loss for a standard of healthy night temperatures, with which to compare the sub-normal ones of phthisis, I requested Mr. Kelly, then clinical assistant at the Brompton Hospital, and a remarkably robust young man, aged 22, to submit himself to experiment. He kindly consented, and was therefore put in a bed in a private ward in the hospital, under the same equable temperature conditions as the patients, at 8 P.M., and remained there till 8 A.M. on the following morning, his axilla temperature being hourly observed by Messrs. Parry and Bernays. Beyond a glass of beer at 12 P.M. and some milk at 3 A.M. he had no food and slept soundly a great part of the night, notwithstanding temperature observations being regularly taken. The heat of the ward varied from 62° to 65°, and he perspired a good deal. Mr. Kelly's night temperatures, combined with Parkes' day records, are shown in Fig. 4 (No. 3). It proves that no special fall occurs at

under the same atmospheric conditions as the consumptive patients, and, like Dr. Parkes' soldier, of the same nationality, renders the experiment more complete.

Dr. Jürgensen made some interesting experiments on the effects of food and starvation on normal temperatures. The experiment of starvation was carried out ruthlessly, for the man was not allowed even water for thirty-three hours. Slight diminution of the temperature, varying in amount from ·15 Centig. to ·2 Centig. followed, but the general course and fluctuations were maintained. On food being largely taken the temperature rose to the normal (*Die Körperwärme des gesunden Menschen*, 1873).

Dr. Finlayson (*Glasgow Medical Journal*, February 1869) made a number of hourly observations on healthy children, but the temperature course in these differs so much from that of adults as to be useless for our purposes.

[1] 'St. George's Hospital Reports,' vol. i.

[2] *Proceedings of the Royal Society*, 1872.

night, and therefore we may conclude that the fall in phthisis is abnormal and probably characteristic of consumptive disease.

It is difficult to assign any special cause for this fall, as observations show that it does not depend on the amount of night-sweats, nor always on the height of the afternoon fever. Sleep is not the cause of the depression, for it takes place in patients still awake. It may be modified and reduced to some extent by taking food, and this is a clear dietetic indication. As we stated at the beginning of this chapter, two influences appear to be at work in the production of the temperature of phthisis—one, an excess of the processes by which pyrexia is produced; and the other, the influence of well-marked collapse on the constitutional powers.

These two agencies are continually struggling for the mastery, and the result of this conflict is the temperature course of the disease. The influence of the first is seen in the rise in the afternoon and evening, well marked in the active forms of all three stages, and regularly recurring day after day, for long periods; the influence of the second is shown in the rapid nocturnal fall and the low temperatures of early morning; the collapse influence is also seen in the subnormal day temperatures occasionally occurring in all stages of the disease, and even where the active processes of lung tuberculisation, of softening, and of excavation, may be taking place. It is, however, chiefly noted in the quiescent forms of the first and third stages.

When low temperatures accompany lung tuberculisation or excavation, it is probable that the *collapse* influence is greater than the *pyrexial*, and therefore masks it.

Where the chart shows occasional fitful variations, these agencies are, perhaps, evenly balanced, and may alternately prevail, as is witnessed in the end temperatures of consumption; and to these deviations noticed in advanced cases may be ascribed the prevailing and erroneous opinion that phthisis has no definite temperature course.

Dr. Burdon Sanderson mentioned to me an experiment which bears on this subject. He inoculated a guinea-pig and a dog with an equal quantity of purulent liquid. The temperature observations were then taken on both animals, and showed a lowering of temperature in the guinea-pig, and a considerable rise in the dog; the poison, therefore, produced pyrexia in the stronger animal, and collapse in the weaker one.

It has always been considered that pyrexia is usually accompanied by wasting, and, in order to test this, the *weights* of the greater number of these patients have been taken before and after the thermometric observations. The following table gives the result, and will, I think, repay study as bearing upon the influence of pyrexia on weight :—

TABLE VI.

WEIGHTS	First stage (active)		First stage (quiescent)		Second stage		Third stage (active)		Third stage (quiescent)		Totals	Per cent.
	No.	*Per cent.	No.	Per cent.	No.	Per cent.	No.	Per cent.	No.	Per cent.		
Gained weight	15	60·	2	—	5	—	7	21·21	13	68·42	42	46·15
Lost	5	20·	3	—	4	—	24	72·73	4	21·05	40	43·96
Stationary	5	20·	—	—	—	—	2	6·6	2	10·53	9	9·89
Unknown	1	—	—	—	1	—	10	··	1	—	13	—
	26	—	5	—	10	—	43	—	20	—	104	—

* Percentages are calculated with the Unknowns excluded.

The thirteen unknowns are excluded in calculating the percentages, but it is doubtful whether they ought not to be included among the losers, as the reason that the weight was not taken, was, generally, that the patients were too ill to leave their beds.

We see that out of 91, 42 gained weight, 40 lost, and 9 remained stationary.

Of the first stage, active, 60 per cent. gained, 20 per cent. were stationary, and 20 per cent. lost.

In the second stage half the patients gained weight.

The greatest loss is, as might be expected, in the third stage, active, where 72 per cent. lost weight; but even here 21 per cent. gained, and when we consider the desperate character of some of the cases, and their collapsed condition, how pyrexia continued in many for weeks and months, it is remarkable that so large a number did not lose weight.

The largest number of gainers is in the 'third stage quiescent,' where the percentage is 68; 10 per cent. were stationary, and 21 per cent. lost, and, considering the number of double-cavity cases, this is noteworthy.

The reason of the large proportion of gain lies in the amount of food taken, combined with rest and comfort.

The quantity of food eaten by even those patients, whose appetites are capricious, is very considerable, owing to the regular stuffing system of the Brompton Hospital, and the number of delicacies supplied to meet every fancy. The scales often discover a gain in the weight of patients in whom it could least have been predicted.

The deductions from this table are—

1st. That, though the tendency of the disease is towards emaciation, gain of weight is possible in every stage, provided there be no diarrhœa or hæmoptysis, and that food can be taken and assimilated.

2nd. That pyrexia is not incompatible with gain of weight provided the appetite be good.

3rd. That while active third-stage cases are the least likely to gain weight, chronic cavity patients are prone to increase in flesh.

4th. That neither tuberculisation nor softening preclude gain of weight.

These conclusions will be found in direct opposition to Niemeyer,[1] who states that he made out by ' numberless examinations, with the thermometer and the weighing-machine, that the decrease and increase in the weight of phthisical patients are respectively in relation to the height of the fever or to its disappearance;' but they confirm Dr. Ringer's[2] remark that ' patients with a very considerable elevation of temperature, but who enjoy a good appetite, do not lose flesh, provided always they are not employed in any active pursuit which entails much waste of tissue.'

Peter[3] has carried out a series of careful and interesting

[1] *Text-Book of Practical Medicine.*
[2] *Op. cit.*   [3] *Leçons de Clinique Médicale,* vol. ii. p. 427.

experiments on the temperatures of different regions of the
thorax in phthisis, specially the portions overlying centres of
tuberculisation, and has arrived at the following remarkable
conclusions :—

1. That tuberculisation causes a local rise of temperature
due to hyperæmia, and that from this hyperæmia may arise
congestion, inflammation, or hæmorrhage.

2. That the temperature of the intercostal space overlying
the pulmonary lesion is always higher than the mean temperature of the body or of that of the corresponding intercostal
space on the opposite side.

3. That the formation of granulations causes a local
elevation of temperature varying from ·5° C. to 1·5° C. above
the mean.

Peter finds the difference between local and mean temperature most marked in caseous or scrofulous pneumonia, where,
although the latter is very high, the former (local) often
exceeds it by ·8° C. to 1° C.; he suggests this as a mode of
diagnosis from ordinary croupous pneumonia. Peter's conclusions have been tested by several observers, and disputed
by many, but confirmed by Vidal of Hyères, and Michel of
Cauterets.

I have also tested the above conclusions repeatedly, and
as I had the privilege of being present on one occasion at the
Hôpital de la Pitié, Paris, when M. Peter took his observations, I was able to follow his methods exactly. But I am
bound to say that I failed in confirming the majority of his
conclusions. I did not find the temperature of the intercostal space overlying the area of active tuberculisation
higher than the general temperature of the body, or again,
higher than the axillary temperature of the same side. In
well-marked cases of one-sided lesions I did occasionally find
a slightly higher temperature in the first or second intercostal spaces overlying the seat of active tuberculisation, or
excavation, than in that overlying the healthy lung, but even
this was not universal. Peter's interesting temperature
results in pneumothorax will be given in the special chapter
on Pneumothorax.

CHAPTER XIII.

THE DIARRHŒA OF CONSUMPTION.

The Diarrhœa of consumption—Three forms—First form arising from indigestion, second form from intestinal ulceration—Its frequency in consumption—
Structures affected—Comparison with ulceration of intestines in dysentery
and typhoid fever—Fatal Hæmorrhage—Perforations—Peritonitis—Contraction of intestinal ulcer—Lesions of mesenteric glands—Tubercle bacilli
in stools—Stages in formation of ulcers—Relative liability of different parts
of the intestine to ulceration—Ulceration and pyrexia—Swallowing sputum
and intestinal ulceration—Third form of diarrhœa arising from lardaceous
disease of the intestine—Prognosis.

DIARRHŒA is often the cause of fatal termination in phthisis,
its exhausting discharges reducing the weight and impoverishing the blood more than the cough and expectoration, or
even than the fever. Three different kinds of diarrhœa are
met with during the course of the disease. 1st, arising
from indigestion; 2nd, from intestinal ulceration; 3rd, from
lardaceous disease of the intestines. A great deal of the
diarrhœa connected with the first stage of phthisis is attributable to the first cause, and many practitioners think,
when they have allayed this, they have arrested ulceration, whereas they have only got rid of dyspepsia. It is
unnecessary to dwell on this form, which as a rule arises
from over-feeding patients, and is accompanied by a coated
tongue, tenderness in the right hypochondrium and the
epigastrium, loaded urine, anorexia, flatulence, and frequent
light-coloured stools.

The second form, arising from intestinal ulceration, is the
commonest cause of fatal diarrhœa in phthisis, and prevails in
a considerable proportion of third-stage patients. Louis found
ulceration of the intestines in five-sixths of his cases, Bayle
in 67 per cent. of his, Lebert in 67 per cent. of his Breslau
ones, and in only 39 per cent. of his Zurich ones. The

Brompton Hospital *post-mortem* book for the year 1880 gave
the following :—Out of seventy-five cases of phthisis in which
the intestines were examined, sixty-one had ulceration of a
tubercular nature, giving a percentage of 81, a higher
percentage than that of either of the above authorities except
that of Louis. In 88 *post-mortem* examinations made on
cases of phthisis during the year 1885 at the same hospital
by Dr. Percy Kidd, intestinal tuberculosis was present in 63,
or in 71 per cent.

The pathology of tubercular ulceration requires a large
number of specimens to display its exact course, most autopsies
only showing the last destructive stages.

Ulcers are to be found occasionally in nearly the whole
intestinal tract below the duodenum, but they are so extensive
and involve so great an amount of the mucous membrane of
the large intestine as often to give it an entirely worm-eaten
appearance, and in many instances the large intestine of a
case of phthisical diarrhœa cannot be distinguished from that
of tropical dysentery, so well portrayed by Sir Joseph Fayrer.[1]
In the small intestine, however, especially in the ileum, we
are able to trace the early steps of the ulcerative process, and
I hope the cases I am about to relate will illustrate these. The
first stage consists principally of an inflammatory hyperplasia of the solitary and agminate glands, as shown by their
prominence when the intestine is opened, though the process is by no means always limited to these structures. The
second stage consists either in their conversion into pustules
by formation of matter, or by their undergoing caseation.
Either phase is followed by the third stage, which is the
evacuation of their contents, leaving behind ulcers with smooth
floors and sharp, clear-cut edges. Up to this point the process
is held by many authorities to be inflammatory, and not
strictly tubercular. After a while grey tubercle is detected in
the floor of the ulcer, which in time undergoes caseation,
causing extension, vertical or lateral, of the original ulcer.
A distinction which has been drawn by Dr. Powell[2] between
this ulcer and the typhoid-fever ulcer is, that in tubercular

[1] *Lettsomian Lectures*, 1881.   [2] *Op cit.*

ulceration the whole of the gland follicles of a solitary gland
or a Peyer's patch are by no means necessarily involved,
and that ulceration is not confined to these structures, but
extends to the mucous membrane beyond. In the typhoid
ulcer the whole Peyer's patch is involved, and the ulceration
is limited to it.

The secondary tubercles follow the lines of the blood
vessels and lymphatics, specially of the latter, and are sometimes visible from the peritoneal surface of the intestines,
demonstrating how deep these ulcers extend. When ulceration has arrived at this stage, each spot presents thickened
irregular edges and an uneven floor. The intestinal wall
around the ulcers is generally thin and wasted, and it is often
easy to recognise them from outside without opening the
intestine.

In time these patches, extending, join neighbouring ones,
and eventually the whole circumference of the mucous
membrane becomes a mass of disease. It is in this condition
that we often find the colon and sigmoid flexure. Before
ulceration, however, reaches this degree, it is not uncommon
to find separate ulcers of great length; Louis found one
measuring eight inches, the floor being formed by the muscular
coat, except in the centre, where even this had disappeared,
and only the peritoneal coat remained. Considerable thickening of the peritoneal coats, especially in the region of the
ulcers, takes place, and thus it is that perforation is comparatively rare. When it occurs, peritonitis is inevitably set up,
but even then it is sometimes of a localised description, the
extravasations being walled off by fresh adhesions. As a rule,
however, the peritonitis is general, and death soon follows the
perforation. Rindfleisch [1] gives a case where several coils of
the intestines were adherent, and perforations took place in
the adjacent portions in five different places. Not rarely, perforations take place between two portions of the intestine
which have become adherent, and fistulous openings are thus
established. Occasionally vessels are penetrated by the

[1] Ziemssen's *Cyclopædia of the Practice of Medicine.*

ulcerative process, and death occurs by hæmorrhage. Hanot [1] states that he saw two such instances in the service of M. Lasègue, in each case the amount of blood lost being very striking. After death the source was found to be a small opening in a vessel in the middle of a Peyer's patch, which was deeply ulcerated. Biermer gives a case of large intestinal hæmorrhage following a tuberculous ulcer of the rectum in a child.

These are the most unfavourable pathological courses of the process : but, on the other hand, the ulcer may shrink, the walls become more indurated, and at length form a puckered cicatrix in the mucous membrane, largely contracting the circumference of the intestine ; and in patients dying of chronic phthisis these scars are occasionally seen.

Accompanying the ulcerative changes we often find considerable swelling of the mesenteric glands, but this is generally the case in the later stages of the disease. Walshe found alteration in these glands in from one-third to one-fifth of his cases of phthisis.

Koch found tubercle bacilli in the intestinal ulcers of phthisis ; and Gaffky detected them in the ochrey stools in great abundance and containing spores ; he also noted some familiar micrococci. Mr. H. H. Taylor also detected them in the stools of two hospital patients of mine who were suffering from intestinal ulceration. Dr. Percy Kidd [2] in nine cases of typically tubercular ulceration of the intestines, found tubercle bacilli present in eight, and in abundance in four. Their presence has been established in mesenteric glands, of which a good example is to be seen in Plate IV (1), where they are to be noted approaching a blood-vessel ; but though abundant in large numbers in cases of recent tuberculosis, and especially in the young, they are difficult to find in cases of long-established disease, and are often entirely absent.

Let us now take some examples of intestinal ulceration to illustrate the above remarks. The first case is one of advanced phthisis, with the common accompaniment of

[1] *Nouveau Dictionnaire de Médecine*, article ‘ Phthisie,’
[2] *Medico-Chir. Transactions*, vol. lxviii.

disease (chiefly lardaceous) of the liver and kidney, giving rise to ascites and anasarca, but intestinal ulceration was proceeding at the same time, causing obstinate diarrhœa.

CASE 34.—John B., aged 29, draper's assistant, was admitted into the Brompton Hospital, October 30, 1879, with a history of cough, loss of voice, expectoration, and wasting for eight months. Night-sweats commenced with his illness, but have ceased lately. Diarrhœa appeared a month ago, but stopped in fourteen days; then œdema of the legs and ankles commenced. On admission, the appetite was good; cough and expectoration moderate; temperature, pulse, and respiration all normal; weight 8 st. 5¾ lb.; pupils rather contracted; urine contained albumen (⅓). *Dulness, flattening over the whole left chest; cavernous tinkle audible from the clavicle to the third rib, and above the scapula, and crepitation heard below these points. Hepatic dulness increased upwards.* In December the patient became excited and wandering in manner and speech. The bowels were costive, the albumen increased to ¾ of the test-tube; the œdema became more extensive. After free purging the head symptoms subsided; the œdema diminished, and the amount of albumen was reduced to ⅜. Diarrhœa returned on January 28, 1880, and at first was very moderate, the motions being two or three daily. It gradually increased to five motions a day, but under a simple regimen of milk and beef-tea the daily number of stools was reduced to three. On March 11 the patient became so weak that a more generous diet, including brandy, was adopted, and the diarrhœa increased to nine and eleven motions a day. Under opiate and astringent enemata it was reduced to two motions a day, but the number speedily increased if the enemata were omitted. On April 2 the œdema increased, and ascites made its appearance; the urine continued abundant, with a large percentage of albumen. *Cavernous sounds became audible in the right lung,* the breathing became more and more difficult, and on April 29 he died, the diarrhœa continuing to the end.

Autopsy, twenty-seven hours after death, by Dr. EWART.—Abdomen: The peritoneum contained a pint and a half of pale opalescent fluid. The mucous membrane of the stomach was thickened and pasty and covered with glairy mucus. The upper portion of the intestine was normal; but from the lower third of the ileum downwards the gut was much ulcerated. No distinct tubercle was recognisable in connection with the ulcerations. These were clearly of follicular origin, and they appeared to result more immediately from the rupture of purulent accumulations (similar to those commonly seen in follicular tonsillitis), some of which were as large as a split pea, or rather larger. The next step of the ulceration was characterised by a clean floor and a very sharp edge. Lastly, in some places, a third step was recognised—viz., that of shrinking and of induration. This was especially noticeable where several ulcers and pustules were arranged in a row in the transverse diameter of the ileum. The ascending and transverse colon were worm-eaten by serpiginous and circular ulcers, situated so close to each other as to scarcely leave any healthy surface. Scattered ulcers were found as far as the rectum. The mesenteric glands were not much enlarged; one was calcified. The liver weighed 3 lb. 4 oz.; it showed fatty and fibroid changes to the naked eye, but isolated streaks and dots of deep brown appeared when iodine was applied. The spleen weighed 15 oz.; it showed a few infarcts and some lardaceous changes. The blood-vessels of the pancreas were distinctly lardaceous. The kidneys weighed 19 oz., and were smooth, pale, with pink pyramids; the cortex showed opaque, yellowish-white streaks, within a semi-transparent basis. Iodine turned several of the Malpighian capsules brown. The kidney was a good specimen of the large white variety, combined with lardaceous disease. Heart: The mitral anterior flap had a small fibrinous vegetation on it. Lungs: Both pleuræ adherent; left anterior lobe occupied by a large vomica with fibrous wall; posterior lobe shrunk and contained tubercular mass in process of softening. The right upper lobe contained a large cavity full of grumous fluid, the rest of the lung being spongy and pervaded by tubercular nodules undergoing rapid softening. Bronchial glands: The sub-tracheal one was large, firm, red, and had a patch of pigmentation, but no tubercle. The larynx showed pinhole ulcerations and erosion of the vocal cords.

CASE 35.—William W——, aged sixteen, single, was admitted into the Brompton Hospital on May 11, 1880, with a history of cough and expectoration for eighteen months, the latter occasionally streaked with blood. Diarrhœa had persisted for two months, and night-sweats for a longer period. Great emaciation had resulted. On admission the cough and expectoration were moderate, the breath short on exertion, and the tongue red and irritable; the bowels relaxed twice daily. Pulse 104; temperature 99° F.; urine, sp. gr. 1015, acid, contains no albumen; weight 6 st. 8 lb. *Dulness and cavernous sounds audible from the clavicle to the third rib on the right side; some dulness above the scapula.* On May 19 the patient was seized with symptoms of acute nephritis; albumen appeared in the urine, and *crepitation was audible at the bases of both lungs, in addition to the above physical signs.* The acute symptoms passed off in ten days, the diarrhœa being somewhat less during their persistence; but albumen continued in the urine up to the patient's death. The diarrhœa returned at the end of June, and gradually increased, in spite of all treatment, from three to ten stools daily, the medicines used being bismuth, bismuth and opium, sulphate of copper and opium, turpentine, enema of opium with acetate of lead, with tannic acid, and also injections of linseed-tea and opium. With the exception of the last named, which reduced the number of motions for a time to two daily, none of the above can be said to have in any way controlled the diarrhœa. Great pain in the umbilicus was noted at the beginning of August. The albumen towards the end of the patient's life amounted to seven-eighths of the tube, and at one time the urine after boiling was solid, and both œdema and ascites were present. The patient sank on August 30.

Autopsy, twenty-four hours after death.—Abdomen contained a small amount of fluid. Intestines showed ulcerations in the jejunum and ileum, and in the large intestine. They were of two kinds—(1) small and round, resulting from simple softening and destruction of tissue; lower down in the intestine were (2) larger ones, possessing a rather thickened floor, in which miliary tubercles were present and visible on the peritoneal surface of the bowel. The neighbourhood of the ileo-cæcal valve was severely ulcerated, and the colon formed one continuous ulcer. From the ileum downwards both forms of ulcer occurred side by side. The liver was fatty and lardaceous, the spleen very lardaceous, kidneys smooth and pale, with pink pyramids. The changes of chronic tubular nephritis were well marked. Malpighian capsules lardaceous, giving a strong iodine reaction. Heart normal, except tricuspid orifice, which

was somewhat dilated. Lungs: Right pleura adherent ; right lung contained
a large smooth-walled cavity, the size of an orange, and a smaller one lined
with a glistening membrane. The lower lobe contained abundance of grey
tubercle, either isolated or grouped ; many of these last were in the process of
softening. Left lung hypertrophied, with some cicatrices at the apex. Bronchial glands enlarged, and contained miliary tubercles. Extensive ulcerations
occupied the whole tract of the larynx and trachea from the tip of the
epiglottis to the second division of the bronchi. The mucous membrane of the
larynx was much thickened, and there was considerable hyperæmia of the
trachea and bronchi. Ulcers were present in the pharynx, on the buccal surface of the epiglottis, on the base of the tongue, and the tonsils were hollowed
out by caseation.

Both these cases are cited to illustrate the morbid anatomy
of ulceration. Case 34 shows well the first steps in the process—the swelling of the glands, their suppuration, and
caseation, Case 35 the different forms of ulcer, and the
secondary infection which takes place in and around them.
As in both instances renal disease existed to a considerable
extent, and exercised a large influence on the symptoms, these
cases cannot be considered fair clinical examples of intestinal
ulceration, but they are undoubtedly good ones of the morbid
anatomy. It is seldom that so much destruction of the
mucous membrane exists, as occurred in Case 35, for here the
pharynx, larynx, and bronchi, in addition to the intestines,
were attacked. No history of syphilis could be traced in
either case. Case 36 is a good instance of perforation and
peritonitis.

Case 36.—Charles C., aged 19, admitted October 7, 1881, into Campbell
ward, with the following history. He was attacked one year ago with pain
in the left side, chiefly on taking breath, accompanied by diarrhœa and
followed by pains in abdomen. The pains subsided, but were followed by
cough, expectoration, night-sweats, and wasting. When admitted, his cough
was moderate ; expectoration scanty ; appetite very bad; tongue red and irritable. Pain was complained of in the lower part of the abdomen. Bowels
open three times a day. Pulse 80 ; temperature 103° F.; respirations 20. The
physical signs showed the existence of cavities at each apex. Bismuth and
opium were given internally, and linseed poultices applied to the abdomen ;
but as the pain in the abdomen persisted, a small blister was applied over the
seat of pain. The diarrhœa ceased, and the cough became less frequent, but
the pain persisted.

On October 19 the pain had extended over the whole abdomen, and there
was tenderness on pressure, chiefly over the right side. Tongue furred; bowels
relaxed ; hands and feet cold. Two days after he died suddenly.

Autopsy.—The abdomen contained abundant purulent fluid, encapsulated
in the lower parts of the abdomen by fibrous adhesions. In the process of un-

ravelling the gut, three perforations were detected in connection with the encapsulated portions of the peritoneum. The intestines presented large ulcers from the middle of the jejunum to the end of the rectum, being most numerous and best marked in the ileum. The ulcers were devoid of raised edges and ree from thickening, except where a few granulations were commencing; the ulcers appeared to enlarge by gradual implication of the neighbouring solitary glands. The mesenteric glands were swollen, some were suppurating and others caseous. The lungs showed both pleuræ strongly adherent at the apices. Both lower lobes were spongy and but slightly congested, but contained fresh translucent tubercles scattered through them. The upper lobes showed congestion of deep venous colour, interspersed with numerous nut-sized cavities containing yellow matter. The cavities, which had apparently been recently formed in the partially indurated tissue, were lined and stood in direct connection with several funnel-shaped bronchi, and were situated at the periphery of the lungs. There were caseous masses in the upper lobes. No tubercle was detected in the upper lobes. The tissue intervening between the cavities under the action of spirit was deprived of its congestion and showed pigmentation and old fibrous changes. The bronchial glands were pigmented; the other organs were healthy.

This is a fair instance of intestinal ulceration leading to peritonitis, at first probably partial, as shown by the encapsulated portions, but afterwards more general. The state of the lungs indicates that an attack of pleuro-pneumonia, probably of both lungs, led to induration of the tissue of the upper lobes, and consequent dilatation of the bronchi. Tuberculosis supervened, and led to the formation of some of the cavities, others being probably bronchiectatic. The infection of the intestines was probably through swallowing the pulmonary secretion, and led to extensive ulceration, perforation, and fatal peritonitis.

Clinical symptoms.—When ulceration is proceeding on a large scale in the intestines the patient complains of thirst and anorexia, the pulse is quick, and there is generally great prostration; the tongue passes through various stages : first it is red at the edges and furred in the centre ; then red all over and glazed ; then red with a brown streak in the centre ; and, lastly, aphtha appears on it. The diarrhœa is remarkable for its persistency ; the motions, especially towards the end of life, are very frequent—sometimes twenty in the twenty-four hours, but this is not always the case, and there may be only three or four daily, the characteristic feature being their steady persistency. They may be checked by treatment, but directly this is omitted, and often in spite of it, the symptom returns; the stools themselves are very loose and ochrey-coloured, often tinged with blood, and towards the end of life contain small sloughs from the large intestine. They are exceedingly fœtid, and, as mentioned before, they contain tubercle bacilli. The patient appears much exhausted after each action.

The aspect in these advanced cases is well marked. Owing to the terrible emaciation the face is drawn, the malar bones appear prominent, a hectic spot is generally seen on each cheek, and the expression of the countenance is that of pain, and pain there generally is in the region of the ulceration, or else in the centre of the abdomen. Tenderness is generally to be found if carefully sought for, but its locality varies in different cases. The right iliac fossa is the common seat in many, and especially in early, instances; but later on it may be detected in the right lumbar, right hypochondriac, epigastric, left hypochondriac, left lumbar and left iliac regions, showing clearly that the whole of the colon and sigmoid flexure are involved as well as the cæcum and lower ileum. Often, too, the detection is rendered more easy by the bowel being distended by flatus. The sigmoid flexure is, in my experience, as often the seat of ulceration as the cæcum, and I have frequently detected this during life, and verified it after death. The wasting arising from the diarrhœa is much greater than in ordinary pulmonary phthisis, and a few days of its persistence will reduce a consumptive, who has previously been gaining flesh, by several pounds.

The question arises, In what part of the intestinal canal does ulceration commence? The experience of *post-mortems* rather indicates that it commences in the small intestine near the ileo-cæcal valve; but how comes it, then, that we usually find the process more extensive and advanced in the large intestine? which will, I think, be generally admitted. The answer is to be found in (1) the long retention of the contents of the large intestine, (2) in the disengagement of foul gases in this region, and (3) in the enormous variations in calibre to which this part of the intestine is subject, all of which circumstances favour the spread of destructive or ulcerative inflammation, if once set up. Ulcers in the large intestine may be considered to be placed under most unfavourable conditions for healing.

The temperature varies according to the amount of the ulceration, provided there are no special exciting phenomena in the pulmonary or other lesions. Where the amount of ulceration is small and not advanced in stage, the temperature is not materially affected, as will be shown presently (Case 37) ; where, however, it is extensive, involving both small and large intestines, there is generally marked pyrexia, often alternated by great collapse. The chief feature, however, is its fitfulness; for one day at 11 o'clock the record may be 97° F., and on the following day, at the same hour, it may show 103° F.; these cases being the least of all subject to the laws governing the pyrexia of phthisis. It is possible, however, to have extensive ulceration without any diarrhœa, for Dr. Douglas Powell[1] gives a case where ulcers an inch and a half long were discovered in the ileum, and during life the patient had no diarrhœa. Dr. Pollock[2] also mentions some exceptional cases of this sort.

Let us now take two instances of the influence of ulceration on temperature.

CASE 37.—Emma K., aged 29, married, was admitted into the Brompton Hospital, October 5, 1875, with a history of cough and expectoration for one year, and of emaciation and night-sweats for nine months. On admission she was much emaciated. Cough very troublesome; expectoration abundant. Tongue furred yellow. Bowels costive. Weight 8 st. 5 lb. *Crepitation sounds over the whole right lung, with dulness over the lower third ; tubular sounds audible in the scapular region.* During the first three weeks of her stay this patient improved, and gained 2 lb., but at the beginning of November diarrhœa came on, which persisted up to her death on December 17, the physical signs indicating that cavities had formed in both lungs. The temperature was taken five times a day from November 26 to December 16, a period of three weeks, and only on five occasions did it reach or exceed 100° F. The following were the averages :—At 8 A.M. 98·2° F.; 11 A.M. 97·9° F.; 2 P.M. 98·5° F.; 5 P.M. 98·8° F.; 8 P.M. 99° F. Maximum, 100·6° F.; minimum, 96° F.

The diarrhœa varied in amount from three to six stools a day, and did not appear to be greatly influenced by treatment, which consisted of astringents applied by mouth and rectum. The patient died December 17.

[1] *Pathological Transactions*, vol. xix.   [2] *Op. cit.*

Autopsy, twenty-four hours after death.—Lungs : Right had apex adherent, and a large cavity in the upper lobe. The rest of the lung was riddled with vomicæ containing purulent matter. In the middle of the lower lobe was a small well-defined band of miliary tubercle. The left lung contained several small cavities full of purulent matter, and a large quantity of transparent miliary tubercle of recent formation. The small intestine for the distance of three feet above the ileo-cœcal valve was very much congested. A small ulcer existed a few inches above the valve, and several small ones in the cæcum. Here some miliary tubercle was also seen. The liver and kidneys were fatty, and the other organs healthy.

This case shows convincingly two things : 1st, that a small amount of ulceration need not cause pyrexia ; 2nd, that tubercle may form in more than one organ, without giving rise to any marked rise of temperature, and in fact with a persistence of sub-normal records. There is no doubt that in this case acute tuberculisation as well as suppurative processes were proceeding in the lungs, and, at the same time, tuberculisation in the intestines, but they only served to depress the temperature, and not to raise it. The next case shows the effects of extensive ulceration on the temperature.

CASE 38.—Herbert N. M., aged 19, a clerk, was admitted into the Brompton Hospital on December 3, 1878. His father and mother died of consumption ; he had suffered from cough and yellow expectoration, loss of flesh and strength for two years. For the last ten months he had had continuous diarrhœa, accompanied by some pain in the epigastrium, after food. Had had night-sweats for three months. At present cough troublesome, expectoration abundant and muco-purulent, tongue furred ; bowels open two or three times daily ; appetite poor ; urine acid, containing no albumen or sugar, sp. gr. 1035 ; weight 6 st. 4 lb. *Flattening under both clavicles, dulness and crepitation above the right scapula, and tubular sounds in the first interspace.* The diarrhœa persisted, in spite of a variety of astringents and diet having been tried, and on January 25, 1879, well-marked tenderness was noted in the umbilical region. *Cavernous sounds were audible at each apex, with crepitation in the lower left lung.* The tongue had become red at the edges ; the number of stools varied from two to four daily, and were loose and ochrey. On January 30 there was much tenderness in the lower portion of the abdomen, most marked in the left iliac fossa in the region of the sigmoid flexure, and to a less degree in the right iliac fossa. There was still some tenderness, quite separate from the rest, in the umbilical region. From these signs and the persistent diarrhœa, ulceration of the cæcum and large intestine, most marked in the sigmoid flexure, with enlargement of the mesenteric glands from secondary absorption, was diagnosed. The patient gradually wasted away under the influence of the pulmonary and intestinal lesions, and sank on February 7.

The temperature was carefully taken five times a day during the last fortnight of his life, and the chart showed records nearly normal in the mornings, with a persistent pyrexia after 2 P.M. The highest recorded was 103·8° F.

This chart, however, proves, as Lebert has shown in other cases, the fitfulness of the temperature in intestinal ulceration. There are several records of 101·5° F. at 8 A.M., once followed by a fall at 11 A.M., and a second rise at 2 P.M. During the last day of the patient's life the temperature fell from 102° F. to 96·8° F., indicating collapse of the vital powers. The following are the averages: At 8 A.M. 99° F.; at 11 A.M 99·9° F.; at 2 P.M. 100° F.; at 5 P.M. 101° F.; at 8 P.M. 101·1° F. Maximum, 103·8° F.; minimum, 96·8° F.

Autopsy, on the following day.—Lungs: The right had the pleura adherent, and contained a large cavity at the apex, surrounded by nodules of tubercle; the lower and middle lobes were congested. The left contained a cavity, with less congestion than the right. Intestines: The small intestine was thin, and the vessels full. Three small ulcers were detected, each about one-third of an inch in diameter, about an inch above the ileo-cæcal valve. The large intestine showed extensive ulceration, chiefly in the lower portions, and most marked of all in the sigmoid flexure, where several ragged ulcers with thin edges were found. The mesenteric glands, especially those lying in the folds of the meso-colon and meso-rectum, were much enlarged. Liver fatty. The other organs were healthy.

This case shows well the effect of extensive intestinal ulceration on the temperature chart, for it is probable the state of the pulmonary lesions was not sufficiently active to have greatly influenced it. The ulcerations were found in the exact positions predicted before death in consequence of the physical examination.

Ulceration of the intestines in phthisis has been attributed by Klebs and Mosler [1] to the swallowing of the expectoration, and the first infection of the bowel, to the bacilli contained therein: an opinion which has much to support it, both from the results of experiments on animals and clinical knowledge. Chauveau [2] fed heifers on food containing tubercular matter, and produced intestinal ulceration with general tuberculosis. Tubercular ulceration does not occur except where cavities have been formed, and are freely secreting, and it generally accompanies (*see* Laryngeal Phthisis) laryngeal ulceration, owing probably to the difficulty of expectoration. The infection from drinking the milk of tubercular cows, already alluded to, shows itself by intestinal ulceration. Ulceration of the intestine is more common in children than in adults, probably on account of the inability of the former to ex-

[1] *Berliner Klinische Wochenschrift*, 1873, p. 509.
[2] *Bulletin de l'Académie de Médecine*, t. xxxiii. 1868, p. 1017.

pectorate ; and it is also most marked in those adults, who
from feebleness or other causes are unable to eject the
sputum.

The third form of diarrhœa, or that proceeding from
lardaceous degeneration of the intestines, is not so marked
in its characters as that arising from ulceration. Its clinical
features have not as yet been completely sketched, but it may
be said to correspond with the colliquative diarrhœa of M.
Peter and other French authors. It prevails in the late
stages of the disease, is not very profuse, but is persistent,
and seems to have some correspondence with the drenching
night-sweats. The stools are much more watery, and diminish
after a profuse night perspiration, the tongue is more furred
than red, there is less pain and no tenderness at all in the
track of the intestines, but the liver is often enlarged and
tender, undergoing similar degenerative changes. Diarrhœa
from lardaceous disease of the intestines is only to be found
(in phthisis) in connection with great discharge from some
mucous surface, or with the profuse expectoration from a large
or from several small vomicæ. Protracted diarrhœa, as that
due to ulceration, may give rise to lardaceous degeneration of
the intestines, so that in many *post-mortem* examinations
we find ulceration and lardaceous disease side by side ; and
the question arises to which of the lesions the diarrhœa may
be attributable, which can only be solved by very careful
reference to the clinical history. In a fair proportion of
phthisical cases lardaceous disease is present without intestinal ulceration, and here we have no difficulty in referring
the effect to the proper cause. After death the appearance
of the intestinal mucous membrane, especially of that of the
small intestine, is pale, and sometimes there is œdema. There
is no marked distinction between it and other mucous membranes until iodine is applied, when a number of small reddish-
brown points appear. These correspond to the villi whose
capillaries and arteries are infiltrated with the lardaceous
deposit, and Dr. Green considers that the diarrhœa is due to
an increased permeability of the infiltrated vascular walls,
which thus allows fluids to pass more easily. Usually the

pancreas, spleen, liver, kidneys, and mesenteric glands are found in the same condition.

The following is an example of diarrhœa arising from lardaceous disease : —

CASE 39.—Kate H., aged sixteen, was admitted into the Brompton Hospital on August 13, 1878. History: was chilled by sleeping in a damp bed in December 1877, and has had cough and expectoration ever since, accompanied by great prostration and marked inanition and night-sweats. Had hæmoptysis, a quarter of a pint in March, and expectoration has been occasionally streaked since. Diarrhœa has prevailed for the last month. Cough is very troublesome, expectoration abundant and yellow ; appetite indifferent ; bowels open three times a day ; pulse 110. *Crepitation is heard over the whole chest, anteriorly and posteriorly ; on the right side loud tubular sounds are audible from the clavicle to the third rib ; on the left side amphoric breathing and tinkling sounds as low as the fourth rib.* The diarrhœa persisted in spite of a variety of treatment, the stools being loose and three to four a day ; the cough continued and the patient gradually lost flesh and died on October 27. On October 2 a careful record was commenced of the temperature five times a day, and was continued up to the death on October 27, a period of twenty-five days. During the first fortnight the temperatures were high, even in the morning ; in the second period there was a tendency to fall, and in the last two days, which I have omitted, as the records were only those of collapse, the temperature fell to 95·8° F., though the diarrhœa did not increase.

Averages	8 A.M.	11 A.M.	2 P.M.	5 P.M.	8 P.M.
1st period (14 days) .	99·2° F.	99·6° F.	100° F.	100·7° F.	101·3° F.
2nd „ (10 days) .	98·5° F.	98·3° F.	99° F.	99·2° F.	100·2° F.
Maximum .	102·4° F.		Minimum .		96·8° F.

The autopsy showed that both lungs contained numerous cavities secreting puriform matter, and in the left was some fresh arborescent tubercle following the line of the bronchioles, some yellow matter filling the lumen of the small bronchi. The intestines, small and large, were intensely lardaceous, giving strong reaction with the iodine ; the liver was enlarged, and both it and the spleen and kidneys were all markedly lardaceous. No intestinal ulceration.

Remarks.—This is a fair instance of diarrhœa arising from lardaceous degeneration of the intestine, as no ulceration could be detected, and the whole intestinal tract, in common with other organs, was infiltrated with lardaceous material.

The appearance of diarrhœa in phthisis is always to be regarded with suspicion, and, if it be proved to be due to one of the last two causes, is of most unfavourable import, and may be considered as foreshadowing a fatal issue ; for, apart from its obstinate character, it wastes the patient more than any other symptom of phthisis, and clearly shows the vast hold the disease has gained on the constitution.

CHAPTER XIV.

PNEUMOTHORAX AND PYO-PNEUMOTHORAX.

Pneumothorax and pyo-pneumothorax—Five forms of pneumothorax: 1.
From emphysema; 2. From abscess or gangrene of lung; 3. From wounds;
4. From empyema bursting into the lung; 5. Tubercular pneumothorax—
Pneumothorax and acute phthisis—Frequency of occurrence—Relative
liability of the two lungs and of different lobes—Number and form of openings—Position of perforation—Symptoms—Perforation accompanied by
great fall of temperature—Differences between affected and unaffected sides
of chest—Physical signs—Displacement of the Heart—Amphoric sound—
Metallic tinkle—*Bruit d'airain*—Succussion—Pyo-pneumothorax—Four examples of recovery from pneumothorax—Localised pneumothorax—Paracentesis in pneumothorax—Double pneumothorax—Diagnosis of pneumothorax—Prognosis.

PNEUMOTHORAX, or the presence of air in the pleural cavity,
is by no means a rare complication of consumption, and may
be due to several causes besides phthisis. We may dismiss
Laennec's simple non-perforative pneumothorax assigned by
him to the exhalation of gas by the pleura, as no proof has
ever been furnished of its existence, and proceed to consider
the other forms, viz.:—

(1) Pneumothorax in connection with emphysema; (2)
that arising from abscess or gangrene of lung; (3) or from
wounds to the thorax; (4) or from an empyema bursting
through the lung into a bronchus; and (5) and lastly, tubercular pneumothorax, with which we are chiefly concerned.

1. Pneumothorax caused by the bursting of an emphysematous vesicle through the pulmonary pleura is rare, but,
lately several well-authenticated cases have been published
including two by Dr. Thorburn [1] and Dr. Austin Flint,[2] both
followed by complete recovery, and Dr. Flint's was the more
remarkable, because the patient, an emphysematous pedler,
twice had signs of pneumothorax after great exertion, which

[1] *British Medical Journal*, 1860.   [2] *Pneumothorax*, 1875.

on each occasion disappeared after prolonged rest. Dr. Whipham's case [1] appears to belong to this class, the patient's signs of pneumothorax disappearing in fourteen days after admission into hospital.

The second and third modes of causation need but little comment, but of the fourth I have seen several examples, and in one remarkable case I was able to verify, after death, the channel by which the empyema had worked its way to the bronchi. [2] In this case the pneumothorax was not detected till the fluid had been removed by tapping, when the presence of air hissing in and out of a perforation into the lung was at once recognised, and after death, near the actual perforation, two well-marked depressions of commencing ulceration in the pulmonary pleura demonstrated the mode in which such perforations form. The *post-mortem* records of the Brompton Hospital contain similar examples.

Pneumothorax may be called an accident of phthisis, and its occurrence depends partly on the pleura not being adherent, and partly on the cavity being very superficial. It has nothing to do with the size of a cavity, as often a very small one, immediately underlying the pleura, if this be non-adherent, is as likely to cause pneumothorax as a larger one situated deeper in the lung.

Pneumothorax more commonly follows in the train of acute phthisis (scrofulous pneumonia) than in that of any other form, principally because in this variety excavation proceeds rapidly, and the pleura is rarely adherent ; and it is unknown in fibroid phthisis, in which adhesion of the pleura and lung contraction is the rule. It is not found in acute tuberculosis, because this form is not marked by the presence of cavities, but it is not infrequent in tuberculo-pneumonic phthisis and in all cases where excavation is rapidly proceeding, the explanation being that in acute excavation the pleura is not sufficiently irritated to form strong adhesions.

[1] *Medical Society's Proceedings*, vol. ix. p. 237. *See* also Dr. S. West's and De Havilland Hall's collections of cases in *Clin. Trans.*, vols. xvii. and xx., and *Brit. Med. Journal*, March 19, 1887.
[2] *Clinical Transactions*, vol. xii. p. 137.

To any one carefully examining the lungs of patients dying of acute phthisis, the marvel is, not that pneumothorax should occur, but that it should not do so more frequently, for it is not uncommon to see what appear to be several abscesses immediately underlying the visceral pleura, and apparently ready to burst. These are caseous masses formed during the rapid disintegration of tubercle, and when perforation takes place, pneumothorax immediately follows. From this we can see that its occurrence does not prove that a large amount of lung has undergone excavation, or even that a large cavity is superficial ; it only means that a caseous mass has rapidly extended, and, owing to absence of adhesions, has burst into the pleura, or that a cavity has extended and opened into the same.

It is not rare to find several caseous masses underlying and ready to penetrate the pleura, and in some instances perforation in more than one spot does take place. In some, two openings have been noted, in others, as many as four, and in one, even six.

I have extracted 51 cases of pneumothorax from the Brompton Hospital *post-mortem* records during the period of eight years (from 1878 to 1886), subsequent to those which Dr. Powell has already published,[1] making, with his 39, 90 in all.

Here are the results :—

With regard to the frequency of death from pneumothorax in phthisis, I find that in 412 autopsies on consumptives pneumothorax was present in 41, or in 10 per cent. This only refers to the patients dying in the hospital on whom an autopsy was performed, and must not be considered to relate to the total number of phthisical patients admitted into the hospital. A number of these die and no autopsy is permitted, and, on the other hand, in a certain number of deaths pneumothorax is not diagnosed during life, but only recognised after death.

This percentage is higher than any other, except Weil's,[2] which was 13 ; the ordinary one being 5 per cent. The per-

[1] *On Consumption, &c.*, p. 142.
[2] *Deutsches Archiv für Klinische Medicin*, 1881.

foration was on the right side in 18, on the left in 32, and on both sides (double pneumothorax) in 1 case.

Dr. Powell's 39 cases give 16 right-sided pneumothorax against 23 left-sided, and adding these to the above, as both are extracted from the Brompton records, we arrive at a total of 34 right-sided against 55 left-sided, and one double, in a series of 90.

This gives a percentage of 38 for the right and 61 for the left, or a greater liability for the left of 3 to 2.

This agrees with the experience of Louis, Walshe, Wintrich, Lebert and Weil, but differs from those of Laennec and Dr. S. West.

In 18 the pneumothorax was limited to a more or less extent by adhesions; in 33 it was unlimited, and involved the whole pleural cavity. The perforation was detected in 48 out of the 50 cases, which witnesses to the care with which the search was made. In Dr. S. West's [1] statistics it was found in 25 out of 43; in Weil's, in 26 out of 33 cases.

The number of openings varied considerably; in 8 there were 2 openings, in one 3, in one 4, and in another 5; in 40 the openings were single. The size and shape also differed greatly. Sometimes it was no larger than a pinhole, and could hardly be traced, and sometimes larger than a crown piece; in some instances it was slit-like and valvular; in others, circular or ovoid, with smooth bevelled edges; in others again it consisted of a large rent with ragged, sloughy walls. Not uncommonly the edges of the perforation were made up of caseous matter evidently undergoing ulceration. The channel connecting the cavity and the pleura showed great variety. Sometimes, and this was the most frequent, it was a simple perforation of the thin wall of the cavity, due to the breaking down of a caseous mass, and it was in these cases that the multiple perforations were found. In two others it consisted of a sinus of some length, opening either between two lobes of the lung or at a point much below the level of the cavity. In two cases the opening was valvular. The position of the perforation was carefully noted in nearly all the cases.

[1] *Lancet*, vol. i. 1884, p. 791.

The upper lobe was the seat of perforation in 17 instances, the middle lobe in 4, and the lower in 16; the sulcus between the upper and middle lobes in 1. In the double pneumothorax case, which is given in full below, the upper lobe of one lung and the lower one of the other were perforated. In the 11 cases of the above series, where the ulcerations were multiple, they opened from the upper lobe in 4, from the middle in 3, and from the lower lobe in 4.

The usual regions for perforation were the posterior and lateral, but it was found to occur occasionally in the anterior regions. These results give a nearly equal liability of the upper and lower lobes to perforation, differing from Dr. S. West's, which are based on a smaller number and give a preponderance of nearly two to one for the upper lobe.

Walshe's[1] statement, that the pleura gives way posterolaterally between the third and sixth rib, is probably not very far from the truth, and includes most of the cases.

The effect of the entry of air into the pleural cavity is generally to cause pleurisy and effusion, first of serum, and later on of pus. The air effused contains, according to Davy, a large proportion of carbonic acid (8 per cent.) and a very small one of oxygen, in fact resembles expired air in composition; but where the expectoration has been fœtid, sulphuretted hydrogen is found (Duncan) in addition to carbonic acid. According to Dr. Powell,[2] where the opening is free and direct, the air pressure within the pleura is nil, but if the opening be oblique or valvular, the air cannot escape and thus accumulates, causing an intra-pleural pressure, varying, in Dr. Powell's twelve cases, in degree from $1\frac{3}{4}$ inch to 7 inches of water. Pleuritic effusion generally follows in a few days if the patient has not already sunk from the shock of perforation; and, curiously, the effusion is often serous, and remains so for some days, though in protracted cases it invariably becomes purulent. It sometimes accumulates in sufficient quantity to close the perforation and to necessitate tapping. The effect of the introduction of air into the pleura, on the lung, depends partly on the presence or absence of adhe-

[1] *Diseases of the Lungs* (4th edition).   [2] *Op. cit.* (3rd edition), p. 133.

sions; and if they exist, on their extent and also on whether
the lung is, or is not, consolidated. If there be no adhesions,
and the granulation is limited to the region of the perforation,
which is of fair size, complete collapse takes place, and the
lung is found, after death, compressed against the spine;
often there is collapse of one portion and the rest is adherent
or consolidated, or both; in this last case the pneumothorax is limited, and consequently its occurrence does not
give rise to such severe symptoms, or grave danger, as a
general pneumothorax.

Localised pneumothorax may occur without the patient's
knowledge, and may continue for years, apparently not preventing its possessor from taking abundant exercise and following active pursuits, as is well seen in Case 43.

The symptoms of pneumothorax are often characteristic:
the patient suddenly complains of sharp pain in the region
of the perforation, of an agonising nature, followed by cold
perspiration and sometimes a sensation of something cold
trickling down the affected side, and by great dyspnœa. The
symptoms of collapse quickly follow, the countenance assumes
an anxious look, the complexion a livid tint, especially of
the lips and cheeks, the extremities become cold, and a clammy
perspiration appears on the forehead. Cough and expectoration nearly cease, the pulse is rapid and feeble, but the marked
feature is the respiration, which increases sometimes to fifty
or sixty a minute, and is the principal index of the sudden
diminution, to half, of the respiratory area. The patient
generally sits up, inclining to the sound side, often holding
the affected one, or he inclines forward with his elbows on
his knees. The voice is feeble and often diminished to a
whisper, but generally the patient is too much occupied with
respiratory efforts to speak at all. The expectoration, when
there is any, contains, as in acute phthisis, abundant tubercle
bacilli. The changes in the temperature, as noted by M. Peter
and myself, are very striking. The first effect of the perforation is to cause a rapid fall of 4° or 5°, as will be seen in
Case 40. This is followed by a rebound to nearly the height
of the previous readings, and later on, by irregular but

pyrexial temperatures. If effusion takes place there is generally a slight rise, but if acute pneumothorax be converted into chronic pyo-pneumothorax, the temperature falls to normal, and often to sub-normal, records. If, however, the issue is fatal, the pyrexial symptoms may continue till death.

The subjoined case illustrates well the fall after perforation :—

CASE 40.—Eliza B., admitted into the Brompton Hospital, May 5, 1874, age 34, married. Lost a sister from consumption at the age of 28. Cough and expectoration came on during pregnancy eighteen months ago, and have persisted ever since, accompanied by emaciation. She had haemoptysis, 3j., eighteen months ago, and to a less extent since.

Night-sweats have existed for two months, and are now very profuse. She has suffered with palpitations and shortness of breath. Catamenia have been

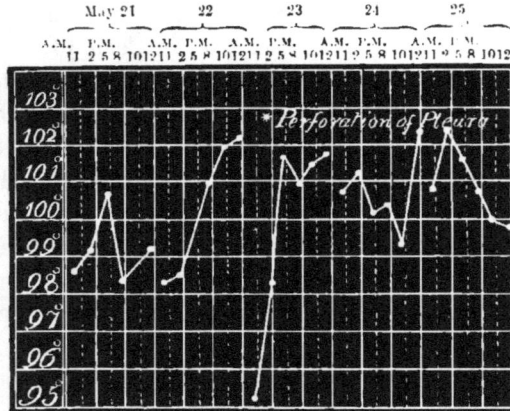

absent since the last confinement. Pulse 130, weak ; respirations 36, temperature 100·6° F.

Crepitation audible over the whole right lung, tinkling cavernous sounds to the third rib on the left side, and crepitation heard over the rest of the lung. The temperature was taken six times a day, and varied from 98·2° F. to 103° F. Night-sweats were profuse and the bowels relaxed. On May 23 she was seized with sharp pains in the left chest, and signs of pneumothorax were detected over that region. The respirations became quickened, rising to 52 per minute, the pulse very feeble, and in one week from that date she expired. The subjoined tracing gives the temperature chart during the periods preceding and after the perforation. It will be observed that a fall of no less than six degrees accompanied the perforation of the pleura, and that afterwards recovery took place, but was followed by irregular temperatures.

Autopsy, thirty-two hours after death. On opening the left pleura air escaped. The anterior and lateral surfaces of the left lung were covered with fresh lymph. About three inches from the apex an opening the size of a fourpenny piece was found on the anterior surface of the lung communicating with a very large cavity. The whole upper lobe was entirely excavated, the size of the cavity being so large, and the walls so thin, that adhesions in the pleura must alone have prevented the occurrence of other perforations. The lower lobe was consolidated, containing several aggregations of miliary tubercle, caseating in some parts and in process of excavation in others; well marked adhesions existed at base and apex. Right lung: Upper two-thirds consolidated, containing a large amount of miliary tubercle and a ragged walled cavity at apex. Lower lobe emphysematous, with scattered miliary tubercle of recent date.

Intestines showed three or four ulcers in the cæcum. Liver friable. Other organs healthy.

A comparison of the temperature of the two sides of the chest—the affected and the unaffected—during the period of perforation, and that succeeding it, is full of interest, and M. Peter [1] has made some observations in the fifth intercostal space on both sides of the chest in a case of left pneumothorax with the subjoined result:—

	5th intercostal space. Left. (affected.)	Right.
At moment of perforation	96·8° F.	97·5° F.
16 hours after ,,	99·1	98·2
3rd day ,, ,,	98·7	99·3
4th day ,, ,,	96·4	98·4
5th day ,, ,,	96·4	98·9

According to these observations, there was a reduction on the left side at the time of perforation, a slight rise after, and during the third, fourth, and fifth days the difference between the two sides amounted to from ·6° to 2·5° F. This is to be explained by the change in the contents of the left pleura, and by the substitution of air for vascular lung, the latter being compressed against the spine. It is possible, too, that an engorgement of the right lung, from the increased afflux of blood, deepens the contrast of the two sides.

The *physical signs* of pneumothorax are as remarkable as the symptoms.

The affected side of the chest is fixed and motionless, the shoulder is raised; there is often bulging of the intercostal spaces, and measurement shows an increase in the circumference. There is absence of vocal fremitus. Percussion

[1] *Clinique Médicale.*

gives hyper-resonance over the whole side, and the hyper-resonant note often extends across the median line into the opposite side of the chest. The heart is often displaced, and frequently extensively so; in the case of a right pneumothorax, towards the left axilla, in the case of a left, to the right of the median line ; this change of the heart's position being noted within a few minutes of the perforation, and therefore having no reference to any subsequent liquid effusion. The abdominal organs are often displaced downwards.

Auscultation often reveals a complete absence of the ordinary respiratory murmur and voice sound, but the presence of a *metallic tinkling* sound on speaking or coughing, which closely resembles that heard over a stomach containing air and fluid. This is held to be due to an echo from the walls of the pleural cavity containing atmospheric air, and is present in cases of empyema where the air is admitted after puncture. The auscultation sounds depend principally on the size of the perforation ; for, if it be large, we hear *amphoric*, or blowing sound, most marked in the region of the opening; if small or valvular, we hear no sound at all, the phenomena of amphoric and tinkling sounds being only present over portions of the chest containing air, not fluid.

Another sign is the coin sound, or *bruit d'airain*, which may be produced when the stethoscope is placed over the affected area, and one silver coin struck against another on the thoracic wall, the resulting loud metallic sound heard through the stethoscope being very characteristic. A sign, dating from the time of Hippocrates, heard when a certain amount of fluid, as well as air, is present, is *succussion*—the splash sound—made by shaking the patient from side to side, and thus bringing the fluid into sharp contact with the thoracic wall, the sound resembling that heard in a half-filled jug under similar circumstances. This sign is only present when the pneumothorax is general, not limited, and it often remains long after the perforation has become closed. Amphoric breathing, tinkling sound, and the *bruit d'airain* may be heard in localised pneumothorax, and their intensity depends on the amount of air present in each case.

The after progress of pneumothorax varies greatly. In some few instances the perforation becomes closed, the air and fluid are gradually absorbed by the pleura, and a cure is thus effected; but this only occurs where the pulmonary lesion is limited and the opening very small; in other cases the fluid effusion increases at the expense of the air, and pyo-pneumothorax takes the place of pneumothorax. Some of these cases cannot be distinguished from those of localised empyema, owing to the fistulæ being closed, and in the majority of cases the actual occurrence of pneumothorax is speedily followed by death, due partly to the shock and partly to dyspnœa from the sudden reduction of the respiratory area.

We will now take some clinical cases to exemplify our descriptions. The first four are fair examples of recovery from pneumothorax in which life was prolonged for many years; all four were accompanied by little or no effusion, and no operative interference to reduce the intra-pleural pressure was attempted.[1]

CASE 41.—*Pneumothorax. Gradual Recovery. Lived 21 years.*—Mr. D. A., æt. 26, surgeon, March 17, 1846; seen by Dr. C. J. B. Williams with the late Dr. John Taylor. Sister died of phthisis. Declares that he was quite well till seventeen days ago, when, riding a very restive horse, he was suddenly seized with severe pains in the whole left front of the chest, catching the breath. This has continued more or less ever since, with short dry cough and quickened breath. Pulse 120, weak. No heat of skin, and other functions natural. *Left chest tympanic and tender on percussion. Loud amphoric breathing, metallic tinkling with voice, and sometimes with breath and heart-beat, which is a little to right of its proper place. Intercostal spaces not depressed in inspiration. Dulness for about three inches in the lower part of left back. Breath-sound puerile in right lung.*

This patient remained in a weak state for twelve months, suffering still from short breath and pain in the left chest with very little cough. He visited Dr. Williams in the spring of 1848, when he had improved in flesh after taking cod-oil, but was still weak and breathless, and with sore feeling in the left chest. Now, however, *the stroke was generally dull; breath-sound very*

---

[1] These cases were published in the first edition of this book in 1871, and therefore preceded both Dr. George Johnson's interesting case of sudden perforative pneumothorax in vol. xv. of the Clinical Transactions (February 1882), and Dr. Samuel West's paper in vol. xvi. of the same Transactions (December 1883), to which the reader is referred for 24 cases in which recovery took place. These latter included five cases of pneumothorax from emphysema and several others produced by over-exertion or strain, but curiously the above four cases are entirely omitted.

obscure, with subcrepitus.   No metallic or amphoric sounds. –To continue oil
with tonic and to paint the side with iodine.

His health improved afterwards sufficiently to enable him to go as surgeon
to an East India merchantman, when further amendment took place, and he
was heard of some years after, conducting a small practice in Kent. His death
was announced in the 'Times,' 1867.

CASE 42.—*Pneumothorax.   Recovery.   Living 10 years after.*—A gentleman, aged 48, first consulted Dr. C. J. B. Williams, December 11, 1861. He
had lost his mother and a sister from consumption, but, with the exception of
a sore-throat four years ago, he had no ailment till last May, when he was
attacked with pain and throbbing in the left side, and with cough. In July
the breath became very short; in September the expectoration was sometimes
tinged with blood, and Dr. (afterwards Sir James) Simpson, on examining his
chest, found *the left side fixed, but clear on percussion, and the heart beating
at the sternum.* Dr. Williams, in addition to these signs, detected *over the
lower half of the left side, back and front, large amphoric blowing sounds
and metallic tinkling with voice and cough, and with many occasional clicks,
also tubular sounds in the upper portion of the lung.   Breathing puerile in the
right lung.* Pulse 100. Skin cool. Oil was ordered, with quinine, phosphoric
acid and tincture of orange, and counter-irritation with cantharides liniment.

April 19, 1862.—After seeing Dr. W. the cough increased, but after being
some time at St. Leonards, taking quinine and oil, he improved wonderfully;
gained 18 lb. in weight. Pulse normal and regular. *Dulness, deficient breath
in lower half of the left chest, but no trace of metallic tinkling or undue resonance.   Large tubular sounds at and above the scapula, and some friction-
sounds in front and at the side, where there has been some pain.   Left chest
contracted, measuring three-quarters of an inch less in circumference than
right.*

July 2.—Is in town frequently, attending at the House of Commons, and
remains well. The pain and friction-sounds have ceased on the left side.

December 20.—Was well till September, when had a severe attack of
measles, accompanied by cough; but was convalescent in a fortnight, and since
then has been free from cough, and with the exception of his breath being short,
is quite well, and able to walk and hunt. *Some breath-sound now heard in left
front, as low as fourth rib, except near sternum, where the heart-dulness and
pulsation remain : dulness and little breath below; bronchophony above both
scapulæ.*—To continue the oil in phosphoric acid tonic.

November 18, 1863.—Quite well, and free from cough and expectoration,
though often exposed to wet and night-air in yacht. Is able to shout as loud
as ever. *Deficient motion and stroke-sound over whole left side, which is still
contracted, but loud and vesicular breathing in front and laterally.   Loud
tubular sounds from the middle of scapula upwards.   Tubular sounds above
right scapula.*

October 1869.—Generally well, and has become stout, but has more or less
cough. Wintered at Caithness. *Still dulness and large tubular sounds audible
at left scapula and above.*

The signs of pneumothorax were unequivocal in this case,
and must have arisen from perforation of the pleura, by
partial disease of the lung in the summer of 1861. Happily

this disease was arrested, the air effused was absorbed, and the lung was gradually re-expanded.

CASE 43.—*Pneumothorax. Complete recovery.* Mr. F., æt. 24, October 2, 1867.—Several half-brothers and sisters have died of consumption. Except a pain in the chest last November, from which he was well in a fortnight, has had no illness till the second week in August, when he again had pain in the front of the chest with a cough, which entirely ceased in two weeks ; and in September was well enough to make a tour in Switzerland, and ascended the Eggischorn with no other inconvenience than short breath and occasional pain in the left side. No cough. In the last three days he has noticed a splashing noise in his chest, but says he feels well. Seems nervous and anxious, and with quickened breathing. Pulse 90 ; temperature normal. *Left chest distended, tympanitic on percussion, and with no breath-sound. Heart pulsation seen and felt to right of sternum, and in epigastrium. Tubular breath and voice at left scapula ; and below this point speaking or laughing is accompanied by a tinkling echo, which can be produced also by succussion. Dulness in lower third of left back.*—To take oil in tonic mixture. Left side to be painted with tincture of iodine.

October 14.—Much the same. Complains only of short breath, and of the splashing noise on every quick turn of his body. Cough and expectoration slight, induced by change of posture. *Signs the same.*

October 22.—Continues pretty well, but breath short, and occasional pain in left chest, which is tender, especially in lower part. Hears noise less. *Left side smaller, with less tympanitic distension. Tinkling and splashing sound higher in axilla and back. More breath-sound above left scapula.*

November 18.—Weaker, and feverish at times, with pain in side. Sometimes faint on exertion, and breath short. *Dulness and absence of breath in lower half of left chest. Stroke clearer, with some breath-sounds above. Bronchophony above right scapula. Heart now in place, or a little higher and more to the left than natural. No tinkling, splashing, or other sign of cavity, nor flattening of walls.*

February 10, 1868.—Has continued steadily taking oil and tonic, and has much improved till the last week, when there has been much more pain in side, with feeling of weakness. No cough, but breath is shorter again ; appetite bad. *Left chest more natural in shape, but lower third contracted ; still dull, while in middle third is less dulness, with obscure breath. Tubular sounds at and above scapulæ. Friction with deep breath in middle front (where has been pain). Heart to left of sternum, with its apex beating above fifth rib.* Nitro-muriatic acid mixture and belladonna plaister to left chest.

June 9.—Quite well, except pain catching breath on exertion felt since leaving off plaisters. Strength good, and up to usual weight, 10 stone. *Same signs, except slight crepitus below left axilla.*—To continue nitro-muriatic mixture, and to apply opium plaisters to left side.

June 8, 1869.—Well all winter, without pain or cough, but breath oppressed by cold. Lately has been a walking excursion, compassing several miles daily, and uphill with a knapsack on back. Weight 10 st. 8 lb. *Lower half of left chest rather duller, and moves a little less than right, but breath-sound heard everywhere except at base, where is crepitus on deep breath.*

September 3, 1869.—Heard that he had been walking 30 miles a day, shooting, rowing, &c., without inconvenience.

CASE 44.—*Pneumothorax. Rapid and complete recovery.*—Mr. M., æt. 48, seen by Dr. C. J. B. Williams with Dr. Stutter, of Sydenham, May 16, 1868.— Lost a brother from phthisis. Has been quite well and gaining flesh lately. Has been in the habit of going quickly up and down a long flight of stairs twenty times a day, till five days ago, and lately had found the exertion cause pain in the right side, and a feeling of oppression. On that day he consulted Dr. Stutter, who ordered a mustard poultice to the side. This relieved the pain, but the breath remained short, and then Dr. Stutter found signs of pneumothorax on the right side. Now complains of nothing but shortness of breath, and a feeling of fulness in the right side. Pulse quiet, urine scanty. *Lower half of right chest tympanitic on percussion down to lower margins of ribs, below which liver dulness reaches down four or five inches in abdomen. Breathing amphoric, with metallic tinkling on coughing in lower half of chest. Above, breath-sound obscure and stroke rather duller than on left side, particularly over the scapula.*—An effervescing saline was given for a few days, and tincture of iodine to be painted on the right chest. Afterwards oil to be taken.

September 1869. —Heard from Dr. Stutter that the signs of pneumothorax soon disappeared and that the patient quite recovered.

August 1871.—Mr. M—— continues well, and has become stout.

CASE 45.—*Localised pneumothorax forming in chronic phthisis.*—Mr. G. H., age 32, consulted Dr. Theodore Williams, September 13, 1883. He had had hæmoptysis twelve years previously, and more or less cough and expectoration ever since. Nevertheless during last winter he hunted five days a week and constantly got wet through.

About one year ago the cough and expectoration increased and the former became most racking, accompanied by severe night-sweats and rapid loss of flesh, amounting in all to 16 lb. Cough is very troublesome, temperature normal, voice slightly affected.

October 14.—He has had severe pain in the right chest, especially affecting the three lower intercostal spaces, and he has expectorated quantities of pus which are found to contain tubercle bacilli in abundance, but no lung-tissue. Cough has diminished, but the expectoration is still profuse. *Slight dulness over the whole right lung, becoming more marked in the scapular region. Hyper-resonance in the axilla and at the posterior base, where tinkling sounds and amphoric breathing are audible.*

August 1884.—Wintered at the Cape of Good Hope and trekked to Bloemfontein, spending much time in the open air, shooting and camping out; has much improved, having gained flesh, but still has cough. Signs of pneumothorax no longer detected, but much dulness exists still over the right lung.

October 30, 1885.—Spent last winter in Ireland, hunting five days a week and living somewhat freely. Has regained most of his weight, but still has cough and expectoration, which is occasionally difficult; *metallic tinkle and amphoric breath-sounds audible over the lower third of the left lung anteriorly and posteriorly.*

July 1, 1886.—Wonderfully well; still cough. *Pneumothorax signs scarcely perceptible.*

There is little doubt that in this case the patient had old tubercular disease, which being rendered active by his frequent exposure to weather, an extensive breaking down rapidly took place in the right lower lobe, and before adhesions could form perforation of the pleura had occurred. As there were doubtless extensive adhesions in other parts the pneumothorax remained localised, and, to judge by his subsequent course of life, did not exercise any marked effect on his general health.

CASE 46.—*Left pneumothorax. Great cardiac displacement, progressing disease of the right lung. Paracentesis. Relief. Death. Autopsy.* —John F., age 28, discharged from the army. Admitted into the Brompton Hospital under Dr. Theodore Williams, November 26, 1885. Has served in India and Africa, but enjoyed good health, with the exception of jaundice, until March 1885, when he caught cold, and this was followed by pain in the chest, severe cough, white expectoration, and occasional vomiting. He was treated for pleurisy, but his breath became short, and he has wasted and had night-sweats ever since. On admission his pulse was 90 and his temperature between 98° and 99° F. *On the left side there was hyper-resonance in front, with absence of vocal fremitus, marked fixity of the side, displacement of the heart, to the right of the sternum, and total absence of breath-sound, with metallic tinkle audible in the second interspace. Posteriorly: dulness over the lower two-thirds, tubular sounds in the interscapular region, hyper-resonance at the apex. On the right side crepitation and tubular sound over the upper third, front and back.*

January 1, 1886.—The patient has continued about the same. At first the temperature rose to 100° F. and reached 101° in the evenings for some weeks, but it is now considerably less, and comparative observations on the two axillæ show the pneumothorax side to be cooler than the other by ·6° F. The right being 99·4° and the left 98·8°. Pulse has risen to 140, and respirations to 40. Patient has vomited several times.

In addition to above physical signs, the *bruit d'airain* is well marked over the posterior regions of the left side, and the splash sound can be heard on shaking the patient. *Crepitation is audible to the fourth rib on the right side.*

February 1.—During the last three mornings has expectorated abundant purulent sputa, especially when lying on the right side. The heart is still felt beating in the fourth interspace on the right side and cardiac dulness extends nearly to nipple in the fourth right intercostal space, showing the organ to be largely displaced. The hyper-resonance on the left side has decreased, and the dulness considerably risen in height.

February 6.—Has continued to expectorate large quantities of un-aerated pus, which is often fœtid; on one occasion 28 oz. and on another 20 oz., with considerable relief.

March 3.—The large expectoration has continued from time to time, varying from 10 oz. to 26 oz.; the patient is much weaker, and after getting up has œdema of the legs and knees. *Cavernous sounds are audible above the scapula on the right side: on the left the physical signs vary with the accumulation of matter, when the dulness increases considerably.*

The urine has been repeatedly examined, but contains no albumen. During the last two months the afternoon temperature has varied from 100° to 103° F.

March 23.—The amount of expectoration has diminished during the last five days, varying from 1½ oz. to 14 oz., but there is evidence of accumulation of fluid in the left pleura. A small incision was made in the eighth interspace to the left of the axillary line, and an aspirator needle inserted and 84 oz. of purulent matter withdrawn, with much relief to the patient's breathing and cough.

A week later the heart had moved considerably towards its normal position, the impulse being felt under the sternum and no longer on the right side ; the left side was hyper-resonant and contained little or no fluid.

After this the fluid re-accumulated, causing displacement of the heart and great dyspnœa. He was again aspirated on April 17, and 72 oz. of greenish offensive pus withdrawn ; this gave some relief, but he became gradually weaker, the temperature fell considerably, the pulse became feeble, his countenance cyanosed, and he passed into a semi-unconscious state, occasionally wandering in his ideas, and died on April 22 from collapse. At the time of his death the heart was in the normal position and the signs of pneumothorax were present.

Autopsy, April 23, 1886.—The right pleura was generally adherent. The left lung quite retracted towards the vertebral border, and the pleural cavity contained a considerable amount of yellow pus. The parietal pleura was thickened and both it and the visceral were rough with lymph. On the upper and lateral aspect of the latter there was a small smooth-edged opening into the lung. The left lung was shrunk into the vertebral groove, the upper part was fibrotic and contained several irregular cavities, one of which communicated with the pleural cavities through the above-mentioned opening. The lower lobe was condensed, and contained a caseous nodule. In the right lung the upper lobe was consolidated with diffused caseous pneumonia, and contained a cavity the size of a bantam's egg at the extreme apex, and a second one in the supra-scapular region. The middle lobe showed caseous masses, and the lower a few fibro-tubercular nodules and groups. Mediastinal and subtracheal glands were enlarged and caseous. Heart enlarged and dilated.

The points of interest in this case were :—

1st. The cardiac displacement, which was more extensive than is generally found in pneumothorax, and is to be accounted for at first by the large accumulation of air, and later on by the purulent effusion, so that towards the close of life the case was more one of empyema than of pneumothorax.

2nd. The activity of the tubercular disease at the right apex, which is not common when there is a plentiful purulent discharge, as occurred in this case, from the opposite side. Considering how completely collapsed the left lung was, it is remarkable that the pus from the pleura was able to find its way into one of the bronchi, and it is probable that the process of erosion during the latter part of the patient's history originated, as much from the pleural cavity, as from the pulmonary one.

3rd. The long-continued and abundant discharge of matter without lardaceous disease being set up.

4th. The lowering of the temperature on the pneumothorax side, thus entirely confirming M. Peter's observations.

CASE 47.—Maud D., age 31, married, was admitted into the Brompton Hospital, under Dr. Theodore Williams, November 24, 1884. She had lost two brothers from consumption, but had enjoyed good health till five months previously, when cough came on with wasting, followed three months later by loss of voice, night-sweats, shortness of breath, which have continued up to the present time. She had hæmoptysis ℥ss. two months ago and ℥ij. a month later. On admission, temperature and pulse were raised.

On the right side: dulness, with some flattening to third rib; crepitation audible to fourth rib and loud tubular sounds over the first and second interspaces; dulness and crepitation over the upper half posteriorly; on the left side crepitation above the scapula and in the interscapular region.

She had a shivering fit five days later, and the temperature rose to 104° F. and the pulse to 102. The cough and expectoration increased, the latter containing abundant tubercle bacilli. The temperature continued high, until the beginning of January, when it fell below 100° F. and remained so, with occasional high records, till the time of death. Excavation proceeded in the right lung and simultaneously in the left ; and on March 12 pneumothorax was detected over the whole left side, the amphoric and tinkling sounds being well marked. On the right side the cavernous sounds were very extensive.

On the 19th her breathing, already rapid, became more oppressed, the pulse was 140, feeble ; she complained of some pain over the right chest, and died apparently of asphyxia in the evening.

Autopsy, March 19, 1885.—On removing the sternum the right lung was seen to be collapsed, but held up in the middle by an adhesion band passing from the junction of the middle and lower lobe to the angle of the fifth rib ; there were a few soft adhesions at the posterior aspect of the apex. The left lung was also collapsed, but firmly adherent at the posterior aspect and border. The pleura was much thickened and contained Oj. of pus. The left lung showed a large perforation in the upper lobe, at the level of the third rib, 2½ inches from the sternal edge. The opening had partly ragged and partly smooth edges of caseous matter, and was twice the size of a crown piece; it led into a large trabeculated cavity with soft caseating walls. The upper lobe, except the sternal portion, which was collapsed, was riddled with cavities. The lower lobe was collapsed, but contained at the upper part some caseating masses. The right lung showed in the upper lobe a large trabeculated cavity with soft and thin walls, consisting in parts only of the pleura, with which cavity it communicated by three perforations at the lower margin of the lobe on its posterior aspect. The external portion of the lower border, laterally, contained traces of spongy tissue, but was in the main consolidated, of a greyish colour, and invaded by softening caseous masses. The middle and lower lobes were partially collapsed, and studded with tubercular groups of various sizes, but the lung-tissue was not, as a rule, infiltrated. The mediastinal glands were pigmented, and in one or two places calcareous. Liver fatty. The intestines showed numerous scrofulous ulcers scattered throughout the colon and one or two small ones in the ileum. The small intestine was slightly lardaceous; mesenteric gland caseous.

This was an instance of double pneumothorax, the fatal issue being due to the occurrence of the right one.

Double pneumothorax is almost unknown,[1] though many cases nearly arrive at it, from the number of the excavations and their comparatively superficial character in both lungs. It is curious to note that the right lung, which was the first attacked, and most advanced in disease, was the last to present perforation of the pleura. In the left lung the disorganising process commenced later, but proceeded more rapidly, and the pleura was eroded earlier, principally from the amount of caseation which took place; the opening in this lung being large and ragged, but single, whereas in the right the cavity communicated by no less than three openings with the pleura.

The exact date of the occurrence of the first pneumothorax is doubtful, and probably preceded the second by a longer period than the physical signs indicated. The second pneumothorax was evidently the immediate cause of death. The case was one of acute phthisis (scrofulous pneumonia), and during residence in the hospital the whole phenomena were developed. Undoubtedly the right pneumothorax only formed within a few hours of death, its advent being marked by the increase of dyspnœa.

The *diagnosis* of pneumothorax is not generally difficult, the characteristic symptoms and the physical signs separating it from most pulmonary affections; but to distinguish between it and a large thin-walled tubercular cavity is often very difficult, as the latter gives rise to hyper-resonance, amphoric breathing, and tinkling sounds similar to pneumothorax. In most tinkling cavity cases, the *position* of the signs at the upper part of the lung and the presence of breath-sounds at the base suffice to separate them from localised pneumo-

---

[1] Laennec gives two cases of it—one under M. Récamier's care, and another under his own.—*Diseases of Chest*, p. 536.

thorax; but where the cavity is so extensive as to convert the lung into a mere sac of the pulmonary pleura, with the bronchi opening into it, diagnosis becomes almost impossible. I remember in such a case all the physical signs of pneumo-thorax were present, including succussion, except the coin sound, which was entirely absent; but in others even this means of differentiation was wanting.

The *prognosis* of pneumothorax may be concluded, from the diversity of the cases in this chapter, to vary greatly. Most of the cases admitted into the Brompton Hospital with pneumothorax are advanced tubercular cases and terminate speedily. In Dr. Powell's thirty-nine instances from the Brompton records the average duration was twenty-seven days, and the longest one year. In Weil's forty-five fatal cases death occurred in nearly a quarter, in the first week, in nearly half, within the first month, and in the rest, life was prolonged for months and indeed for years. One patient lived two years and another two years and three quarters. Traube gives an instance of a woman who had pneumothorax for at least seven years and recovered. Our own cases just cited show that among the upper classes, at any rate, life may be prolonged for many years, even in tubercular pneumothorax, and that in some cases complete recovery may take place. Prognosis may be said to depend on the extent of the pneumothorax and the amount of disease present in the other lung. Pneumothorax arising from other causes has a most hopeful prognosis.

CHAPTER XV.

ALBUMINURIA OF CONSUMPTION.

Its insidious onset—Relation to excavation—To diarrhœa—Lowering effect on temperature, even of acute tuberculisation—Influence on duration of consumption—Pathology of kidney—Lardaceous Granular—Large white kidneys—Bamberger's, Southey's and Brompton statistics—Condition of Malpighian vessels—Urine analyses—Reduction of urea—Cause of lowering of temperature—Hammond's experiment—Bernard and Barreswil's—Stolnikoff's experiment—Uræmia—Prognosis.

THE supervention of albuminuria in phthisis is far from rare; and, until recently, no attempt has been made to explain its pathology. It has always been regarded, and justly, as a very serious complication, which, if it do not immediately terminate life, places the patient in the category of the condemned; the patient lingers on, but some of the most promising cases of arrested phthisis have succumbed to this lesion. It is difficult to predict beforehand the kind of case likely to be associated with albuminuria, though our histories show that there has generally been some exhausting discharge, as diarrhœa or profuse expectoration. Nevertheless, it also arises in patients in whom neither of these symptoms exists; and some observers have connected it especially with fibrosis of the lungs. The approach of albuminuria is so gradual that it is frequently overlooked; and often the first indication which the patient has of the mischief is œdema over the ankles or the shins. The urine is then tested, and for the first time found to contain albumen. Sometimes, however, the preliminary symptoms of renal mischief are present, such as nausea and pain in the region of the kidney. Careful palpation will discover tenderness in the loin, as was noted in some of the patients whose cases are here given; but in most cases the symptoms are so obscure as to be overlooked.

When albumen has appeared in the urine, it is rare for it wholly to disappear; and though it may escape a superficial examination, it will generally be found, by carefully testing the urina sanguinis —*i.e.* the urine passed on rising in the morning and also that passed late at night. In two private cases, where very careful nursing was carried out, I remember the albumen disappeared from the urine for a period of some months, but returned later on, when both patients soon sank. I have known it disappear for even a longer period, but this is quite the exception.

Another feature is, that it rarely occurs except where cavities have formed; in all our sixteen cases there were distinct signs of excavation in one or both lungs. It by no means followed that the discharge from the cavity was large, provided such discharge existed, and that the cavity was one of long standing.

The 2 following cases are given in full to show the influence of albuminuria on the temperature of phthisis, while 16 are tabulated (*see* Table I.). The patients were twelve males and four females, of ages varying from 16 to 46, the average being for males 27, for the females 37, and, with one exception, they were all hospital cases. Hereditary predisposition was present in 3, syphilis in 3, diarrhœa in 11, and dropsy, either in the form of anasarca or of ascites, or both, in 14. The first 2 cases are especially interesting, from the albuminuria having supervened while the patients were under observation in the hospital. The temperature was always taken in the mouth for a period of five minutes.

CASE 48.—Edward G., aged 18, single, a worker in stained-glass windows, was admitted into the Brompton Hospital, December 31, 1878. He had had measles and purpura some years ago, but of late enjoyed fair health till nine months previously, when cough came on, with night-sweats, followed by diarrhœa, the latter lasting two weeks. Cough and night-sweats had continued up to admission, and in the last three months he had lost much flesh. He had never had any swelling of the legs. On admission, his aspect was pale, his cough hard and troublesome, expectoration greenish and copious. tongue red; he had slight diarrhœa; purpuric spots were visible on the thighs; the urine was acid, of specific gravity 1020, and contained no albumen or sugar.

Some dulness and coarse crepitation from the clavicle to the fourth rib, and above the scapula on the right side; loud cavernous sounds, audible over the upper third of the left front and back. Tenderness on pressure in the left iliac fossa. During the three weeks following admission the diarrhœa continued, being only partially controlled by medicine, the number of stools varying from three to four. He also had epistaxis and some discharge from the ear. From January 31 to February 13 a careful record of the temperature was taken five times a day, and was kept, with the number of stools, to ascertain the natural course of the pyrexia. Line A in the subjoined diagram gives the curve of the averages of these fourteen days. The diarrhœa gradually subsided, reaching an average of one stool a day on February 12; and while the 8 A.M. record of temperature varied from 97·80° F. to 99° F., the 11 A.M., 2 P.M., and 5 P.M. records ranged from 100° to 103°, on only three occasions falling below 100°. The 8 P.M. notes show the time of daily subsidence.

February 13. He has improved; the cough was easy; the bowels were open once a day; there was no tenderness of the abdomen. As the fourteen

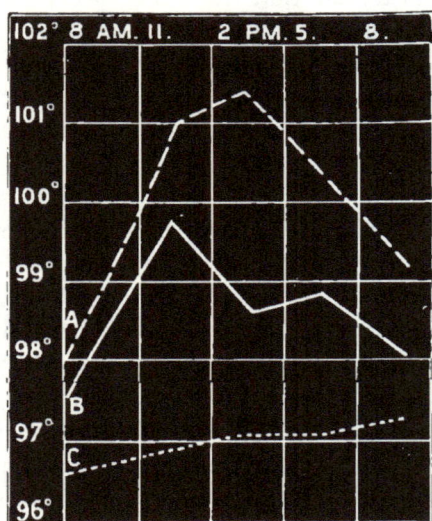

days' temperature observations had clearly shown the pyrexial characters of the chart the record was discontinued.

March 6. —Yesterday, some œdema of the lower extremities was noted, which to-day diminished. The patient complained of nausea and of pain in the region of the left kidney, and some tenderness. The abdomen was somewhat distended. The urine was acid, of specific gravity 1023; it contained albumen (about one-third). Observations in temperature were resumed five times a day on March 7, and were continued up to the patient's death.

He became more and more dropsical. Ascites was detected on March 19, and two days later there was general anasarca. On March 20 fine crepitation was audible over the bases of both lungs, in addition to cavernous gurgle at both apices. The patient died April 3.

The time during which the temperature was taken after the appearance of the albuminuria up to the death was twenty-seven days; and for the purpose of comparison with the pre-albuminuric period it has been divided into two periods of fourteen (n) and thirteen (c) days each, and the subjoined are the means of the three periods:—

—	8 A.M.	11 A.M.	2 P.M.	5 P.M.	8 P.M.	Max.	Min.
From January 31 to February 13, fourteen days (pre-albuminuric)	98·1°	101·1°	101·4°	101·4°	99·4°	103°	97·2°
March 7 to 20, fourteen days (albuminuric)	97·6°	99·5°	98·8°	98·8°	98·1°	100°	96·8°
March 21 to April 2, thirteen days (albuminuric)	96·8°	96·9°	97·1°	97·1°	97·3°	100·2°	95·6°

The fall after the albuminuria appeared is very striking, and this feature of temperature is well seen in the maximum and minimum columns, but it appears to affect the temperatures taken at all times of the day. The third period, when the dropsical symptoms were fully developed, showed still further lowering; the minimum, 95·6°, being reached shortly before death. The urine varied in quantity from 15 oz. to 28 oz., the specific gravity from 1018 to 1024, and the daily excretion of urea from 135·5 grains to 337·6 grains, the amount apparently depending on the amount of urine passed; in other words, the more urine, the more urea. The total quantity of urine was small; the amount of albumen was generally about two-thirds of the test-tube.

The diarrhœa, consisting of two to three loose stools, accompanied the pyrexial period. In the second period it increased, once reaching five stools a day, and averaging two stools. In the third period it almost subsided.

Necropsy, on the day following death, showed both lungs to contain large ragged cavities, of comparatively recent origin, at the apices. The rest of these organs were thickly disseminated with recent tubercle. The liver was small, weighing three pounds, and soft. The spleen was soft and lardaceous. The kidneys were in the early stage of lardaceous degeneration, giving marked reaction with iodine. The intestines were unfortunately not examined. Dr. Ewart kindly made a microscopic examination of the kidneys, and reports as follows:—'A very plain reaction was obtained with solution of iodine. The Malpighian tufts and the peripheral zones of the secreting tubules were the parts mainly affected by the reagent. The tubes were somewhat enlarged, and choked by the proliferation and the swelling of their secreting cells. Fatty granular degeneration had made some progress, but it had not reached an extreme degree, such as to mask the presence of the lardaceous state; in many places the cells were fused into long lardaceous lumps, which assumed a mahogany colour with iodine.'

This case is of interest as illustrating the relations of tubercular disease and lardaceous changes, and their relative influence on the temperature. The patient was suffering from acute tubercular phthisis of the lungs, and probably of the intestines. The lungs were rapidly undergoing tuberculi-

Q

sation with accompanying excavation, and the temperature-chart gave the usual curve of acute third stage. The lung-changes seem to have proceeded unchecked up to death ; but the sudden supervention of kidney-disease, which, owing to the early examination of the urine and the marked symptoms, was immediately detected, caused a cooling of the body ; and, from that time forth normal and sub-normal temperatures appeared. The lardaceous disease was probably due to the extensive discharge from the lungs and intestine, and appears, from Dr. Ewart's report, to have already, to some extent, been succeeded by fatty degeneration.

Case 49.—William W., aged 16, a coppersmith, was admitted into the Brompton Hospital on May 11, 1880, with the following history. He had enjoyed good health till eighteen months ago, when, after a chill, cough with expectoration came on, which persisted up to the present time, the expectoration being occasionally tinged with blood. Shortly afterwards, night-sweats and rapid emaciation supervened, and for the last two months there had been diarrhœa.

On admission, cough and expectoration were moderate. The tongue was red and irritable. The bowels were relaxed—two motions a day. The breath was short on exertion. Pulse 104 ; temperature 99° F. The urine was acid, of specific gravity 1015 ; it contained no albumen. His weight was 6 st. 8 lb. *Dulness and cavernous sounds on the right side to the third rib, and some dulness above the right scapula.*

May 19.—The diarrhœa was checked by bismuth ; the night-sweats were profuse ; the voice was affected.

May 25.—The patient had improved till May 24, when he vomited four times. His face was flushed ; his skin hot ; pulse 120, small and compressible. Temperature (evening), 103·1° F. Pupils were slightly dilated ; tongue coated white in the centre, with red irritable edges.

May 27.—Vomiting had ceased. The urine contained albumen. *Crepitation audible over the lower third, front and back, of the right side, and in the scapular region.* Three days later, the diarrhœa returned ; and on July 10 a careful record of the temperature, the quantity of urine, and the amount of urea and albumen was commenced and carried out up to the patient's death, on August 30, with a short intermission, by my clinical assistant, Mr. Wightwick.

The diarrhœa had been continuing more or less since admission, on May 16, and was treated successively by bismuth ; bismuth and opium ; opium pill ; opium and sulphate of copper ; enema opii ; enema opii, with acetate of lead ; turpentine enema ; by combinations of glycerine of gallic acid with tannic acid ; by cupping (dry) ; and finally, by injection of half a pint of decoctum lini and twenty minims of laudanum. All of the remedies failed to check it, with exception of the last mentioned, which for a short time reduced the number of stools to two a day. The patient never had pains in the abdomen till August 4, when he complained of some in the region of the umbilicus ;

neither was there any œdema till August 18, when it appeared on the feet and
ankles, although for a considerable period the urine had shown as much as
seven-eighths of albumen in the test-tube. The pupils throughout the illness
were much dilated, in spite of the large quantities of opium he was taking,
till the day before death, when they became remarkably contracted.

The patient died August 30.

The record of the temperature was omitted for twelve days between August
7 and 19, which enables us to compare the two periods of the temperature,
which are given below. The last gives somewhat lower records than the
former. The first period of twenty-nine days, from July 10 till August 7,
gives as averages, at 8 A.M., 98·5°; at 11 A.M., 99·2°; at 2 P.M., 100°; at 5 P.M.,
99·6°; at 8 P.M., 100·6°; maximum, 102°; minimum, 98°. The second period
of twelve days, from August 19 to 30, is as follows:—At 8 A.M., 98·3°; at 11
A.M., 98·9°; at 2 P.M., 99·6°; at 5 P.M., 99·4°; at 8 P.M., 99·5°; maximum,
100·6°; minimum, 98°. All these temperatures were taken when albuminuria
was present; but the lowering of the later ones was no doubt partly due to the
supervention of the uræmia, and partly to the general sinking of the temperature at the approach of death.

The quantity of urine varied very little, averaging from thirteen to twenty
ounces, a very small amount. The albumen steadily increased up to the
patient's death. At first (July 10), it was only one-fourth of the test-tube,
then one-half, then three-fourths, gradually increasing to seven-eighths; and
on August 19, the urine became quite solid on boiling, and the determination of
urea had to be relinquished. The urea, on the other hand, diminished from
201·4 grains on July 10, to 54·2 grains on August 7, a reduction of nearly
three-fourths; the quantity of urine remaining the same. The specific gravity
increased in the same ratio as the albumen, and in the inverse ratio to the
urea. It rose gradually from 1014 to 1035. The diarrhœa steadily persisted,
varying from three to ten motions a day, increasing, on the whole, as death
approached.

Necropsy, on the day after death, by Dr. Ewart. The abdomen contained a
small amount of fluid. The intestines were largely ulcerated, the ulcerations
extending high up to within six inches of the duodenum, and consisted of two
kinds: (1) small and round, the result of simple softening; and (2) some
situated lower down in the intestine, of good size, with a thickened floor, in
which miliary tubercles were visible through the peritoneal cont. The neighbourhood of the ileo-cæcal valve was severely attacked, and the colon was one
continuous ulcer. The liver weighed 2 lb. 13 oz., and showed signs of fatty,
fibroid, and lardaceous changes. The spleen weighed 4½ oz., and was lardaceous. The kidneys weighed 13½ oz.; they were smooth, pale, and with
pink pyramids. The changes of chronic tubal nephritis were well marked.
The Malpighian capsules gave the iodine reaction. The heart weighed 5½ oz.
The tricuspid valve was dilated. The pleuræ were closely adherent on the
right side. The right lung contained abundance of grey tubercle, either
isolated or in small groups. In other parts, small cavities were forming from
the gradual disintegration of these groups, and some collections of pus were
visible. The upper lobe contained a large smooth-walled cavity, of the size of
a small orange, and a smaller one lined with a glistening membrane. The
left lung had cicatrices at the apex, but was otherwise healthy. The bronchial
glands were enlarged, and contained miliary tubercle. The larynx, trachea,

Q

and bronchial tubes showed extensive ulceration from the top of the œsophagus to the commencement of the bronchi, with some thickening of the mucous membrane. Ulcers were also present in the pharynx and epiglottis (buccal surface), and at the base of the tongue. The tonsils were hollowed out by caseation.

In this case, during the last two months of the patient's life, acute tuberculisation and excavation were proceeding in the right lung, and tuberculisation and ulceration throughout a considerable portion of the intestines, a combination which, as I have shown elsewhere,[1] causes a high pyrexial chart of a fitful character. Nevertheless, the maximum was only 102° F.; the average hardly reached 100° F., and in the last period sank below it, showing the marked influence of the blood-poisoning in cooling the system.

Judging from the symptoms, the nephritis appears to have come on about May 24, a fortnight after admission, so that the whole series of clinical phenomena arising from it occurred under observation, and a great part are contained in our records. These consisted of—(1), steady increase in the diarrhœa; (2), a constant quantity of urine secreted; (3), a rapid increase in the amount of albumen, and apparently (4) a proportionate decrease in the amount of urea; (5), a steady increase in the specific gravity. Considering the small amount of urea excreted in the twenty-four hours (54·2 grains), and the absence of coma, we may conclude that a great part of the urea was excreted through the bowels.

The *post-mortem* examination showed lardaceous changes in the liver, spleen, and kidneys; though, in the latter organ, there was superadded much blocking of the tubes from nephritis, and to this were no doubt due the diminished quantity of water and urea, and the larger amount of albumen.

The other fourteen cases, with two above given, have been already published,[2] and are too lengthy for quotation, but the annexed table (Table I.) gives a good general idea of their character and complications, some of which have already been mentioned. Albuminuria was present in all. Of the sixteen patients thirteen died while under observation, and twelve

[1] *Lancet*, 1861.   [2] *British Medical Journal*, December 22. 1883.

No.	Name	Age	Stage and Extent of Disease	Renal and Intestinal Symptoms	Result	Period of Albuminuria (when known)	Total Duration
1	E. G.	18	Phthisis III, active. Excavation and Tuberculisation in both lungs	Diarrhœa. Dropsy. Œdema of lungs—	Died	4 weeks	12 months
2	W. W.	16	Phthisis III, active. Cavities in right lung. Tuberculisation. Intestinal Ulceration	Diarrhœa, Ascites, and Anasarca	Died	12 weeks	21½ months
3	Alfred B.	26	Phthisis III, active. Cavities in both lungs. Tuberculisation	Diarrhœa. General Dropsy—	Died	8 weeks	28 months
4	Mary A. P.	46	Phthisis III, active. Cavities in both lungs. Caseous Pneumonia. Intestinal Ulceration. Peritonitis	Diarrhœa. Œdema. Peritonitis. Death by Coma—	Died	Uncertain	5 yrs., 1 month
5	Thomas T.	35	Phthisis III, active. Cavity in right lung. Tuberculisation proceeding in both lungs	Œdema of lungs and legs	Died	Uncertain	9 months
6	John B.	29	Phthisis III, active. Cavities in both lungs. Tuberculisation in right lung. Intestinal Ulceration	Diarrhœa. Œdema—	Died	29 weeks	14 months
7	George M.	38	Phthisis III, active. Cavities in both lungs. Syphilis. Ulceration of Larynx and Intestines	Diarrhœa. Dropsy—	Died	Uncertain	12 months
8	Frederick G.	19	Phthisis III, active. Cavities in both lungs. Tuberculisation. Ulceration of Larynx and Intestines	Diarrhœa. Œdema—	Died	Uncertain	30½ months
9	James R.	27	Phthisis III, active. Cavity in left lung. Tuberculisation of both lungs	Œdema—	Left hospital	Uncertain	Not known
10	J. T.	30	Phthisis III. Cavities in both lungs. Recent Tuberculisation. Pleurisy	Diarrhœa. Hydrothorax. General Anasarca—	Died	Uncertain	3 yrs., 2½ mnths.
11	J. M. W.	28	Phthisis III, active. Cavities forming in both lungs. Tuberculisation	Diarrhœa. Dropsy—	Died	Uncertain	2 years, 3 weeks
12	Catherine T.	38	Phthisis III, active. Cavities in left lung. Tuberculisation of right lung	Uræmia. Coma—	Died	1 week	2 years, 6 weeks
13	Eliza Ann T.	46	Phthisis III, active. Cavities forming in right lung. Fresh Tuberculisation in left lung. Œdema and Emphysema of lungs	Anasarca. Pulmonary Œdema—	Died	4 weeks	14 months
14	Mary P.	18	Phthisis III, active. Cavity in left lung. Tuberculisation of right lung	Diarrhœa. Œdema—	Left hospital	Uncertain	Not known
15	Mr. B.	29	Phthisis III, after Syphilis. Pleurisy with fresh Tuberculisation	Diarrhœa. Œdema—	Died	6 weeks	7 yrs., 9 months
16	Reuben B.	28	Phthisis III, active. Cavity in left lung. Tuberculisation in right lung	Œdema—	Left hospital	Uncertain	Not known
			Averages ('uncertain,' 'not known' excluded)			6 weeks	29 months

post-mortem examinations were made, from the results of
which, and the physical signs, the stage and extent of the
tubercular disease has been compiled.   All appear to be
active third-stage cases ; in eight, cavities had formed, or
were forming, in both lungs.   In the rest, excavation had
taken place in one lung, and tuberculisation was proceeding
in the same, or the opposite lung.   In six, at least, there was
ulceration of the intestines, and probably this was the case in
two more.   Ulceration of the larynx occurred in five.   This
indicates not only extensive and acute disease, but from the
amount of excavation present, a large discharge of pus from
the lungs, which, combined with the diarrhœa noted in eleven
cases, rendered the presence of lardaceous disease of the
organs not improbable.   From the fourth column we see
that dropsy in some form or other existed in all but one.
Two patients died with marked symptoms of uræmic poisoning.
The rest succumbed in the usual way by either apnœa or
syncope.   In the last column, the duration of life in these is
given, which, excluding the three ' not known,' amounts as an
average to twenty-nine months, scarcely two and a half years,
a figure far below the proper one, even for this advanced stage
of phthisis.   The true influence of albuminuria on the duration
may be seen in the last column but one, which shows in six
cases the short time the patient survived the onset of this
complication.   This period varied from twenty-nine weeks to
one week, the average being six.

The influence of the supervention of albuminuria on the
temperature has been already shown in the two cases given,
but the collective result of the observations on these sixteen
cases, taken five times a day, fully confirm the conclusion
that the effect is a lowering one, as the subjoined chart will
demonstrate.

The total averages of the temperature are : at 8 A.M.,
97·7° ; at 11 A.M., 98·5° ; at 2 P.M., 99·1° ; at 5 P.M., 99·4° ;
at 8 P.M., 99·8°.   These figures show a gradual rise from
2 P.M. to 8 P.M., as is usual in phthisis, but the temperatures
are below the febrile standard.

For comparison, I have placed on the chart the curve

(continuous line) of the averages of temperature of acute third-stage phthisis, founded on forty-three cases without albuminuria. Below it is the curve (broken line) of the sixteen albuminuria third-stage cases. This demonstrates a steady lowering of about 1° at each period of observation.

It will be noted that the temperature-curve of the albuminous cases is exactly the same as that of the other; only, owing to the cooling influence of the blood-poisoning, it is carried on 1° lower.

The maxima and minima also showed the tendency to extremes common to active third-stage cases, the range vary-

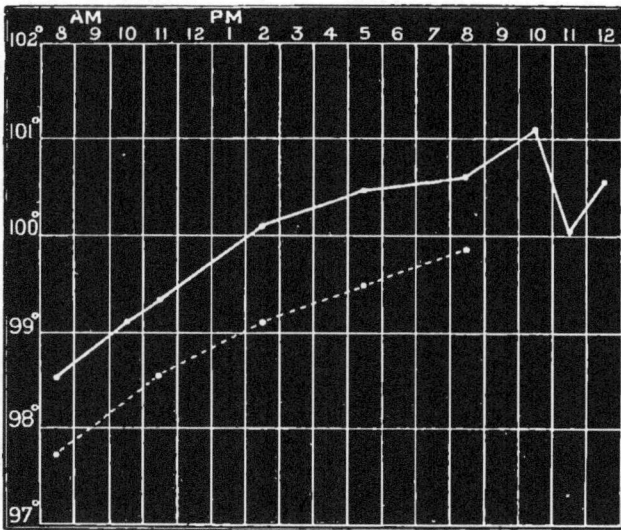

ing from 94° F. to 104·2° F.; the average maximum of the sixteen cases being 100·7° F., and the average minimum being 96·5° F.

Dr. Whipham [1] communicated a case of acute tuberculosis, where lardaceous disease of the liver, spleen, and kidneys had supervened, in which the temperature-course was normal or sub-normal. The case was interesting as confirming the above conclusion, and also for the diagnosis of the case, from

[1] *Medical Society Proceedings*, vol. vi. p. 352.

the presence of bacilli in the sputum, when the physical signs were obscure.

The principal facts relating to the *urine* of these patients has been summarised in Table II., in which some of the results of the *post-mortem* examinations are also entered, and it will considerably facilitate our comprehension of the urinary phenomena if we turn to this first. Of the sixteen patients, thirteen died while under observation, and in twelve necropsies were obtained ; but we need not here go into the state of the lungs in each case, but will proceed at once to the consideration of the kidneys.

The kidneys in seven of these twelve cases were undoubtedly lardaceous; in three more instances they were described as either large white kidney, or mottled kidney, or chronic nephritis, and it is not improbable that all these cases might, after more careful examination, have been classed as lardaceous. In the remaining two cases, the kidneys were, in one, in a state of interstitial nephritis, and in the other, they were granular with some tubal inflammation. In five of the seven cases of undoubted lardaceous kidney, there was similar disease of other organs, viz., of the intestines, liver, spleen, or pancreas, the spleen being most frequently involved. This gives a preponderance of probable lardaceous disease—viz., ten out of twelve, *i.e.* five-sixths of the cases ; but, even taking the undoubted cases only, more than half were affected with this form of disease.

In the fatal case in which no necropsy could be obtained, the character of the urine, and the presence of diarrhœa and œdema, pointed to lardaceous disease ; and of the survivors, Case 16 was probably lardaceous ; while, in Case 14, the large reduction of the urine and of the urea pointed rather to nephritis, and it is possible that Case 9 belonged to the same category. It was only by very careful examination that lardaceous disease was detected ; and often, in doubtful cases, microscopic examination settled the question. According to Ecorchet [1] phthisis is present in 60 per cent.

---

[1] Hanot's article in the *Nouveau Dictionnaire de Médecine et de Chirurgie Pratiques.*

TABLE II.—*Illustrating Condition of Urine and Morbid Anatomy of Kidneys.*

Case	Name	Age	Urine					Morbid Anatomy	
			Average Quantity (ozs.)	Average Specific Gravity	Albumen	Urea (grains)	Rate per lb. body weight (grains)	Kidneys	Other Organs
1	E. G.	18	23	1021	⅓ to solidity	236·8	2·27	Lardaceous (incipient)	Liver fatty, fibroid, and lardaceous.
2	W. W.	16	17	1014 rising to 1010		201 to 54		Lardaceous, nephritis	Spleen lardaceous
3	A. B.	26	21	1031	¼	169	1·37	Lardaceous, large white	Liver and intestines lardaceous
4	M. A. P.	46	36	1026	½	180	1·63	Lardaceous	Spleen lardaceous
5	T. T.	35	40	1018	⅓	216		Lardaceous	Liver fatty and lardaceous. Spleen
6	J. B.	29	40	1015	½	228	2·03	Lardaceous, large white	and pancreas lardaceous
7	G. M.	38	27	1012	⅙			Large white and mottled. Nephritis. Weight 22 oz.	
8	F. G.	19	17		¾	166		Lardaceous	Liver and spleen lardaceous
9	J. R.	27	30		abundant			Left hospital	
10	J. T.	30	30	1027	abundant			Large mottled	
11	J. M. W.	28	44					Chronic nephritis, probably lardaceous	
12	C. T.	38	48	1012	1/10			Not lardaceous, interstitial disease (cyst), and nephritis. Weight, 14 oz.	Spleen lardaceous
13	E. T.	48		1011				Interstitial disease with tubal inflammation (granular). Weight, 8½ oz.	
14	Mary P.	18	7	1039	¾	82	·98	Left hospital. Nephritis	
15	Mr. B.	29		1014				Died. No *post mortem* examination	
16	R. B.	28	38	1019		207·9		Left hospital	

of the lardaceous kidneys.   Dr. Southey states,[1] as the result of his hospital experience, that phthisis was associated with 8·9 per cent. of the granular kidney, with 12·9 per cent. of the large white kidney, and with 37 per cent. of the lardaceous kidney.

Professor Bamberger, in a valuable contribution based on 2,430 cases of Bright's disease, from the records of the Allgemeine Krankenhaus in Vienna, gives a list of the various diseases to which 1,623 of his cases of Bright's disease were secondary; and in this it appears that in 15·7 per cent. of the whole number, it was secondary to tuberculosis, phthisis, and scrofula.

Bamberger found that, while lardaceous disease of the kidney and other organs was a tolerably frequent attendant on chronic phthisis, the parenchymatous and tubular forms of nephritis were most common.   Wishing to investigate this subject from the other standpoint, I endeavoured to ascertain the number of cases of phthisis in which lardaceous disease of the kidney was found after death, and for this purpose examined the records of the Brompton Hospital for one year and a half, during which period Dr. Ewart had been curator, and had examined most of the kidneys microscopically.

In 133 necropsies in cases of phthisis, the kidneys gave a marked reaction with iodine in 50 cases, or in 37 per cent.   In 19 more a slight reaction was obtained, chiefly confined to the Malpighian tufts, giving a total of 69, or 52 per cent.   This is a higher percentage of lardaceous disease in phthisis than any previously given ; but it must be remembered that the Brompton Hospital has special advantages in point of numbers for investigations of this sort.

The microscopic observations confirm the admirable ones of Dr. Dickinson,[2] demonstrating that the vessels of the Malpighian tufts were those first attacked, and that, later on, the epithelium of the tubuli contorti was involved in the staining.   In many instances, fatty changes were also taking place in the epithelium.

Urine.—In the lardaceous cases, the average quantity of

[1] *Lumleian Lectures.*   [2] *Pathology and Treatment of Albuminuria.*

urine varied from 17 to 40 oz. in the 24 hours. In Case 14, of which the diagnosis was chronic nephritis, it at one time fell to 2½ oz., the specific gravity rising to 1047. In this case the urea fell to between 74 and 93 grains, and the patient nevertheless showed no head symptoms. The specific gravity averages of the lardaceous cases varied from 1012 to 1031, a much higher estimate than Dr. Dickinson's, who gives it as 1006 to 1015. In Case 2, where great obstruction occurred in the tubes, it rose gradually from 1014 to 1040, with such an enormous increase of albumen that the urine appeared to be a mere solution of serum. In Case 13, one of granular kidneys, it fell to 1011.

The albumen was very abundant, and seldom amounted to less than half the test-tube. In Case 2, the urine eventually became solid on boiling, and precluded further investigation.

In the 'granular case' it averaged one-tenth, but never exceeded one-sixth. In Case 7, one of large white kidney, presenting no lardaceous reaction, the albumen was one-half the test-tube, the specific gravity being only 1012, and the kidneys, after death, weighing 22 oz. Casts were found in six cases, granular in two, granular and hyaline in one, hyaline with oil-globules in one, simple hyaline in one, and blood-casts in one.

The urea was determined by Russell and West's method. The maximum in the cases of lardaceous kidney was 337 grains in the 24 hours, which occurred in Case 1 ; the minimum was 54 grains, which occurred in Case 2 : the normal amount being about 500 grains *per diem*, showing a great decrease from the ordinary standard.

The most remarkable diminution occurred in Case 2, where, in one month, the urea diminished from 201 to 54 grains, the quantity of urine remaining the same, and the dietary not greatly varying. In this case, diarrhœa was present, and increased somewhat towards the end of the period ; but the increase was not sufficient to reduce the quantity of the urine passed, though probably some of the urea was excreted by the bowels. Our lardaceous cases, therefore, show a decided diminution of water and urea, but the presence of a

very large amount of albumen. Unfortunately, time did not
allow of the quantitative analysis of the other constituents of
the urine in these cases. In column 7 of Table II. I have
entered the rate of urea-excretion per pound of the patient's
weight, according to Parkes's plan, who estimated that a healthy
man should excrete daily $3\frac{1}{2}$ grains for every pound of his
weight. This shows that the highest rate was $2\frac{1}{4}$ grains, and
two of the patients excreted 1·6 grains and 1·3 grains per
pound respectively, and one patient, rather less than 1 grain
per pound—a very small quantity.

The question arises as to the cause of the lowering of the
temperature in the above cases, and we naturally turn to the
state of the blood for an explanation. In albuminuria,
generally, there is a marked decrease in the number of red
corpuscles and of albumen, with a relative increase in the
water, fibrin, and leucocytes; hence the great pallor and the
tendency to dropsy. As the red corpuscles are the carriers of
oxygen to the tissues, it would appear that the diminution in
their number, which, according to Dickinson, amounts to one-
fifth, may considerably interfere with perfect oxidation; but
in many cases of disease great pallor is combined with high
temperature, and so this will not afford a complete explanation. Nor will the presence of an increased amount of urea
alone do so; for it has been shown by Hammond's[1] important
experiments that, provided the kidneys remain intact, a considerable quantity of urea can be injected into the blood of
dogs without greatly disturbing the circulation, and that, when
symptoms of uræmia appeared, they disappear under the
excreting action of the kidneys. Even when the kidneys have
been removed, or the renal arteries tied, and thus the urinary
constituents retained within the blood, uræmia does not take
place, if, as has been pointed out by Bernard and Barreswil,
the urea or the products of its metamorphosis can be discharged
by the stomach and intestines. The state of the kidneys in
Case 2, blocked by tubal nephritis, and the small quantity of
urea and water excreted by them, without any symptoms of
coma, with the obstinate, but doubtless necessary diarrhœa,

[1] 'Uræmic Intoxication,' *American Journal of Medical Sciences*, 1861.

present a remarkable similarity to this set of experiments. Hammond, however, shows that urea, injected into the blood in quantities beyond such an amount as can be excreted, produces toxic effects.

It is probable, however, that the condition known as uræmia is due to the retention within the blood, not of one urinary constituent alone, but of many; and an experiment of Stolnikoff[1] bears on this. He found, in experimenting on rabbits, that if, before injecting putrid material into the blood and setting up septicæmia, the renal arteries were compressed, and thus the conditions of nephritis reproduced, no pyrexia followed, but rather a lowering of the temperature. He also found that, when nephrectomy was performed on animals in a feverish state, the fever sank.

The temperature of chronic Bright's disease, whether nephritic or granular, is usually low, and often sub-normal: and so it is not surprising that the state of blood which lowers the temperature in this form of disease should reduce the high temperatures of pyrexial phthisis. We may conclude, therefore, that the lowering of the temperature in cases of phthisis associated with albuminuria is due—partly, at least—to the retention within the blood of the constituents of the urine. This gives rise to uræmia, which is not sufficiently intense to cause coma.

Prognosis.—Though our cases disclose a very unfavourable outlook for the complication of albuminuria with phthisis, we must remember they were hospital ones, and private patients appear to live longer, and cases have been quoted where life has been prolonged even for years. Dr. Quain[2] cited a case of phthisis of fifteen years' standing, with albuminuria of five years' duration; and one of extreme albuminuria and extensive chronic phthisis, with but little influence on the general health. In one of my patients, who has had phthisis for nineteen years, and a considerable-sized cavity for ten years, albuminuria supervened eight years ago, and he still survives, although in a very precarious condition.

[1] *Centralblatt für die Med. Wiss.*, April 24, 1880.
[2] *Proc. Roy. Med. Chirurg. Soc.*, 1879.

Another patient, of twelve years' duration, has had slight albuminuria for one year, and is still able to travel, though living the life of an invalid. A very interesting case of 32 years' duration, terminated by albuminuria, will be found in Chapter XX. These are the longest cases I can recall; and they must be regarded as the exception, and not the rule. Generally these patients sink a few months—or it may be weeks—after the albuminuria sets in ; so that this symptom must, in all cases of phthisis, be regarded as of very serious import.

CHAPTER XVI.

ACUTE TUBERCULOSIS.

Acute Tuberculosis—Similarity to infective fevers—Connection with Chronic
Consumption—Sketch of clinical history and *post-mortem* appearances—
Tubercular Meningitis—Varieties of Acute Tuberculosis—Primary and secondary—Absorption from caseous centres—Cases of Acute Tuberculosis—
Obscurity of symptoms of Meningitis—Examples of Acute Tuberculosis
arising from caseous bronchial glands, vertebral abscess and measles—
Diagnosis from Typhoid Fever, Capillary Bronchitis and Hysteria—Prognosis—Distinction from Scrofulous Pneumonia—Acute Tuberculo-pneumonic Phthisis—Prognosis.

WE will now proceed to describe the principal varieties of
phthisis, beginning with the acute forms and ending with the
more chronic ones, which are therefore more amenable to
treatment.

Acute tuberculosis is the most fatal of all the forms of
phthisis, and while it is characterised by the outburst of
tuberculisation in the various serous membranes and in the
parenchyma of the lungs, and other organs of the body, it
displays so little of the localised features of phthisis, and
so many of the general symptoms of the infective fevers, as
to be classed by some writers, not as a variety of phthisis,
but as a fatal complication of the disease. Acute tuberculosis,
however, only differs from other forms in the area of the
system involved: the pathological products are identical with
those of the more chronic forms of phthisis, and it is the
completeness of the self-poisoning, as we may call it, *i.e.* the
omnipresence of tubercle in the system, which gives to this
form its characteristic features.

A patient is attacked with pyrexia, with cough and scanty
viscid expectoration, with extreme oppression and overwhelming weakness, the symptoms much more resembling those of
continued fever than those of consumption. The tongue

becomes red, with brownish centre, the bowels are costive ; the
appetite is entirely wanting; the skin becomes hot, the temperature ranging from 103° to 106° F., sometimes even higher,
and less marked by remissions than in scrofulous pneumonia ;
the urine is scanty and high-coloured ; the pulse is rapid, varying from 120 to 140. Sordes appear on the lips and teeth.
But the most prominent feature is the marked dyspnœa, the
respirations being 40, 50, and 60 a minute ; and even the
countenance testifies to the obstructed state of the breathing.
The patient wastes rapidly, and is often delirious. The cough
is hacking and generally unaccompanied by expectoration, or, if
that be present, it is very slight in amount and simply mucous.

Physical examination of the chest shows marked resonance over the anterior surface, and a lesser degree of the
same over the posterior surface. The natural respiratory
sound is everywhere replaced by bronchitic rhonchi and fine
crepitus, the latter, though fine, being more scattered in
character than the crepitation of croupous pneumonia. In
spite of all remedial measures, the breath grows shorter and
shorter, the face and nails become livid, the prostration
increases, clammy sweats break out, the sounds in the lungs
become more and more liquid, the patient grows weaker, and
dies, often of apnœa, owing to the great pulmonary obstruction.
When death supervenes in this way, the temperature falls
considerably some time previously, and a morning record of
only 95° to 96° F. is often observed, though the evening one
keeps up to the last.

Frequently, however, other organs besides the lungs are
attacked, though generally secondarily to the lungs ; the
brain is the most common one. The patient complains of
severe and persistent headache, is seized with vomiting,
becomes excited and delirious, and often confused in his
memory and ideas ; has twitching of the extremities, squint-
ing and double vision, though not necessarily any loss of
power. The urine contains phosphates, the bowels are costive.
The pupils are somewhat contracted. Ophthalmoscopic examination shows the fundus oculi to be congested, and sometimes detects granulations. Then follow the symptoms of

effusion into the ventricles : the pupils are dilated, the patient becomes comatose, passes motions unconsciously, and cannot void his urine, which has to be drawn off, or, on the other hand, he passes it unconsciously, and he gradually sinks, the temperature falling rapidly before death, and the pulse becoming very weak. Of the cerebral symptoms the muscular twitchings are the most certain and characteristic previous to the coma. The temperature is not always high, and cannot be depended on for diagnostic purposes. Sometimes pain in the abdomen and diarrhœa indicate the presence of miliary tubercle in the intestines and peritoneum.

Post-mortem examination in cases of acute tuberculosis shows the lungs disseminated with fine miliary tubercle, grey and white ; the granulations generally soft and easily squeezed out, showing their recent formation. Sometimes they are aggregated and present centres of caseation ; and sometimes, when the outburst is extensive, the lungs seem to be packed full of these nodules, so as to offer the appearance of consolidation. In these cases the intervening tissue is hardly visible, and section of the lung shows a greyish surface with numerous yellow centres. When, however, it is visible, we find a great variety of conditions in different patients. In some, it is in the stage of simple engorgement : in others, in that of red hepatisation ; in others, again, it is highly emphysematous. The pleura is the seat of fine glistening granulations, and the bronchial glands are swollen and hardened, while sections under the microscope show the appearance of sections of miliary tubercle, varying according to the date of the lesion (*vide* chap. III.). Where the symptoms have indicated abdominal complications, we find fine granulations scattered throughout the peritoneum and its various processes and involutions, especially on the under surface of the diaphragm, in the great and small omenta, and on the peritoneal surfaces of the liver, pancreas, and spleen. In the intestine we may find fine miliary tubercles in the solitary glands and in Peyer's patches standing forth in bold relief from the mucous surface, but much of the formation of tubercle in the intestines is secondary to ulceration, and the result of absorption. Intestinal ulceration is often secondary to lung excavation and laryngeal ulceration. Tubercle of the intestine (not peritoneal) cannot be said to be a frequent accompaniment of acute tuberculosis. The relation of the tubercle bacilli to the changes of miliary tuberculosis are well shown in Plate II. 1, where, in a lung section, the crowding of the organisms, and the extensive cell proliferation they give rise to, are well demonstrated.

Where tuberculosis has involved the meninges of the brain, the appearances are very characteristic. In addition to flakes of lymph on the hemispheres, fluid in the ventricles, and softening of the adjoining structures, we find granulations, varying in size from a pin's head to a millet-seed, sprinkled over the surface of the pia mater, and its various and numerous processes, the choroid plexus, &c., invading the optic disc. The hemispheres are gorged, and there is considerable arachnitis and effusion of lymph at the base of the brain, and especially in the neighbourhood of the circle of Willis, the fissure of Sylvius, and the longitudinal fissure, and following the course of the blood-vessels.

Hanot [1] and Jaccoud both describe several varieties of acute tuberculosis dependent on the predominance of the pulmonary, abdominal, or cerebral symptoms; but we think that, considering the same lesion, viz., miliary tubercle, is always present, further division is unnecessary.

Acute tuberculosis is sometimes primary, sometimes secondary. It supervenes on various fevers, on old scrofulous diseases of various organs, or on chronic phthisis.

In these cases it appears to be due to absorption of tubercle bacilli from some old caseous centre, very commonly a bronchial gland, of which examples will be given. The first one displays well the series of brain symptoms in tubercular meningitis supervening in a third-stage case of phthisis.

CASE 50.—Anne Charlotte D., aged 17, single, by occupation a teacher, was admitted under Dr. Theodore Williams on May 1, 1877.

History.—Subject to winter cough for years, which has been continuous now

---

[1] *Nouveau Dictionnaire de Médecine et Chirurgie.*

for six months, and accompanied by yellow expectoration, night-sweats and emaciation. For the first four months she suffered from dyspnœa.

On admission she complained of great lassitude. Her aspect was pale, but she was not very thin. Cough hacking, rather troublesome, with little or no expectoration. Tongue coated. Bowels regular. Appetite fair. Catamenia absent for four months. Pulse 80. Weight 7 st. 2 lbs. *Dulness existed over the whole left chest, most marked over the upper half, which was somewhat flattened. Cavernous sounds were audible from the first to the third ribs. Scattered crepitation was heard below and over the whole posterior region on the left side. Tubular sounds in the first interspace on the right side.*

The evening temperature was taken on May 3 and 4, and each time found to be 101° F. The temperature was noted five times daily in the mouth, viz., at 8 and 11 A.M., and 2, 5, and 11 P.M., from May 11 to May 28 (inclusive), a period of 18 days. The course was somewhat irregular and on the whole febrile, the evening observations ranging from 98° to 100° F., but rarely passing 100°; the morning, especially at 8 A.M., falling to 98°, 97°, 95° F., and on one occasion to 93° F.

During the first month of her stay in the hospital, with the exception of diarrhœa, lasting two days, and slight hæmoptysis, she improved, gaining 3 lbs. in weight and becoming stronger. On May 22, when the temperature was still low, the blood was examined microscopically, and found to contain large granular protoplasmic masses, varying in size from three to seven times that of a leucocyte. The leucocytes and the red globules were normal.

July 17.—Patient gaining weight and cough less troublesome ; expectoration slight. She complained of severe headache.

More flattening on the left side. Cavernous sounds were audible in front as before, but also posteriorly. Coarse crepitation was heard over the upper half and gurgling at the angle of the scapula. The blood was examined ; no granules could be detected. The patient complained of great pain in the lumbar region of the spine. There has been slight menstrual discharge.

July 23.—The patient still complained of headache and had vomited several times. The temperature had risen slightly, being at 8 A.M. 100° F. Her manner was strange and variable ; sometimes apathetic, and at other times excited and violent, and occasionally she was incoherent. She complained of pain in the cervical region, but slept a good deal. Tongue white and indented, appetite bad, bowels open, pulse 100, urine passed involuntarily. No photophobia. Pupils normal and responded to light. Ophthalmoscopic examination showed nothing abnormal in the fundus oculi.

July 24.—She had been delirious for three hours in the night, muttering and singing ; was now semi-comatose and recognised no one. No photophobia, pupils normal, jaw clenched, and dysphagia apparently present. Uneasy twitching of the hands. Six leeches were applied to the back of the neck and a soap injection administered. The leeches drew well, and the patient also had epistaxis. The bowels acted freely. Pulse 120 ; respirations 45.

July 26.—Better, and took nourishment by mouth freely. She was quite conscious and recognised everybody. The blood was again examined and contained granules and granular matter in abundance.

July 30.—She has been sleepless and constantly muttering to herself and picking at the bedclothes. Last night she was much excited and bit her hand. Consciousness decidedly less. The conjunctivæ were less sensitive ; the pupils variable, but for the most part dilated. Jaws firmly closed. The patient could not be roused from stupor. Bowels open. She took food. Pulse 140. Ordered four leeches to the neck.

July 31.—Free bleeding from the leeches was followed by a short sleep. The patient was extremely restless and constantly moaning. She vomited once. She took food and the bowels were open. Ordered ice cap to the head.

August 1.—She had convulsive twitching of the face and neck with some lividity. She was sensible at times to-day, slept very little, and her face was always turned to the left. Bowels not open. Abdomen tympanitic. She vomited this morning.

August 2.—She was moaning and almost entirely unconscious. The pupils were dilated and responded very little to light. The urine was drawn off, acid, of a specific gravity 1025 and contained no albumen or sugar. The conjunctivæ were hardly sensitive.

August 3.—She appeared more conscious; has had several convulsive attacks: these consisted of a drawing down of the angles of the mouth, and a contraction of all the muscles attached to the hyoid bone. The respirations varied greatly in number, and were attended by a moaning and sometimes by a gurgling sound. During the spasm respiration almost ceased. The chest movement was very slight; little or no air passed in and out. The spasms lasted about fifteen seconds; the muscles then relaxed and breathing became rapid and shallow, with flapping of the dependent cheek.

August 4.—More comatose; the convulsive attacks are more frequent; the pupils are dilated; a shallow ulcer has appeared on the right cornea. The urine is passed involuntarily, but the bladder is not distended. The labia majora are enlarged. She died at 3.50 P.M. of coma.

Autopsy.—Six hours and half after death. Permission could only be obtained to examine the brain. The calvaria were thin; the convolutions flattened, the superficial vessels congested; both ventricles contained slightly turbid fluid, not exceeding an ounce in quantity; the septum lucidum and commissures of the third ventricle were softened and almost diffluent; the puncta vasculosa were very distinct. The pia mater at the base of the brain contained numerous fine granulations, especially the processes in the longitudinal and Sylvian fissures. The granulations were about the size of a pin's head, and crowded close on each other; the arachnoid was thickened and very opaque. In the neighbourhood of the pons Varolii there was a thick layer of plastic lymph completely concealing the basilar artery and circle of Willis and all the structures contained therein. Processes of the lymph could be traced in the Sylvian fissures; the lymph masses in the transverse fissures joining formed a well-marked process, which extended for about two inches along the middle wall of the cerebellum, on its upper aspect.

The patient exhibited a larger amount of cerebral disturbance than I have usually noticed in tubercular meningitis supervening on chronic phthisis. This may be accounted for by the hemispheres and medulla being early affected, as shown by the delirium and stupor preceding the convulsive twitching of the extremities and spasms of muscles which are generally referred to affections of the membranes. The loss of consciousness was gradual, and great improvement followed the first application of leeches, which unfortunately did not recur when the experiment was repeated. The temperature is worth noting, for though Dr. E. Long Fox,[1] in his admirable 'Clinical Observations on

---

[1] *St. George's Hospital Reports*, vol. iv. p. 87.

Tubercle,' states that the temperature in meningitis seldom rises above 102° F., many physicians hold that the supervention of tubercular meningitis necessarily raises the temperature very considerably. In this case we have during the period of greatest excitement, a record often but little above normal, and sometimes below it.

In most of these patients the supervention of meningitis is very obscure, and twitching of the muscles, with a certain degree of stupor, are the only symptoms to draw attention to the important change. In one case under my care, the first symptom was bladder paralysis, which continued to be the sole indication for some days of brain mischief, though impairment of sensation and convulsions followed at a later date. Generally headache and vomiting appear early, followed by more or less delirium, but the muscular twitching is the symptom most to be relied on before the signs of effusion appear. In some cases irregularity of the pulse, which is often slow, and phosphates in the urine, are noticeable, but these symptoms are by no means always present.

Acute tuberculosis, when secondary, as it generally is, may be considered to arise from some infective centre in the system, whether a caseous mass in the lungs, or in some gland, or caries of some bone, as of a vertebra[1], or of the bones of the skull after scarlet fever, and it greatly resembles pyæmia in its mode of origin and its spread through the circulation, though it differs very greatly in many features from that disease.

We will now take some more striking examples of acute tuberculosis.

CASE 51.—*Meningitis. Acute tuberculosis. Death in twenty-one days.*— Robert I., aged 13, admitted under Dr. C. J. B. Williams, January 13, 1848. A diminutive boy, four feet in height, employed in spinning tobacco. Quite well till sixteen days ago, when, after some failure of appetite, he began to suffer from pain in the head, which increased so much in two days that he was confined to bed, and screamed and moaned all night. Vomiting then came on, and has continued since, with obstinately costive bowels. Urine scanty and high-coloured – the day before admission he had passed none for twenty-four hours. Has been repeatedly leeched, and has vomited all the medicine given. Eyes suffused; pupils dilated. Strabismus of right eye, said to be habitual. Head not very hot; but carotids pulsate more strongly than the radial arteries. Pulse 108. Slight twitching of arms. Head to be shaved. Eight leeches to temples. Three grains of calomel every four hours; ten drops of tincture of cantharides in a draught three times a day.

[1] See some cases by Dr. Southey, *Brit. Med. Journal*, October 1877.

January 14.—Blisters were applied to the nucha last night. He is conscious to-day, but is dull ; moaning, and drivelling at the mouth. Pulse 150, weak. No urine voided since admission.   Catheter was passed, but with negative results.

January 15.—Catheter brought away three ounces urine, being the first for thirty-two hours : alkaline, phosphatic, not albuminous. More has since been passed in bed. Patient more sensible, but moans much and takes hardly any food. Sordes on lips and teeth.

January 17.—Almost unconscious, except when attempts are made to move him, when he screams.

January 18.—Pupils now unequally dilated, right most. Pulse hardly perceptible. Very noisy in the night ; and now talks incoherently. Right side slightly convulsed. Extremities cold.

Died on January 19.

Autopsy five hours after death. Body moderately emaciated. Dura mater vascular. Convolutions flattened. A good deal of serum escaped on removing the brain, and at the base ; especially on the tuber annulare and crura cerebri, there was a considerable thickness of lymph infiltrated with serum. Opaque patches, not granular, were seen in the membrane between convolutions of cerebellum. That at the base is partially so. Cineritious matter darker than usual, but few red points in the centrum ovale. Serum also in the lateral ventricles, the whole quantity of serum being about two ounces. Lungs pale and light, weighing, right four and half, left three and a half ounces. Small granular tubercles were found scattered through both lungs. One caseous tubercle of the size of a pea in right lung, and a bronchial gland on the right side also was enlarged with cheesy matter. Several tubercles in the spleen. Mesenteric glands simply enlarged.

This case is remarkable for the entire latency of the pulmonary tubercles which must have existed before the commencement of the fatal attack of meningitis and for the completeness of the train of cerebral symptoms.

CASE 52.—*Acute tuberculosis. Meningitis. Caseous bronchial glands.*— John B. aged 15, admitted under Dr. C. J. B. Williams, April 22, 1843. Lived in ill-ventilated house and unhealthy part of town. Father spits blood. Last winter had whooping-cough, and not well since. Two months ago had a rash, which some said was scarlatina. Three weeks ago suffered from pain in chest, cough, and short breath ; hot skin and costive bowels. Has been very ill ever since.

Is now dull and heavy, sleeping much ; skin very hot. Pulse 120. Bowels very costive ; fæces very dark. Cough slight. Nose has bled several times lately. Pupils dilated. Much emaciated. Urine scanty, strong-smelling *Dulness and tubular sounds in both scapular regions, most in right.*—Head to be shaved. Pulv. hydrarg. subchlor. et pulv. Jacobi, ter die. Cras mane olei ricini ʒij.

April 24.—Nose has bled again. Head still hot. Veins of neck full. Cough worse. Bowels open ; fæces dark. A leech to each temple.

April 25.—Very restless, crying out with pain in the head. Cold lotion.

April 27. More quiet, but moans in sleep. Heat less. Lips parched, and teeth covered with sordes. Pupils still much dilated. Takes no food, and passes excretions under him. Pulse 180 and more.

April 28. Died.

Twenty hours *post-mortem*. Body much emaciated. Skin of back of neck and head suffused with blood. Both lungs studded throughout with miliary tubercles, mostly yellow and crude. Masses of yellow tubercle in bronchial glands. One at the root of the right lung much enlarged, with cheesy matter partially softened, and communicating with the right bronchus by an opening with vascular margin and elevated edges. A few tubercles under the pericardium covering the heart. One in the left ventricle and another in the septum. Liver studded with miliary tubercles, on the surface white and granular ; in the substance, larger, yellow, and seeming to enclose the ducts, some of which were enlarged, and containing green bile. Peritoneum covering under surface of diaphragm patched with granular tubercles. Spleen more thickly studded than any other organ with tubercles, of size of hemp seeds. A few minute transparent granules on the surface of intestines, omentum, and meso-colon. A few pale tubercles in the kidneys.

Convolutions of brain much flattened on upper surface, and membranes very vascular. Granular tubercles on the right hemisphere, near the falx, collected in a patch posteriorly, and here the pia mater adhered firmly to the brain. Carotid arteries and branches studded with granules. About three ounces of bloody serum at base of skull.

Here the caseous bronchial glands seem to have been the cause of the general eruption of tubercle.

CASE 53.—*Acute tuberculosis from vertebral abscess.*—James S., aged 18 ; admitted March 11, 1843. Footman. Lived well, and enjoyed good health till last Christmas, when, for a short time, suffered from rheumatism in the hips. Two months ago, after a chill, began to cough, which increased, with pains in his limbs, headache, and rapid loss of flesh and strength. The pains have lately been bad in his back and chest, especially left side. Expectoration scanty in yellow clots ; once brought up a mouthful of blood. Pulse frequent ; skin hot ; sweats at night ; tongue furred ; face pale, with slight flush on each cheek ; moderate emaciation. *Dulness and crepitus, fine and coarse, throughout left side ; most in upper parts. Some crepitus also on right side.* —A blister to left side. Mixture containing antimony and hydrocyanic acid.

March 13.—Breath and cough somewhat better, but complains much of pain in left hip and back. Pulse 104. Heat great ; very weak. Sputa viscid and opaque. Died next day.

Examination, eight hours after death.—Left lung very red and much congested, with many clusters of miliary tubercles, most in the upper lobe. The lung was closely adherent to the posterior walls of the chest, and on removing it purulent matter was observed to pour out from an opening in the posterior mediastinum. This was traced to the body of the third dorsal vertebra, which was carious ; the disease did not penetrate to the spinal canal, but there was a small abscess of a size of a pea outside the theca of the cord. The matter from the mediastinum also poured out from the posterior and lower part of the upper lobe of the lung, in which was an abscess about the size of a small egg. The right lung generally adherent. Its substance was partially consolidated by clusters of grey tubercles. At the apex were some opaque and partially excavated masses ; and around each of these the lung texture was deep, red, and consolidated, but soft. Posterior portions of this lung all much congested, but crepitant. Liver large, weight 3 lb. 13 oz. ; much congested, as was also the spleen.

CASE 54.—*Acute tuberculosis. Measles.*—P. S., aged 23, admitted under Dr. C. J. B. Williams, January 4, 1848. Formerly farm labourer; lately policeman. Health good till two and half years ago; had syphilis, for which he was salivated, and was quite well in a few months. Thirteen days ago was chilled whilst on duty, and began to cough and expectorate, and felt very weak. He vomited much dark green matter the day before admission, and that morning an eruption came out over the whole body. This rash, which is obviously measles, is now fully out, with suffused eyes, flushed face, and urgent cough and short breath. Pulse 88; tongue furred.— Ordered antimonial saline : calomel and a senna draught.

January 5. - Expectoration more viscid, partly opaque. Eruption was out in morning; now more faded. Pulse 100; soft, reduplicated.

Breath-sound obscure, especially in right back. Mucous rhonchus, most on left side.

January 6. - Breathing more embarrassed. Face much congested, almost livid. Expectoration in ragged opaque masses, floating in brownish liquid. *Left back dull; crepitation on both sides.* Cupped between the shoulders to ʒiv., and later in the day, on right side to ʒviij.

January 8.—Breathing was much relieved yesterday, but to day became as bad as ever; and this morning had ʒxj. blood drawn in full stream from the arm, with some relief, but he is very weak.

At the visit carbonate of ammonia and nitre mixture were substituted for the antimonial saline. At night the dyspnœa increased, and a blister was applied.

January 11. In the last two days breathing easier, and face less livid. Urine abundant; s. g. 1029, with plenty of lithates.

January 13.—Breathing easier in morning, but always worse at night, preventing sleep. Pulse frequent, jerking, sometimes irregular, and weak ; urine 24 oz.; s. g. 1021, contains a little albumen. *Breath-sound weaker on both sides, superseded by irregular crepitation. Tubular sound above right scapula.*

January 15.—Breathing more laboured, with pain in right side. Lividity increased. Pulse frequently irregular. *Crepitus has risen higher, and breath-sound is weaker.*

Died on January 16.

Post-mortem, 46 hours.—Both lungs contain numerous miliary tubercles scattered through their texture ; none hard, but some are grey, transparent, and offer some resistance. Others are red and soft, yet firmer than the tissue around. At the apex this distinction is most obvious, where are red bodies as big as a pea, sinking in water, while the intervening tissue is comparatively free, but the lower lobes are much congested, and the right partially hepatised posteriorly : even here the slightly firmer miliary tubercles can be felt scattered through it. A patch of recent lymph was found on the posterior surface of right lung. Other organs healthy.

The remarkable points in this case are, the rapid production of miliary tubercles in twenty-four days, the suppression of the measles rash, and increase of the pulmonary oppression ; and the soft plump character of the tubercles in their most recent state.

The *diagnosis* of acute tuberculosis is not always simple, as it is often mistaken for typhoid fever, and sometimes, on account of the early physical signs, for capillary bronchitis. From the former it is separated by the gradual development of the lung phenomena, from the latter by the higher temperature and the progressive emaciation. Cases of commencing tubercular meningitis are sometimes mistaken for hysteria, occurring as they often do in young females; but the head symptoms are early supplemented by the febrile ones and the signs of effusion and brain pressure, which clearly separate the former.

The *prognosis* is generally hopeless. The duration varies from a few days to a few weeks. This is true of acute tuberculosis, but I do not intend to include under this heading the duration of chronic consumptive cases, in which tuberculosis suddenly supervenes, as here the date of the tuberculosis, and not that of the primary symptoms of phthisis, should be regarded as the beginning of the fatal illness.

The mode of death varies greatly in different patients. Where the eruption of miliary tubercle in the lung is rapid, death takes place by apnœa through pulmonary obstruction; in others, where diarrhœa comes on, death occurs by exhaustion; and where tubercular meningitis supervenes, coma closes the scene.

Acute tuberculosis contrasts in the following points with scrofulous pneumonia, which will shortly be described: 1. In the dyspnœa. 2. In the absence of expectoration. 3. In the physical signs. 4. In the lesser tendency to breaking down of the pulmonary lesion. 5. In the infection of other organs, such as the meninges of the brain, and the peritoneum.

The second acute variety is *acute tuberculo-pneumonic phthisis*, which may be described as a link between the first two varieties—scrofulous pneumonia and acute tuberculosis—as it presents some of the clinical and pathological features of each. It resembles the latter in so far that rapid tuberculisation takes place in the lungs, and sometimes involves other organs, as, occasionally, the intestines; it resembles the former in the fact of a pneumonic process being also present, and excavation soon following on consolidation. The pathological differences it presents when compared with the other acute varieties, are : 1. The tubercle is aggregated and tends to rapid caseation. 2. The cavities are formed by the breaking down of tuberculous masses in the lung, and not, as in scrofulous pneumonia, through a breaking down of the infiltration products. Moreover, whilst excavation is taking place in one part of the lung, fresh tuberculisation may be proceeding in another, and thus total disorganisation of the pulmonary area soon takes place.

After death, we find in the lungs large tracts of catarrhal or croupous pneumonia, surrounding small and large racemous masses of aggregated tubercle. These masses are generally more or less yellow in colour from the process of caseation, and present in some parts excavations of evidently recent date. Scattered miliary tubercle is generally present, though the aggregated form is more commonly met with. In this variety we find occasionally ulceration of the intestines, and the presence of tubercle in the solitary glands, but the serous membranes are free. The clinical symptoms may be described as those of chronic tubercular phthisis in an exaggerated form. The cough is very troublesome, and is early accompanied by large expectoration ; anorexia is generally present ; the tongue varies, but is sometimes red and irritable in appearance, and diarrhœa often occurs. The pulse is rapid, varying from 100 to 130, and the respirations are hurried and count between 30 and 40.

The temperature is distinctly pyrexial, with occasional but rare periods of collapse, the high figures being maintained more uniformly towards the end than is the case in scrofulous pneumonia, the reason probably being that excavation is accompanied by fresh tuberculisation even during the last weeks of the patient's life.

The *prognosis* is very unfavourable, though occasionally, the tubercular mischief seems, for a time at any rate, arrested by the formation of cavities. Re-absorption of the virus appears to take place easily from the cavity wall, and fresh eruption in the lung is only too common. As acute tuberculo-pneumonic phthisis is only a link between the two other acute varieties, it will not be necessary to give examples of it.

CHAPTER XVII.

ACUTE PHTHISIS OR SCROFULOUS PNEUMONIA.

Intensity of Pyrexia and Rapid Emaciation—Acuteness of Excavating Process
Rarity of Miliary Tubercle - Absence of Fibrosis Frequent Occurrence of
Pneumothorax--Limitation of Disease to Lungs - Examples.

ACUTE phthisis, or scrofulous pneumonia, is characterised by
a well-known group of pyrexial and *consuming* symptoms.
The patient, generally young, may or may not have had
cough or hæmoptysis previously. He is then attacked by
what appears to be pneumonia of one or both lungs. The
pulse becomes rapid, from 120 to 140, hard at first; the
temperature rises to 100°, 101°, 103° and 104° F., and even
higher; the skin is pungent and burning to the touch, and
especially to the ear of the auscultator, but nevertheless
there are chills and sweats at night. The tongue becomes
red and beefy in appearance. There is marked anorexia,
with nausea, vomiting, diarrhœa, and thirst. The wasting is
rapid and progressive; the cough is continuous and distressing; the expectoration abundant, opaque, shreddy, and
purulent, and on microscopic examination is found to contain
lung-tissue in large quantities and abundant tubercle bacilli.

The flesh wastes and the strength ebbs away. On examination of the chest at first we find nothing, then appears
crepitation, under one or both clavicles, followed by dulness,
which soon extends over the whole of one or both sides,
varied occasionally with spots of cracked-pot sound.

Auscultation reveals coarse crepitation with gurgling in
parts. Bronchophony is general at first, but gives way to
islets of pectoriloquy, or to cavernous whispering sounds,
and the physical evidence is soon complete that the lung is
quite disorganised. In many cases the wasting continues,

though the temperature may fall, the pulse and respiration continuing rapid ; and in these patients the duration of the disease is a short one, often not exceeding a few weeks. Where, on the other hand, the appetite improves, it not uncommonly happens that, after the destructive lung pro- cesses have abated, leaving those organs mere wrecks, the temperature falls, and even assumes the collapse type ; the pulse also diminishes in frequency and becomes weak, though it rarely resumes the normal standard. In this feeble state, with the greater part of the lungs excavated or infiltrated with caseous or tuberculous masses, and his body worn to a skeleton, the patient may even fatten and gain strength, if his appetite be good. The temperature may fall to the normal in the morning, though there is generally an indication of an afternoon rise ; the pulse generally remains quick, and the respiration somewhat hurried. In this precarious condition, cases of acute phthisis sometimes last on for a considerable time, though as mere shadows of their former selves, and liable at any time to be carried off by the various accidents, as they are termed, of the disease, viz., hæmoptysis, diarrhœa, or pneumothorax. Generally, however, the duration is a short one, and the patient sinks, worn out by the cough and febrile process.

On *post-mortem* examination we find the lungs to be the principal seat of the morbid changes, and these are for the most part of an inflammatory nature. Whole lobes are in- filtrated with caseous pneumonia having a greyish-white appearance, some portions caseating and others undergoing excavation, and thus are formed those large thin-walled cavi- ties which often give rise to nearly the same physical signs as pneumothorax and occasionally, through perforation of the thin wall, pass into it. Sometimes the greater part of one or both lungs is in a state of croupous pneumonia, with caseating centres scattered throughout, showing a strong tendency to coalesce and to reach the pleural surface, where they are often to be detected before breaking down has taken place. Often miliary tubercle is entirely absent, as also fibrosis, though the former, if the case is prolonged at all, is pretty sure to appear as a secondary infective product. In the most rapid cases the pleura is non-adherent, and hence the tendency to pneumothorax.

Let us now take a typical case of scrofulous pneumonia, or galloping consumption, which ended in pneumothorax.

CASE 55.—Mary Ann M., aged 23, single, a servant, born in Gloucestershire, was admitted into the Brompton Hospital on June 29, 1871. Her sister had died of phthisis. She had enjoyed tolerable health till six months before admission, when cough came on, accompanied for the last three months by muco-purulent expectoration, slight loss of flesh and appetite, and by night-sweats. Lately she had had pains in the left side, and some dyspnœa on exertion. No hæmoptysis. Appetite bad; tongue very red, slightly furred in the centre. Catamenia absent for six months; bowels regular; pulse 133, weak; respirations 31. Temperature (morning) 100·4° F.

No physical signs.

At the end of a week I examined her chest and still discovered nothing. The temperature was taken twice a day, and varied between 100° F. and 103° F.; the pulse was 140, the tongue continued red, and the patient wasted visibly. Cough very troublesome.

July 13.—Patient still very feverish. *Crepitation audible under the right clavicle.*

August 1. Cough very troublesome; expectoration scanty; temperature 103° F.; pulse 132; respirations 48. *Crepitation now heard under both clavicles and at the base of the left lung.* Tongue beefy.

August 21.—In the last fourteen days the temperature had fallen somewhat, and had ranged from 100° to 102° F. Pulse from 116 to 128; respirations from 36 to 48. Emaciation considerable; *cracked-pot sound and gurgling detected under the left clavicle. Crepitation audible under the right.*

A week later *cavernous sounds were detected under the right clavicle, and scattered crepitation over the lower portions of that lung.* By degrees the cavity on the left side appeared to enlarge, the patient assumed all the appearance of advanced disease; night-sweats became profuse and wasting well marked. Nourishment, in the form of beef-tea, egg-flip, and milk, was taken regularly, and there was entire absence of diarrhœa.

On October 1 she complained of sharp pain under the left mamma, which did not yield to local applications, and on examination signs of pneumothorax were detected. The tongue became aphthous; she showed inability to take food, and gradually sank, and died on October 11, having survived the appearance of the pneumothorax ten days.

My clinical assistant, Mr. William Rose, kept a regular and careful record of the pulse, respirations, and temperature for two months, and this showed that the pulse varied from 104 to 140, the average being 120, the respirations from 32 to 48; the temperature, taken always twice daily, frequently oftener, from 99° to 103·4° F., the former point being only once recorded, when death was at hand; as a rule the range was between 100° and 103° F., the evening being slightly higher than the morning temperatures. The treatment was chiefly directed to subdue the fever and support the patient. For the first purpose salines, quinine, digitalis, and sulphurous acid were given, and with little or no effect. The appetite continued good till aphthæ set in, and large quantities of nourishment were taken, and failed to arrest the progressive emaciation.

The necropsy was made forty-eight hours after death. The body was greatly emaciated; rigor mortis was slight. The left pleura was adherent at the upper third; the lower two-thirds contained about a pint of offensive purulent fluid. The left lung communicated with the pleural cavity by two openings on its anterior surface, one an inch in diameter and very ragged, and the other somewhat smaller. The pleura also contained several loose bands of adhesion. The lung itself was in a state of caseous pneumonia, riddled in parts with cavities, the upper lobe containing one extending throughout its whole length, and communicating by the above-named openings with the pleura. The right lung was adherent at the apex. The upper lobe was full of caseous pneumonia, and contained a small cavity at the top. Caseous masses were found in the middle lobe, along its upper border, extending seemingly from certain centres. The lower lobe was much congested and contained a few miliary tubercles. The heart weighed 7 oz. and was fatty, but the valves were healthy. The liver was enlarged, of nutmeggy consistence, and weighed 3 lb. The spleen was lardaceous. The kidneys and other organs were healthy.

This was a fair instance of scrofulous pneumonia, and exemplified the striking features of this variety of consumption, which may be summed up as follows :—

1. The intensity of the pyrexia and rapid emaciation.

2. The acuteness of the disorganising processes, due probably to the swarms [1] of tubercle bacilli present, excavation quickly succeeding consolidation.

3. The large part played by inflammation and the rarity of miliary tubercle.

4. The absence of fibrosis.

5. The occurrence of pneumothorax owing to two causes : (1) the size of the cavities formed, and (2) the absence of pleural adhesions.

6. The tendency of the disease to localise itself in the lungs, and the freedom of other organs, such as the intestines and the serous membranes, from secondary infection.

The following cases are also instances of caseous or scrofulous pneumonia, in two of which the disease came to a standstill for a long period :—

CASE 56.—*Acute Caseous Pneumonia.*—S. N., aged 18; admitted into

---

[1] This case was taken previous to Koch's discovery, but I have detected swarms of tubercle bacilli in four somewhat similar examples of scrofulous pneumonia.

University College Hospital, under Dr. C. J. B. Williams, November 23, 1839. Cellarman. Always weakly. In last six months overworked in wine-cellars, damp and draughty, and kept up at night at the bar. Three weeks ago became suddenly weak, with loss of appetite, thirst, and violent cough, and mucous expectoration. Ten days ago pain of left side came on, and patient since been confined to bed. Has lost much flesh and strength. Expectoration was streaked; now is viscid and opaque. Pulse 84; respiration 27. Urine high-coloured. *Dulness in upper left, most front. Cracked-pot stroke under clavicle, and loud tubular sounds. Breath-sound bronchial, with some dulness in left back, except the base, which is quite dull, and œgophony is heard in mid-region. Breath puerile in right lung.*— 18 leeches to the right chest, calomel, James's powder, and opium every night. Senna draught in morning; nitrate and tartrate of potass, in camphor mixture, three times a day.

November 26.—Much relieved in breathing and pain, especially since blister on the 23rd. Still much cough and expectoration, but latter less viscid, and no blood. Pulse 120. *Less dulness in left back, and more breath-sounds.*

December 5.—In last week cough relieved by eruption on side, produced by tartar emetic ointment. *Left front of chest still dull and collapsed.* Pulse 96.

December 17.— Cough easier, but weakness increasing.—Ordered iod. potass. in infus. cascarillæ.

December 24. —Cough increased, with rusty sputa, and pains on both sides, with increasing weakness. *More crepitus and dulness in posterior regions.*— Ordered antimony and henbane mixture, instead of cascarilla, &c.

December 31.—Has been better in every respect, and feels stronger. Cough and expectoration diminished, but the latter still rusty. Squills substituted for antimony.

January 14.—In last few days more cough, rusty expectoration, and increasing weakness. *Loud amphoric breathing, pectoriloquy and gurgling in left front.* Pulse 120. Occasional night-sweats.

Continued to get weaker, with harassing cough and copious yellowish clotty expectoration, and died on March 11.

Examination 48 hours after death.—Great emaciation. Left pleura firmly adherent throughout. A large cavity in anterior part of left lung capable of holding half a pint of fluid. Its anterior walls were little more than the adherent pleuræ; in other parts the surface was irregular, with some bands stretching across, and contained muco-purulent matter. The upper and posterior part of this lung was in a state of grey consolidation, here and there mottled, and with small excavations communicating with the large cavity. The base of the lung was firmly adherent to the diaphragm by a large mass of organised lymph, which contained in its interior opaque patches of yellowish-white colour, some tough and some softened (yellow tubercle). Right lung in first stage of pneumonia, with several patches of yellow tubercle, some crude, some soft. No adhesions in right pleura. Mesenteric glands much enlarged in parts, with patches of crude yellow tubercle. Mucous membrane of larynx and trachea rough, red, and partially thickened with numerous isolated pits, apparently ulcerated follicles ; and several were found also in the bronchi of the left lung.

CASE 57.—*Phthisis. Acute Caseous Pneumonia.*— Fred. R., aged 38, tailor, admitted into University College Hospital, March 20, 1840, under Dr. C. J. B.

Williams. In good health till last five months, when he was out of employment and living badly. He then began to cough, and soon to expectorate, and lose flesh and strength. A fortnight ago the cough became much worse, and he brought up about a tablespoonful of blood, which has recurred nearly every morning. There is also pain in left side; thirst and increasing weakness. Pulse frequent. *Whole left side more or less dull, and with crepitus superseding breath-sound. Crepitation also in upper right.* Was cupped ten ounces, and ordered tart. antimony.

April 2, 1840.—Pain and cough better. Vomited several times after medicine. Sweats at night. Pulse 108. Ipecac. wine and tinct. camph. co. substituted for tartar. emetic.

April 7.—Cough has again become severe, and breath more tight. Has been blistered and again put on antimonial treatment. *Left side dull (except at apex), with mixed crepitus and bronchophony. Dulness and coarse crepitus at right apex.*

April 21.—Has continued to get worse in spite of all remedies; breath very short, cough urgent, and now the sputa have become partially rusty. Pulse 120. Veins much distended. Distress great. To be bled to six ounces. Ammonia and wine.

Some relief followed from the bleeding, but the weakness increased, with inability to expectorate.

Died on the 25th.

Percussing the chest below the left clavicle elicited the cracked-pot sound, which could be traced to the mouth of the subject, showing that it was caused by succussion in the cavity communicated to the air in the trachea.

Both lungs adherent to the ribs and diaphragm, most firmly posteriorly and below; but some recent lymph was found at left base. In the substance of both lungs were consolidations of various sizes- those in the left pervading many lobules, and of the density of liver; some small as miliary tubercles, harder than the others. Parts of the general consolidation were yellow, opaque and softened at points. At the summit of the left lung was a cavity of the size of a small orange, and some smaller ones below. At the summit of the right lung were several solid grey masses, with small yellow patches of softening in some. The lung-tissue between the masses was generally red and congested, and in parts hepatised. Bronchial membrane generally much injected. Many dilated air-cells in the marginal lobes.

The consolidations of the lungs in this case presented every gradation between recent acute hepatisation and the grey nodules commonly called tubercular, and the change to the opaque state (caseation) was seen in parts of both.

CASE 58. --*Scrofulous Pneumonia. Cavity. Death from Hæmoptysis.* - Mr. J. V., aged 28, seen by Dr. C. J. B. Williams, April 9, 1867.—A brother had hæmoptysis. Devoted to tropical horticulture; has travelled in Australia and China. Cervical glands occasionally swollen. In the end of February caught severe cold with cough, which has been increasing with tinged expectoration, pain in the left chest, high fever, and extreme weakness and wasting. Pulse 120; skin hot, Temp. 104° F. *Dulness, deficient breath, crepitation on deep breath, in whole left, except apex. Reedy bronchophony in mammary and scapular regions.* Blister to left side; nitric acid, calumbo, and glycerine morning and midday; effervescing saline with opiate, evening and night.

May 31. – Fever, hard cough, and viscid expectoration continued a fortnight, requiring repeated blistering and continued saline. Then expectoration became purulent, cough looser, and temperature lowered. Cod-liver oil was then added to the morning doses, and improvement soon followed. *Cavernous sounds in mammary region. Some breath above.*

October 15. –Has steadily continued the oil with phosphoric acid and hypophosphite of soda; quinine and iron being added at times. Fever and sweats have long since subsided ; the cough and expectoration are much moderated ; appetite is good ; ten pounds have been gained in weight, and he can walk a mile. *Dulness and collapse of left front, with croaky cavernous sounds, most marked in mammary region. Moderate dulness and sub-crepitus with obscure breath posteriorly and above. Tubular sounds above right scapula.*

June 2, 1868. Wintered at Hyères. Able to be much in the open air, and steadily improving the whole time, taking oil with phosphoric acid and hypophosphite of soda, and sometimes quinine. Weight has increased from 9 st. 9 lb. in October, to 10 st. 7 lb. in April. Cough and expectoration gradually diminished ; and have been trifling in last three months.

Collapse and dulness with cavernous sounds in left front, most marked at third, fourth, and fifth ribs. Clearer, with a little breath above and in upper dorsal region; more dull and obstructed below. Tubular sound still in upper right.

September 23.—Quite well through summer, except shortness of breath and slight cough with opaque expectoration. *Cavernous sounds are rather higher up, and heard in the scapular region also, but less. Heart drawn up about an inch. Rather less obstruction in lower dorsal region.*

May 5, 1869.—Again wintered at Hyères, regularly taking oil and hypophosphite. Weight has risen to 11 st. 1 lb., and he has walked four or five miles daily.

June 25.—On his return home he attended to business in nursery grounds and hothouses, and lived more freely, and at end of May coughed up 6 oz. of blood, and a less quantity for three days, with pain in right side, fever and viscid expectoration. *Fine crepitation in lower two-thirds of right lung.* After a blister and a few days with antimonial saline, this attack of pneumonia subsided, leaving him weak, with loss of 13 lb. weight. *Now breath pretty clear throughout right lung, but large tubular sounds above right scapula. Cavernous sounds on left side appear to be still further drawn upwards and backwards, and the heart occupies the front up to the third rib.*

August 25. ·Has gradually improved, having lost all pain and hardness of cough, and has regained some strength and flesh. *Signs in left chest much the same. There has been some crepitus at right apex, but that is no longer heard.* Is bilious, and requires occasional omission of oil, and a dose of blue pill, &c.

October 11.–Has been distressed, and called into exertion by the death of his father. More cough, expectoration, and weakness. Has required blistering with acetum cantharidis. *More crepitus in left lung, but no extension of cavity.*

May 31, 1870.--Wintered again at Hyères, taking oil with strychnia and phosphoric acid and hypophosphite mixture ; and has gradually improved, gaining 12 lb. in weight, and some breath and strength, but not up to the point of this time last year. *Dulness and collapse of left front as before, but the cavernous sounds are more croaky and muffled. A little crepitu also above right scapula.* Still occasionally bilious, as at the present time. Treatment accordingly.

August 1.—Has had a slight recurrence of hæmorrhage, the bowels being confined at the time. Since has frequently taken sulphate of magnesia in mornings, and has improved, and breath and cough are better. *More breath-sound at left apex. Cavernous sounds croaky and dry in scapular region. Crumpling crepitus below. No crepitus at right apex, but heart-sounds are loud there.*

In the following week, during Dr. Williams's absence from town, hæmorrhage came on, became profuse, and death followed in two days.

This case at first appeared to be a hopeless one of galloping consumption, and was pronounced to be so by a physician who had been consulted. The great and continued improvement afterwards was very remarkable, and might have been more permanent, but for the patient's too early return to active business and his subsequent heavy trial. Although at first the disease appeared to be scrofulous pneumonia, the subsequent affection of the apex of the right lung was probably due to the formation of miliary tubercle.

CASE 59.—*Scrofulous pneumonia. Cavity. Arrested.*—Mr. F. W., aged 24. January 14, 1867. Mother scrofulous. Is reported to have been quite well till October, when he caught a severe cold, followed by pain in left side, cough and expectoration, which lately has become very clotty and opaque. Has lost much flesh and become extremely weak, with fever and profuse sweats at night; much annoyed with piles. Pulse 110.

Dulness and obstructed breath through whole left chest. Large tubular sounds and coarse crepitation in mammary region, and less above. Tubular sounds above right scapula.—To take oil with phosphoric acid and strychnia. To paint with the tincture of iodine. Morphia linctus at night.

July 12.—Has vastly improved in all respects. Gained 21 lb. in weight. Has long lost fever and sweats. Cough and expectoration moderate. In last two months quinine has been added to tonic. *Still much dulness and obstruction in left chest ; but no crepitation. Dry cavernous sounds in left mammary region. Tubular above.*

October 10.—Except a bilious attack, requiring a few doses of aperient and the suspension of the oil, &c., for a week, improvement has been constant. Gained 8 lb. more. Has lately suffered from bleeding piles. *Cavernous sounds in left front and upper back. Breath still obstructed below.*—To take electuary of senna with sulphate and bitart. potass. Continue oil and tonic.

March 30, 1868.—Has wintered at Hyères, and generally well, having been regularly out for exercise. Had gained 8 lb. up to January. Then suffered from piles, and left off the oil for five weeks, and lost 9 lb. in weight. Now has a fistula. Cough and expectoration still, but moderate. *Much less dulness and obstruction in left side. Cavernous sounds obscurely heard in mammary and subclavian regions. More tubular sounds above scapula, with weak breath below.*

October 13. - Has been well and active and gained 6 lb. Cough ceased for three weeks in August. Fistula still open. *Cavernous sounds heard only on deep breath, which causes a 'squash,' and some crepitus posteriorly.* Taken oil once daily.

April 24, 1869. -At Hyères, and out all the winter, though often wet. Gained 1½ lb. and much strength. No cough, except in mornings, with a little dirty expectoration. Fistula has discharged more, till touched with caustic. *Obscure pectoriloquy with croak on deep breath in upper left. Partial obstruction and dulness below. Tubular sounds above right scapula, but surrounded by good vesicular breath-sound.* Enlarged bronchial gland.

September 23.—Has been well and active in his business (builder), till six weeks ago ; caught cold and has had more cough and expectoration, which is now quite yellow. Has lost 3 lb. *Signs much the same.*

March 31, 1870. Wintered well at Hyères, although a bad season. Has taken three tablespoonfuls of oil with tonic once daily. Gained only 1½ lb. *No material change in signs.*

August 6. Continues well ; about the same in flesh ; but stronger and in active business. Pulse 72. Little cough or expectoration. *Cavity smaller, and with only crumpling crepitus around.*

February 14, 1877.—Last year had profuse hæmoptysis, but recovered, and cough has been troublesome since. Left chest is much contracted, measuring two inches less than the right. The right lung is drawn across the median line and the heart's apex raised. *Crepitus audible at base.*

This case had all the aspect of acute phthisis from scrofulous pneumonia, until arrested by treatment. The fistula had probably a salutary influence, and thus the patient acquired sufficient health to return to his work as a builder and contractor, and to live at least ten years from the first symptoms. The cavity contracted, causing displacement of the adjacent organs.

CASE 60.—*Scrofulous Pneumonia. Cavity. Arrested.*—A medical man, aged 30, whose paternal aunt had died of consumption, saw Dr. C. J. B. Williams on October 14, 1859. In 1856 he had a bad attack of scarlatina, and has not been strong since, but remained in laborious practice till March 1858, when he was attacked with severe caseous pneumonia of the right side, followed by copious purulent expectoration, sweats, emaciation, and other signs of acute consumption. Generally improved, and wintered at Malaga, but suffered much from cough and expectoration. Occasional hæmoptysis to the amount of ℥iv. After a time he crossed to Tangiers, and there improved a little, but remains very weak and thin, and lately has been suffering from nausea and hoarseness. 'Cannot take oil.' *Collapse, deficient motion, dulness, large tubular (cavernous ?) sounds in right front ; less dulness, some breath-sound audible, but mixed with crepitus, in back.*—Ordered oil in tonic of phosphoric acid, strychnia and tincture of orange and the use of a cantharides liniment. To winter at Hyères.

October 24, 1863. -Wintered at Hyères, steadily continuing oil and tonic, and much in open air. Has improved so much in all respects that he has returned there every winter since, practising as a physician. Passes summers in North Wales with further improvement in flesh and strength. Breath better, but still short on exertion, and has always had some cough, expectoration, and hoarseness. Last winter, after going out at night, had another attack of the right lung, which weakened him much ; but he gradually recovered, and again gained much ground during the summer in North Wales, although the weather was wet. *Dulness, collapse, deficient motion and breath throughout right chest. Loud tubular or dry cavernous sounds above. Coarse crepitation below.*

June 2, 1867.—Still living at Hyères in winter, and in North Wales in summer. Has steadily improved ; is stouter than he ever was, and has little

s 2

cough or expectoration. Last winter expectorated some calcareous matter. Still collapse, deficiency of motion, dulness of whole right side of chest, but dulness diminished in anterior portion. Some breath-sound, with sub-crepitus is heard, especially under clavicle, at scapula, and in lateral region. Strong percussion gives a raised pitch note in upper half. Tubular and cavernous sounds to be heard nowhere, except tubular above the right scapula. Left lung quite healthy.

1871.—Remains very well in health, although he has been much tried in mind and body of late. Oil has been taken at least once a day regularly. Had charge of an ambulance during the second siege of Paris.

1887.—Has continued at Hyères, and the lung symptoms appear to have remained in complete abeyance, but he has had symptoms of renal calculus and pyelitis. This has greatly enfeebled him at times, but he has with great fortitude persevered in the practice of his profession, though confined to his room at times. He has quite lost his cough and expectoration. Twenty-eight years have elapsed since his first symptoms of phthisis.

CHAPTER XVIII.

SCROFULOUS PHTHISIS. CATARRHAL PHTHISIS.

Scrofulous Phthisis Affections of joints, bones, &c.—Fistula -Caseous glands — Rindfleisch's large cells- Action of tubercle bacilli—Correlation of scrofulous discharging surface and pulmonary symptoms—Fitful Temperature course—Rapid loss and gain of weight -Hereditary character of Scrofulous Phthisis—Tendency to excavation—Prognosis—Two typical cases of Scrofulous Phthisis — *Catarrhal Phthisis* - Two modes of origin — First, Catarrhal Pneumonia—Second, Chronic Bronchitis Symptoms of Tuberculisation supervening—Localising of Bronchitic Sounds -Masking influence of Emphysema- Tendency to excavation - Prognosis.

SCROFULOUS PHTHISIS is applied to a large number of patients in whom consumptive disease of the lung is preceded by, or accompanies, scrofulous affections of various joints, caries of the sternum, ribs, and vertebræ, or lumbar and psoas abscesses, otorrhœa, fistula in ano, or, as is most common, enlarged and caseating glands, cervical, bronchial, axillary, and mesenteric. Rindfleisch explains the non-absorption of scrofulous matters by the presence in exudations of this character of relatively large cells with glistening protoplasm, and that the emigrated leucocytes, which pass from the blood-vessels of the inflamed part into the adjoining structures or into the lymphatics in scrofulous persons tend to grow larger on their way through the connective tissue. They appear to absorb albuminous substances, and in this very swelling slowly degenerate and die. The large size of the cells has been verified by Godlee, Schüppel, Green, and others. There is little doubt that these cells swell in consequence of irritation set up by the tubercle bacilli in their neighbourhood; these, being few in number, are not strong enough to produce complete destruction of the cells. Cases of scrofulous phthisis show an early infection of the lymphatic system and a remarkable correlation appears to be established between the external gland or discharging surface and the condition of the lungs. If the glands are suppurating, or if the fistula be open, or if the carious bone be freely discharging, the lung disease will remain quiescent and progress may be made towards arrest; but if, on the other hand, any of the above discharges should be checked or cease, the lung disease bursts forth into fresh activity, making considerable advance and extension. Nor is this wonderful when we remember that according to Koch's and other researches, tubercle bacilli are present in all scrofulous diseases and are to be found in certain discharges, as that from a strumous kidney, or from a fistula. The cough of scrofulous phthisis is generally accompanied by expectoration often streaked with blood, and hæmoptysis is not rare. The temperature course in these cases, where active lung changes are proceeding, is remarkably fitful, showing evening exacerbations of 102° F. to 104° F. and morning depressions of 96° F. to 95° F., and night-sweats are usually very profuse. Patients of this type lose and gain flesh with great rapidity, owing probably to the pyrexia and fitfulness of the appetite. It is not uncommon to lose or gain a stone weight in fourteen days.

In other cases the temperature pursues a quiet, almost sub-normal course, the symptoms being slight and the discharge persisting. In time these patients, however, pass into the general class of chronic tubercular phthisis, and at an early date manifest their markedly constitutional character. Ulceration of the intestine, setting up uncontrollable diarrhœa, is a very common termination of scrofulous phthisis, and after death it is not rare to find lardaceous disease of various organs, such as the liver, kidneys, and spleen, arising from the prolonged suppuration from the strumous lesions.

Scrofulous phthisis is strongly hereditary : it prevails chiefly among children under fifteen, as shown by Dr. Pollock,[1] many of these presenting the well-known strumous aspect, the clear complexion, enlarged glands, chronic in-

[1] Elements of Prognosis.

flammation of the eyelids, or discharging ears, and are attacked early with hæmoptysis, accompanied by cough and wasting. The course of the disease, probably on account of the relief afforded by the various discharges, is slow, and the patient lasts on for a considerable period; but, as might be expected, the development of the individual is slow and often stunted. It is very rare indeed to find active tubercular disease proceeding at the same time as extensive suppuration from an external gland.

The *post-mortem* appearances do not differ from those of ordinary phthisis, except in the more general complications of the lymphatic glands.

Prognosis.—It is rather difficult to give a certain prognosis in these cases, because, although improvement is often rapid, in unfavourable cases deterioration is equally so, and the future depends as much on the state of other organs as on that of the lungs. Where the appetite keeps good, and the discharge is not excessive, and the amount of lung involved is small, recovery may be expected; and I have records of cases of this type living and thriving ten, fifteen, and twenty-two years after the lungs were first attacked.

Happily most scrofulous phthisis is of this class, and does well; but in a certain number even slight changes in the state of the primary lesions are accompanied by marked pyrexia, wasting, and progress of the lung disease. The prognosis in these cases is unfavourable, for the fever generally persists as the appetite fails, and extensive excavation, causing increased pulmonary discharge, follows, and the patients become gradually wasted and sink from exhaustion.

The chief characteristics of this form of phthisis are—first, the relation of the lung symptoms to the state of the discharging surface; secondly, the tendency to excavation, but also to cicatrisation of cavities; thirdly, its distinct constitutional character, as shown by various organs becoming involved; fourthly, the rapid changes in weight, large losses and gains being very common and dependent for the most part on the state of the patient's appetite; fifthly, the great extremes of the temperature curve.

The subjoined are good illustrative cases of this form of phthisis :—

Case 61.—*Phthisis arrested. Scrofulous Abcesses. Death Ten Years after from Subacute Pleurisy.*—Mr. G., aged 17. Fair complexion ; family phthisical ; mother afterwards died of scrofulous pneumonia.

First seen by Dr. C. J. B. Williams November 4, 1848. In May returned from Germany, and a fortnight after caught cold and spat a little blood. Cough continued after, and in August brought up large quantities of blood. Was starved, leeched, and cold vinegar kept constantly to the chest. Is very much reduced in flesh and strength, and breath is very short. Pulse 120 ; skin hot. States that he has been under the care of an eminent physician, who declares his case to be hopeless, and that he cannot live six weeks.

Dulness, defective motion, and erepitus superseding breath-sound throughout right lung. Moist cavernous sounds, and most dulness in upper part. Some dulness and mixed erepitus at left apex also.

Cod-liver oil prescribed, with mixture of nitric and hydrocyanic acids, tincture of hop and orange peel : morphia linctus for the cough, and a blistering liniment.

In a week considerable improvement. Pulse reduced to 100, appetite good, cough and expectoration less, strength increasing.

Cavernous sounds now drier, and extend from right claviele to fourth rib.

July 10, 1849.—Continued to improve in all respects, although shut up in London through the winter. Has gained much flesh and strength, but breath still short on exertion. Cough moderate, with opaque expectoration. Has continued oil regularly. *Collapse, dulness, and small dry cavernous sounds below right claviele to third rib. Obscure vesicular breath below and in the back, with clearer stroke. Tubular sounds, but no erepitus at left apex.* A few days ago, after walking in the heat, fainted, and has spit some blood since.

He afterwards gradually improved in flesh and strength, and the cough became little troublesome, with only morning expectoration. The next winter was spent in the neighbourhood of London, and the following year he went to Torquay, where he continued to improve, riding on horseback. Last year one of the knees became painful, and the joint swelled much. It proved to be a scrofulous abscess, which opened and has been discharging since ; and another has since formed in the thigh, connected with diseased bone, and has discharged enormous quantities of matter. He was consequently so much reduced, that, when brought to London, Sir B. Brodie thought he would not live ten days. However, by resuming the oil with quinine, and partaking liberally of bitter ale and a very generous diet, gradual improvement followed in flesh and strength, and the discharge from the wounds diminished. The cough, which had entirely ceased, has lately returned, with some opaque expectoration.

Note of physical signs not kept.

May 31, 1854. Abscess in thigh quite healed, the knee-joint anchylosed. His health is fair, with little or no cough. The last three years have been passed in London or at Fulham. He has continued the oil and a liberal diet. Lately, after the exertion of throwing a large dog into a river, he brought up an ounce of blood.

Right front still flattened and rather dull, but much less so than formerly, with some puerile vesicular breath-sound. Tubular sounds only heard close to sternum. More dulness and deficiency of breath in dorsal regions, and slight crepitus in parts.

[A sad but remarkable expisode connected with this case deserves to be recorded. The patient's mother, who had most devotedly nursed him and dressed his wounds during his long and trying illness, was attacked with scrofulous pneumonia, which in a few weeks ended in excavation and death. Previously she had been quite healthy and strong.]

1858.—He continued in good health till the beginning of this year, when, travelling in Spain, he was attacked with ague at Madrid, and returned to England, broken in health and very bloodless. On May 7, Dr. R. D. Harling found him suffering from pleurisy of the right side, with effusion, which resisted all treatment; and as the extensive effusion threatened suffocation, it was determined to attempt relief by paracentesis. This was performed by Sir Prescott Hewett on May 17, with complete temporary relief, the fluid drawn off being clear serum, forming a fibrinous clot on standing. The dyspnœa, however, soon returned; and on the 20th the operation was repeated, a smaller quantity flowed, and was now somewhat turbid, but not purulent. In four days the breathing had again become as difficult as ever; but the patient refused to submit to another operation, and died on the 26th.

Unfortunately, *post-mortem* examination was not permitted, and the precise condition of the chest at last remains doubtful; but, after the evidence of advanced consumption at first, the wonderful recovery from this and from extensive scrofulous suppuration subsequently, the enjoyment of tolerable health for four years are all most remarkable; and no less so the fact, that the fatal attack brought on by adequate external causes did not present the purulent or consumptive character of his former disease.

CASE 62.—*Scrofulous Phthisis with Fistula. Long sea voyages. Recovery.* —Mr. ——, aged 21. Towards the end of 1860 had an abscess near the rectum, which has left a small fistula. After much anxiety during that winter, in April 1861, had left pleuropneumonia, which was treated by calomel and opium, salines, and repeated blisters. In June the fistula was still discharging. Much loss of flesh and strength. Still cough and purulent expectoration, and occasional recurrences of pain in the left side. *Dulness, deficient motion and breath in lower two-thirds of left chest. Loud bronchophony at and within left scapula; crepitus on deep breath.*—Ordered the oil with nitric acid tonic. A saline with opiate at night. Cantharides liniment to the side.

November 9.—Rapidly improved and gained flesh. Had lost cough, but it has returned in last week. Fistula still open. *There are now heard coarse crepitation in left back, and cavernous sounds at scapula. General dulness rather less.*—To continue oil tonic and liniment.

May 28, 1862.—Wintered in Madeira, and pretty well; but after fast riding once spit three ounces of blood; and in April caught cold, with increase of cough and expectoration. Taken oil regularly, and not lost weight. *Still much dulness and obstruction in left chest. Obscure cavernous sounds in scapular region. Large tubular sounds above right scapula.*

June 6, 1863.—Passed winter in Natal and Capetown, returning by way of Brazil. Has taken oil and tonic regularly. Quite free from cough and expectoration; and is in good flesh and strength. Discharge from fistula slight. *Signs of cavity more obscure, and some improvement in breath and percussion sounds.*

May 21, 1864.—In October went the voyage to Australia, and after spending a month there has just returned. Quite well the whole time, except slight hæmoptysis at Capetown. Gained 7 lb. Sometimes morning expectoration. *Loud tubular sounds at left scapula ; some breath-sound, weak in some parts and rough in others, on that side.*

May 14, 1866.—To Australia and back this winter, and to the Cape and Bombay last winter, taking the oil and tonic all the time ; and quite well. Gained 5 lb. in weight. *Left chest still rather duller than right, especially in lower part behind, which is contracted. Tubular sounds at scapula ; peculiar dry vesicular breath-sound below, especially on full inspiration (flaccid emphysema).*

May 12, 1868.- Last year took a voyage to the Cape, but this year has remained in London, quite well ; stronger and stouter than ever. Takes oil only at times. Fistula still open, but discharges little, and gives no inconvenience.

This gentleman has been so well in the last three years, that his going away during the winter was a matter of inclination rather than of necessity ; and the last winter was passed in London without inconvenience.

1871.—Heard of as quite well.

Catarrhal Phthisis.—By catarrhal phthisis I designate a class of cases which apparently commence in catarrh of the bronchi induced by cold or damp, creeping down into the alveoli, and thus originating catarrhal pneumonia, followed later through bacillar invasion, by implication of the alveolar wall.

It is the catarrhal pneumonic phthisis variety of many authors, and a large number of the patients of the Brompton Hospital are included under it. Out of one thousand private cases whose origin was investigated by Dr. C. J. B. Williams and myself, no fewer than one hundred and eighteen, or nearly 12 per cent., were of this form. Family predisposition was present in a few instances, but in a large number it was entirely absent, and as the patients belonged to a class, for the most part exempt from the depressing influences of bad air, or scanty or improper food, they may be taken as a proof of catarrh as a cause of predisposition in a considerable number of cases of phthisis.

Two principal modes of onset are to be noted in this form of the disease. The first is as follows :—A patient has an attack of pneumonia, generally one-sided and not necessarily accompanied by great pyrexia, and for the most part limited to one portion of the lung. He recovers from the attack ; but the dulness persists, and there is also some bronchophony and absence of vesicular murmur. He regains his general health, but does not lose his cough or expectoration, the latter becoming purulent; and sometimes the breathing is slightly impaired; though the existence of this last symptom of course depends on the number of alveoli blocked and the amount of surface impaired. The cough and expectoration continue, and the latter is found to contain bacilli, and these symptoms are followed shortly by loss of appetite, night-sweats, and wasting. There may also be rise of temperature and pulse, but this is not necessary. The chest is examined, and over the area of consolidation are detected signs of active disintegration; coarse *râles* and some cavernous sounds are audible. In a word, excavation has taken place in the old spots of catarrhal pneumonia. And here I may draw attention to the rule that the tendency of pneumonic consolidations to resolve seems to be in the inverse ratio of the fever and constitutional disturbance which accompany their development. After an attack of croupous pneumonia, accompanied by high temperature, involving the whole surface of both lungs and threatening the patient with fatal results, resolution, when it comes, is for the most part speedy and complete; whereas a slight and limited pneumonia, involving only the lower lobe of one lung, and hardly giving rise to any constitutional disturbance, is slow to disappear, and sometimes does not disappear at all, but remains—to percussion a dull spot, the nucleus of future mischief—perhaps to undergo disintegration and excavation. The subjoined is a good illustration of this form of catarrhal phthisis.

CASE 63.—Mary N., aged 39, married, with no children, was admitted, under Dr. Theodore Williams, into the Brompton Hospital on July 18, 1871. Her brother had died of some form of chest disease. Six weeks ago she had pneumonia of the right lung and was laid up for some weeks, and since that period has had cough. Four months ago, she stated, she caught a fresh cold, and since that date the cough had been much worse, the expectoration mucopurulent, and for the last two months streaked with blood. There had also been night-sweats, emaciation, and loss of colour. On admission her cough was troublesome, with abundant expectoration. The patient complained of pains in the left chest, most marked in the interscapular region. The appetite was moderate; the tongue clean; the bowels relaxed; the catamenia irregular; pulse 92; respirations 40; afternoon temperature 98·7° F.; weight 7 st. 1¾ lb.

Dulness over the whole of the right side of the chest back and front, with bron-chophony ; some flattening and deficient extension visible over the upper third of the right front ; cavernous gurgle heard below the clavicle ; and coarse crepi-tation from the third rib downwards to the base the sounds being more liquid in the upper than in the lower portions. Posteriorly somewhat finer crepitation was heard everywhere. Sonorous rhonchus in the left mammary regions.

The patient was treated with cod-liver oil, alkaline gentian mixture in the day, and effervescing ammonia draughts at night, and a generous dietary ; and after three months in the hospital, was discharged ' improved,' having gained 7½ lb. in weight. The cough and expectoration were less. *Crepitation had diminished in extent and intensity, but cavernous sounds were audible over a large portion of the right lung.*

This case exemplifies well the catarrhal pneumonic form of phthisis. The inflammatory origin was clearly proved (1), by the extent of the dulness and bronchophony; (2), by the unilateral character of the lesions; (3), by the history. The absence of any considerable amount of shrinking of the side distinguished it from phthisis originating from interstitial pneumonia, and of which we shall speak hereafter.

A second form of attack is the following: The patient, generally middle-aged, has been subject for years to winter cough and expectoration: and in many cases there is present a certain amount of pulmonary emphysema. In the summer months the patient is, as a rule, quite free, but some one winter, owing to being exposed to more unfavourable circum-stances than usual, he suffers a more severe attack of bronchitis, involving a larger surface.

After the severity of the symptoms has subsided, the cough remains ; the expectoration becomes more purulent and abundant, and is found to contain tubercle bacilli ; the pulse is quick ; and not uncommonly there is a rise of temperature in the evening. The patient loses a certain amount of flesh. The physical signs are at first those of bronchitis ; sonorous and sibilant rhonchus scattered over the whole surface of the lungs, succeeded at a later date by bubbling sounds, the percussion being resonant or hyper-resonant. These signs disappear from the greater part of the lungs, and remain fixed in one position, either over a whole lobe, or under the clavicle, or at the angle of the scapula, or below or above the scapula. The localising of the physical signs of bronchitis is always suspicious, and more especially so if any degree of

crepitus accompany the expiration, thus distinguishing it
from the pneumonic subcrepitant rhonchus. Next comes
slight dulness on percussion, chiefly to be detected by comparison with the note of the opposite side; or a whiffing
sound in situations overlying the principal bronchi, below the
clavicle, in the supra-scapular, interscapular, and scapular
regions, showing that the bronchial sounds are conducted to
the ear by some solid medium. Flattening of the chest wall
follows. The next step is increase of crepitation; then
the ominous croaking rhonchus; then gurgling, and often
the cracked-pot sound on percussion. A cavity has formed,
and these latter degrees of the destructive process are easily
proved by careful examination of the sputum, which rarely
fails to detect both tubercle bacilli and lung tissue. All these
signs may be traced in selected cases; but they are not
always present, and even if present, are not always detected:
for, on account of the emphysema present, many of these
are masked, and consequently overlooked. We often find
ourselves listening to gurgle and cavernous breathing where
a few days before we heard only sonorous rhonchi and mucous
râles. The great point in watching the physical signs is to
observe whether or no they become localised; for if they do,
we may know full well that the hour of tuberculosis has
come.

In many of the instances under my notice, there has never
been any physical signs of extensive consolidation during life;
but breaking down of the lung has occurred very rapidly and
cavities have formed in one or more portions; and though
the physical examination may teach us much, the only way to
be sure that this process is going on is to examine the sputum
frequently and carefully, both for tubercle bacilli and for lung
tissue.

Most cases of catarrhal phthisis gradually become those
of chronic tubercular phthisis, and consequently the *post-
mortem* appearances do not differ materially from those of the
latter variety. In early instances, however, as Rindfleisch
shows, where only certain lobules are invaded by catarrhal
pneumonia, centres of caseation may be traced in the track of

these lobules, and surrounding these, or following the line of the peribronchial lymphatics, are groups of miliary tubercle, thus showing the path of the bacilli.

The principal features of catarrhal phthisis are the tendency to excavation, and the comparative absence of fibrosis ; and to these are due its less favourable course and more limited duration, compared with that of other chronic varieties and especially the fibroid. Patients may remain with their lungs in a tolerably quiescent state for years ; but when, from some fresh exciting cause, more tuberculisation or excavation is induced, there seems to be no limiting power, the lungs become rapidly disorganised and the patient sinks.

CHAPTER XIX.

FIBROID PHTHISIS.

By the term fibroid phthisis I would signify cases of phthisis in which fibrosis is the main feature; and therefore I do not enter on the question of its being primary or secondary, though for the most part it was secondary in the instances I have seen. Fibroid phthisis would include all that Sir Andrew Clark[1] intended when he communicated his admirable paper to the Clinical Society, and in addition, cases where tuberculosis has supervened. There are two principal modes of origin of fibrosis: firstly, from attacks of pleurisy and pleuro-pneumonia or interstitial pneumonia; secondly, from chronic pneumonia resulting from long continued irritation of the lungs, through the inhalation of various kinds of dust and dirt, such as prevails among fork and knife grinders, colliers, and button-makers. In this group tubercle bacilli can be detected in the sputum. The history of symptoms in the first group may be sketched as follows:—

[1] *Clinical Transactions*, vol. i.

A patient has an attack of pleurisy with effusion, from which he recovers with absorption of fluid ; but percussion shows dulness over the whole side and somewhat feeble respiration. He recovers, but experiences dragging pains in the side, a dry, hacking cough, somewhat paroxysmal in character, with little expectoration, and the breathing, always short, becomes still more so on exertion. There is emaciation and often night-sweats. These symptoms increase, and in a few months later we find marked immobility of the affected side, still dulness throughout, and now considerable shrinking; the circumference of this side, whether taken at the level of the second rib, or of the nipple, or of the ensiform cartilage, measuring one or two inches less than the healthy side. On auscultation, we notice the breathing to be very deficient in some parts, and in others bronchial, and sometimes cavernous in character ; but generally there is everywhere absence of the vesicular breathing. Careful percussion of the opposite side of the chest shows the line of resonance to extend beyond the usual limit, passing to the edge of the sternum, and often an inch or two farther, demonstrating that the contraction of the affected lung has caused the healthy one to be drawn across. Other organs are also displaced. If the left lung be affected, the heart is tilted, not necessarily upwards as where a cavity is contracting, but outwards. The stomach rises, its note being audible as high as the fourth rib. The heart is not only displaced, but is uncovered by the retreating lung ; and the right auricle and ventricle are clearly distinguished by their pulsations, while the right lung is drawn across to the left side to the extent of one or two inches. If the right lung be affected, the left may be drawn over ; and the area of resonance may extend as far as the inner half of the right clavicle, and a line drawn thence sloping towards the middle of the sternum. The heart is transposed, and its impulse may be traced in the fourth interspace on the right side. The liver rises up to the fifth rib, and shrinking of the chest wall takes place as on the other side. The expectoration has been occasionally but not often found to contain tubercle bacilli. The pulse is generally slow, and the temperature seldom rises above the normal, and is often sub-normal. When the temperature rises above 100° F., it signifies that something besides fibrosis is going on. The cough is troublesome and often induces vomiting, and the expectoration becomes more and more difficult, and in time, on account of retention, fœtid. Meanwhile the dyspnœa increases to an enormous extent, the other lung becoming involved in the change; signs of obstructed circulation appear; dropsy of the extremities takes place, and rapidly increases; the urine becomes albuminous and of low specific gravity; and the patient dies either of dyspnœa or of blood poisoning, his death contrasting strongly with the ordinary termination of consumptive disease. After death, we find a lung contracted to the size of a man's fist, with enormously thickened and adherent pleura and widely dilated bronchi, with interlobular septa much increased in size and encroaching on the lung-structure, which seems to be replaced by a hard fibrous tissue, mottled with grey in parts, deeply pigmented, and resembling cartilage in its resistance to the knife. Embedded in this structure are found caseous and cretaceous masses, or again, excavations of various sizes; the walls of these and of the dilated bronchi being rigid and inelastic from the presence of the fibroid material, and thus affording some explanation of the difficult expectoration and consequently troublesome cough. Careful examination of microscopic sections of this fibroid tissue of the lung has entirely failed, according to Watson Cheyne and Percy Kidd, to detect tubercle bacilli, a conclusion confirmed by Sir Andrew Clark and others. In the caseous masses enclosed within the fibroid tissue, Watson Cheyne found tubercle bacilli. Besides these changes, we may find the other lung the seat of tuberculosis, though this is not usual; but commonly the bronchial glands are hardened and deeply pigmented. There is often lardaceous disease of the liver and spleen, and also of the kidneys.

The second form of fibroid phthisis results from the irritation to the lungs caused by prolonged inhalation of gritty or metallic particles, and prevails among knife and fork grinders, potters, colliers, button makers, pearl cutters, flax dressers, and

T

others, the changes in whose lungs have been well described
by Drs. Peacock and Greenhow.[1]   The *post-mortem* lesions
generally consist of deeply pigmented tissue, more or less
consolidated, containing large quantities of the particles
inhaled, whether these be coal, grit, or iron dust, some of
these being traced even into the epithelium of the bronchi
and the bronchial glands.   The lungs are largely invaded
by fibroid tissue and the pleuræ generally thickened and
adherent.   Tubercle bacilli were detected by Watson Cheyne[2]
in caseous masses in three cases of potters' phthisis, but none
in the fibrosis nor in that accompanying a case of miners'
phthisis in which the lung had undergone this condition.

The symptoms resemble those of the first form, in the contraction of the lungs and shrinking of the chest wall, but
owing to the great irritation to the bronchi and alveoli from
the presence of foreign particles in the lungs, the expectoration
is more abundant and contains a large number of tubercle
bacilli, as I found in a case of this kind under my care in the
Brompton Hospital.   This form appears to be more limited
to the lungs and not to affect other organs to the same extent
as the first form of fibroid phthisis.

Such is fibroid phthisis : but we must bear in mind that
its principal element is *fibrosis*, which has been described in
Chapter ii., and which must be looked upon as a secondary
product.

Thence we have a large development of fibroid tissue,
causing a thickening of the alveolar walls, and afterwards
filling up and obliterating the alveolar spaces themselves ; and
besides this, there is an enormous increase of the inter-lobular
connective tissue in all its various ramifications, and
especially in the parts surrounding the large vessels and
bronchi as they pass to the lobules.

Though the fibroid tissue is apparently the same in whatever part of the lung it may occur, it is not so in reality, for
Dr. Green[3] has shown that while the fibrosis of the inter-

---

[1] *Pathological Transactions*, vols. xii., xvi., xvii., and xx.
[2] *Practitioner*, April 1883, p. 294.   [3] *Pathology of Phthisis*.

lobular septa does not caseate and break down into cavities, that of the alveolar growth does, because the former is amply supplied with blood-vessels and the latter is destitute of them; and this gives an important means of distinguishing between the two. It is needless to say that cases of fibroid phthisis differ widely, according to the amount of lung involved, the presence or absence of cavities, the state of the bronchial tubes, and the amount of lung available for respiratory purposes, all which conditions materially influence the symptoms; and we must bear in mind that most cases of phthisis, especially chronic ones, have the element of fibrosis present in them, and that while, as in the subjoined case, it may strangle the patient by its overwhelming growth, in most instances it is the limiting agent to the spread of tuberculosis, and therefore highly to be desired.

We will now consider a typical case of the first form of fibroid phthisis :—

CASE 64.—Isabella C., a married woman, aged 29, a dressmaker, was admitted under my care April 18, 1871, and gave the following history. She had lost a cousin on her mother's side from consumption :—

For three years she had had cough and expectoration, and about two years ago noticed that she was emaciating. Hæmoptysis to the extent of ℥iv. came on soon after the beginning of the illness, and the same amount three months later, and again after two months, in February 1869, ℥ij. were coughed up; but none since then. She had had night-sweats, which ceased before admission, and she once, for two weeks, suffered from diarrhœa. Lately she had complained of increasing shortness of breath, which had become very troublesome.

In September 1869, she became an out-patient under Dr. Douglas Powell, who found flattening, dulness and cavernous breathing in the upper right chest anteriorly, and tubular respiration and moist crepitation over the right posterior base. He observed that the left lung was enlarged, and extended to the right of the sternum, and he detected crepitation at the apex on the posterior surface. Dr. Powell's diagnosis was a cavity in, and contractile disease of, the right lung and granulations in the left.

In October 1869, she was admitted into the hospital under the late Dr. Cotton, when the right lung was found to be in the same condition, but in the left, harsh blowing respiration had taken the place of the humid crepitus. She remained in the hospital for three months, taking phosphorated oil alternately with quinine and other bitters, and gained 7½ lb. in weight; the physical signs remaining the same. After this she continued better for some time, but the symptoms returned, when increasing shortness of breath caused her to apply to the hospital, where she was admitted under my care.

April 1871.—Her cough was very troublesome, expectoration frothy, tongue clean, bowels regular, catamenia absent for two years; countenance pallid, and showing great emaciation, especially of the thorax. The right side of the chest was considerably smaller than the left ; the shoulder greatly depressed, and the upper portion of the chest fixed. Marked depression was visible in the upper right front as low as the third rib, this space being decidedly resonant. *Cavernous breathing was heard under the clavicle, though chiefly below the sternal end. From the fifth rib downward there was absence of breathing, and marked dulness, not modified by change of position. Posteriorly, cavernous breathing was audible in the right inter- and supra-scapular regions.* The heart's sounds *were more audible over the lower portion of the sternum than in the cardiac regions, and the apex could not be felt ; it was therefore concluded that the heart was displaced towards the contracted side.* The diagnosis was : fibroid disease of the right lung, with some excavation of the upper portion, and displacement of the left lung, liver, and heart towards the affected side.

The patient was ordered cod-liver oil and quinine, but she did not improve, and on April 27, the evening temperature, which had been normal, rose to 100° F. ; 23 days later the morning temperature rose to 101° F., and the record then remained between 100° F. and 101·6° F., with the exception of one day, till May 4.

The patient became weaker, and the breath shorter. Rhonchus and harsh breathing were audible in the left lung.

On May 15, the morning temperature fell to 98·4° F. ; the evening to 99° F., the pulse being 100, respirations 40. After this the morning and evening temperatures varied between 98° F. and 98·5° F., and on May 20 the breath became very difficult ; crepitation being audible at the base of the left lung. Dyspnœa increased, and the patient died on the 25th.

The *post-mortem* examination was made by my then clinical assistant, Mr. William Rose, in the presence of Dr. Powell and myself, and the following record is taken from our notes :—

The body was greatly emaciated, there was marked concavity from the right clavicle to the second rib, and general shrinking of the whole right side of the chest. On removing the cartilages, the right lung could not be distinguished at first, but was afterwards found contracted to a third of its normal size, occupying the axillary and lateral regions, and not extending lower than the fifth rib.

The heart was drawn over to the right side of the sternum, the apex being slightly to the left of the ensiform cartilage. The liver, which was much enlarged, was drawn up, and, owing to the retraction of the right lung, had become superficial below the fifth rib. The left lung stretched across the median line as far as the costo-sternal articulations on the right side, this being the case as low as the third rib. The right lung, weighing 20 oz., was universally adherent to the chest wall, the pleura, with intervening adhesions, being on an average half an inch thick, and thickest over the diaphragm, where it contained in its interior a great deal of fibro-gelatinous material. The texture of the lung itself was dark grey, traversed by whitish fibrous bands proceeding from the pleura, interlobular septa, bronchi, and vascular sheaths. In the upper lobe were the remains of an old cavity into which some bronchial tubes opened, their mucous membrane seeming to terminate at its entrance.

The cavity appeared to have undergone contraction, and was divided by numerous fibrous septa, stretching across it, separating it into smooth-walled loculi which communicated with each other. The left lung was much enlarged and congested, the edges being highly emphysematous. It contained some cheesy nodules, around which were miliary tubercles of recent date. In the portion drawn over to the right side was a small cavity the size of a walnut, which corresponded to the area of the cavernous respiration heard under the right clavicle.

The heart was large, the right ventricle slightly dilated, valves healthy ; the liver and spleen were both lardaceous ; kidneys healthy.

In this case the shrinking of the right lung had caused several remarkable changes. The chest wall was contracted, the liver pressed upwards, the heart displaced to the right of the sternum, and the upper part of the left lung into the right chest. Owing to the portion of the left lung, which had been drawn over, containing a cavity, it gave rise to cavernous sounds under the right clavicle, from which it was naturally concluded that the cavity existed in the *right lung*. Whereas strange anomaly, it really was in the *left lung*, though in the right chest.

This portion of the lung would doubtless during respiration have reached further to the right side, and thus given rise to cavernous sounds over a larger area of the right chest than appeared in the *post-mortem* examination. The fact of the remains of an old cavity in the right lung might explain the cavernous sounds heard by Drs. Cotton and Powell at an earlier period, and it might be suggested that these physicians listened to the sounds from a different cavity to the one I directed my attention to, but this is improbable from the fact that Dr. Powell even then detected the presence of the left lung to the right of the sternum, and this displacement would hardly have occurred without a retraction of the right lung, including its excavated upper lobe, to the axillary and lateral regions of the chest, where, if anywhere, the cavernous sounds from this source would have been audible. The condition of the pleura indicated that the disease originated in an attack of pleurisy, and that the contractile tissue had slowly extended since, gradually obliterating the lung, and giving rise to increasing dyspnœa. Death was probably caused by congestion, and eruption of miliary tubercle in the left lung, by which the remaining breathing surface was further reduced in extent. The rise of temperature between April 27 and May 4 was undoubtedly due to the tuberculosis. The case may be considered a typical one of fibroid phthisis, the dyspnœa and the changes in the shape of the thorax being very well marked and the life-history and *post-mortem* records filling up almost every link in the evidence.

I have notes of a somewhat similar case, where the *left* lung was the seat of fibroid phthisis, and caused a remarkable amount of displacement of the neighbouring structures. In this patient the left side measured two inches in circumference less than the right; the spine was curved, the convexity being towards the right, the left shoulder was depressed, and the physical signs denoted an old cavity at the apex of the left lung, and nearly complete absence of breathing over the rest of the side. The right lung was drawn half-way across the left chest; the heart's apex tilted to the left, and the stomach raised as high as the fifth rib; the right lung was healthy. The patient died from abscess of the brain, and *post-mortem* examination exactly confirmed the diagnosis of the state of the lungs, indicated by the physical signs.

The left lung contained a small cavity at the apex, and consisted of a hard fibroid mass, impermeable to air, and occupying hardly a third of its proper area, the rest of this space being filled up by the encroachments of the adjoining organs, as had been described during life.

This patient lived four or five years from the date of the first symptoms, and might have lived much longer but for the cerebral disease, which seemed quite unconnected with the pulmonary lesions.

The *prognosis* depends on the amount of lung involved in the fibrosis and in the rate of its contraction. Where half a lung, or at the most one whole lung, is permeated by alveolar fibrosis, and where this process is limited by the white bands of interlobular tissue, where, in fact, matters remain at a standstill, there the patient may live for years, provided he limits his exertions in proportion to his breathing powers, and does not place himself under circumstances to induce a fresh inflammatory attack, or necrobiosis in the old fibrotic lesion. The patients who trace their disease to pleurisy, seem to have the best chance of life, for these cases last

longer than those arising from pure interstitial pneumonia.
I have records of two cases, one lasting ten years and the other
fifteen, each of which commenced in pleurisy. In both of these
excavation took place, but their health remained tolerably good
until the opposite lung was attacked. In other instances,
fibrosis has proceeded rapidly, permeating the whole lung,
and by obstructing the vessels, has induced dilatation of the
right side of the heart and dropsy of the abdomen and lower
extremities, the kidneys usually becoming affected towards
the close of life. It is remarkable how general the rule
appears to be that when fibrosis predominates over the other
morbid processes, the life-history of the case becomes at once
changed; the typical consumptive character is lost, obstruction to the respiration and circulation, with accompanying
dyspnœa and dropsy, close the scene, contrasting sharply with
the termination of catarrhal phthisis or scrofulous pneumonia.
It will be seen that the prognosis, therefore, depends on the
amount of lung involved. A good instance of fibroid phthisis
is subjoined :—

CASE 65.—*Fibroid Phthisis, with cavity, arrested fifteen years. Emphysema.*
—A merchant, aged 45, first consulted Dr. Williams July 13, 1855. Was well
till eight months ago, when he caught a severe cold, with cough and yellow
expectoration, which have continued with much loss of flesh, and lately with
very short breath. Took oil in the winter, but soon sickened of it. *Extreme
dulness on left side of chest, mostly in upper front, where there are large
tubular sounds, and some liquid rhonchus, almost gurgling. Less dulness
behind, but tubular sounds and mucous rhonchus.*— Ordered oil in a tonic of
nitric and hydrocyanic acids and strychnia, and counter-irritation with acetum
cantharidis.

April 19, 1856.—Wintered at Hastings, and very much improved under the
above treatment. Now walks six miles. *Still marked dulness and obstructed
breath in left front; but voice less tubular. Coarse crepitation in parts.*

November 28, 1856.—Passed summer well, but six weeks ago brought up
half-a-pint of blood and was freely leeched. Has taken oil regularly,
except during six weeks, and is quite stout. Breath still short. Cough
has increased in last ten days. *Physical signs the same, except the addition
of loud tubular breath above right scapula.*—To continue oil, but in tonic of
nitric and hydrocyanic acids, with iodide of potassium and tincture of orange.

April 28, 1857.—Continues well, and fatter than ever. Breath short, but
walks six miles. Just now has headache and increase of cough.

December 7, 1859. Passed last winter at Hull, and pretty well; but breath
short, and had morning expectoration. No oil for one year. Has lately had
more cough and slight hæmoptysis. *Dulness, cavernous croak and voice-sound
in upper left chest.*

February 21, 1861.— Has taken oil, but was shut up at Hull all the winter. Lost flesh, and lately appetite and strength. Cough increased, and occasionally streaked expectoration. *More cavernous sounds in upper left, front and back.* —Ordered strychnia with oil.

March 29, 1862.— At Hull through the winter, and not confined to the house, but breath shorter, and losing flesh since August. Three weeks ago had pain in left side, increase of cough and expectoration, which was more opaque. Symptoms relieved by blisters, and patient has resumed the oil and strychnia since. *Large tubular sounds and crepitation in upper left front and back. Cavernous sounds below clavicle.*

March 24, 1864.—Again recovered, and has been generally well. Has little cough, but breath very short. Weighs twelve stone. Attends to business. No oil for one year. *Less dulness, no cavernous sounds, but breath weak and subcrepitant. Tubular voice, and little breath at and within left scapula. Large tubular sounds above right scapula.*—Ordered nitric acid, tincture of nux vomica, and glycerine.

March 7, 1865.—Fatter and pretty well; but breath always short. Six weeks ago had hemiplegia of left side, and confused state of mind; but, after leeching and blistering, has recovered. *Moderate dulness, crepitation and croaky sounds over whole left chest. No large tubular sounds, except above right scapula.*

June 1, 1866.—Pretty well; but lately palpitation, and pain in left arm. *Breath-sounds feebly audible throughout left lung, only tubular in back.*

April 15, 1868.— Was pretty well; but during last year breath has become shorter, and palpitation has increased. Has little cough, with only transparent expectoration. No oil for two years and a half, and has spent the last winter at Hull. *More dulness in left front and upper back, and large tubular or cavernous sounds above left scapula and immediately below left clavicle; dulness and obstruction sounds in lower part of left lung, with some sibilus; more tubular sounds in upper right chest. Heart's apex drawn up, and beating at left mamilla; action weak.*

March 4, 1869.—Wintered at Hull pretty well, with little cough, but breath very short on exertion, and some palpitation. *Still moderate dulness and defective breath and motion throughout left chest, but no cavernous sounds, and tubular only at and above scapula. Heart-sound and impulse, high and feeble.* —To take three drops of liquor arsenicalis and two grains of hypophosphite of soda in a gentian and glycerine mixture, twice a day.

February 26, 1870.— Improved much in strength, and somewhat in breath during the summer, complaining only of the latter. Lately has suffered from fluttering at heart. Four months ago, for three days, had difficulty of articulation, but it passed off; only he complains of his memory failing. Lately closely confined to the house, and has indigestion, and become paler. Urine highly coloured and scanty. *Physical signs much the same. Liver rather full and tender.*—To take iodide of potassium and digitalis, with tincture of calumba, and a few mercurial pills.

May 26, 1870.—Health and appetite better; but breath very short, and urine becomes scanty when he leaves off medicine. About three months later the legs began to swell; and he died suddenly in a faint.

The remarkable degree of dulness in the left chest found on first examination makes it probable that the disease was inflammatory, causing a hard consolidation. This was afterwards partly softened and excavated, and partly absorbed and contracted. The general inference from several examinations (some of which, for want of space, have not been given) was, that emphysematous dilatation of the air-cells took place in the consolidated front of the left lung, rendering the stroke-sound more clear, whilst the breath and voice were obscured; but the signs of old cavity and consolidation were still heard in the scapular region. In the last two years the lung disease was stationary, and there was neither cough nor expectoration; but the heart showed signs of weakness, and death is to be referred to this cause. *Post-mortem* examination was not permitted. More than fifteen years elapsed since the first attack.

Laryngeal Phthisis is the form of phthisis in which, in addition to the pulmonary lesions, the larynx is the seat of tubercular attack and undergoes infiltration of its tissues with inflammatory exudations, such changes being set up by the tubercle bacilli. The larynx becomes the seat of extensive ulceration, the lesions found in the lungs being simply those of acute or chronic tubercular phthisis.

Phthisis, when accompanied by aphonia, is not necessarily laryngeal phthisis; many cases of aphonia in consumption are instances of hysterical aphonia, or of weakness of the laryngeal muscles, chiefly of the adductors, and in these, as a rule, the voice returns under the use of galvanism. Aphonia arises in other cases from catarrh of the larynx, from syphilitic disease, and various kinds of tumours; but these forms of aphonia are to be distinguished from the distressing one caused by actual tubercular disease of the larynx by laryngoscopic examination and by the course of the symptoms.

It is very rare indeed in phthisis to have the larynx alone affected, though it is possible that it may be the first organ attacked by the bacilli. Dr. Morell Mackenzie[1] states, that though he has detected disease in the larynx before it has been discovered in the lungs, he has only three times met with this lesion in the *post-mortem* room, without finding corresponding pulmonary disease.

I have never seen a case which I could clearly identify as laryngeal phthisis, where the lungs were not involved and generally to a large extent, though owing to the pain and difficulty the patient experiences in performing the acts of coughing and deep respiration, and owing to consequent deficiency of breath and voice-sounds in portions of the lungs, ausculta-

[1] Reynolds' *System of Medicine*. vol. iii.

tion is rendered unsatisfactory, and hence important lesions
are often overlooked. It may be safely stated, however, that
the majority of cases of laryngeal phthisis are associated
with advanced stages of pulmonary disease.

Laryngeal tuberculosis, according to the Brompton Hospital
Records, is found in no less than 54½ per cent. of all the *post-
mortem* examinations on consumptives, and is frequently
accompanied by intestinal ulceration even in cases not specially
marked by bowel symptoms. This coincidence points to both
being due to difficulty of expectoration, leading in one case
to lodgment of the bacilli-laden sputum in the larynx, and
consequent infection of that organ, while, if it, at a later date
be swallowed instead of being coughed up, it may and does give
rise to fresh tubercular centres in the stomach and intestines.

The *symptoms* are at first those of chronic laryngitis.
Aphonia gradually comes on in the form of increasing hoarseness, which becomes permanent, and the voice diminishes to a
whisper; and this, in spite of the struggles of the patient to
make himself audible. This aphonia may be due to loss of
power to adduct the cords, and not necessarily to ulceration of
the cords themselves.

The cough, which at first was probably the ordinary expectorating cough of chronic phthisis, becomes more troublesome, likewise paroxysmal in character and metallic in sound,
and less amenable to treatment. The expectoration does not,
as a rule, increase in amount, but contains more tubercle
bacilli, the secretion from the larynx being never very large.

The respiration is difficult, especially during swallowing,
and distinct fits of dyspnœa occur. The dysphagia soon
comes on, and is the great cause of distress, as it accompanies not only the swallowing of solids but also of liquids
even in sips. Pain, referred to the larynx, is much complained
of at this stage, and both this and the dysphagia are due to
the action of the constrictors of the pharynx, rather than to
any ulceration processes.

The patient also often complains of pricking along the
track of the Eustachian tubes, extending to the ears. And
his distress is sometimes very remarkable: he dreads swallow-

ing, and often starves himself to avoid it. To alleviate the
pain, he will sometimes, during the process of swallowing, hold
his head with his hands, but this does not prevent the convulsive cough and urgent dyspnœa coming on. As the disease
proceeds, these symptoms increase, the larynx becomes
tender to external pressure, the temperature rises considerably and is somewhat fitful in character, but preserves on
the whole the well-marked features of phthisis, *i.e.* the alter-
nate collapse and pyrexia. The pulse quickens and the respiration is, as a rule, only accelerated at certain periods. The
aspect of these patients is truly cachectic, the wasting is very
rapid. and night-sweats are generally present.

Many of the changes in the larynx can be traced by the
laryngoscope; we first, as a rule, see a general hyperæmia of
the mucous membrane, but this in time gives way to extensive swelling, involving the whole tract, which varies in
colour according to the part, but is generally pale in colour,
with occasional streaks of red.

According to Dr. Morell Mackenzie[1] there is nothing in the
early stages to distinguish the affection from chronic laryngitis,
but when exudation takes place the ary-epiglottic folds stand
out as pyriform masses, and, owing to the inter-arytænoid fold
being involved, approximation of the vocal cords is prevented.
This pyriform swelling he holds to be pathognomonic of the
disease.

Dr. Marcet[2] notes the presence, shortly after the swelling
of the larynx, of a white milky secretion, oozing out between
the arytænoid cartilages, due probably to commencing ulceration, and according to him, characteristic of the lesion. At a
later date ulceration becomes visible, although the process is
probably going on in parts of the larynx not visible by the
laryngoscope. The changes in the epiglottis often obscure
our vision of the laryngeal phenomena, for this structure
sometimes swells and becomes folded over. obstructing the
view; but when this is the case, we may conclude that active
changes are proceeding in the larynx.

The *morbid anatomy* of laryngeal phthisis presents two dis-

---

[1] *Op. cit.*   [2] *On Diseases of the Larynx*, p. 95.

tinct features—infiltration of tissues and extensive ulceration. *The first* is seen in the œdema and thickening of the interior of the larynx, especially of the ary-epiglottic folds, but more commonly in the swelling of the epiglottis, which becomes extensively thickened, and loses its normal contour and elasticity.

The second varies greatly in extent, from a superficial ulcer of the mucous membrane, to a deep-seated erosion, involving the cartilages. The favourite localities for ulceration appear to be, first: the vocal cords, then the inter-arytænoid fold, and next the epiglottis, especially on its under surface. Sometimes this and the upper part of the larynx present a worm-eaten aspect, from the number of small ulcers present, which has been assigned to the cascation and removal of the numerous glands situated in these parts: but generally the ulceration is far too extensive to be explained in this way.

Again, the sacculus laryngis may be found to contain an ulcer reaching to the cartilage, or what is considered somewhat characteristic, a number of pin-hole ulcers are found on the interior of the larynx above the vocal cords; the large crater-like ulcers are also usually found above the vocal cords, and at their anterior angles. The epiglottis sometimes undergoes more change in this disease than any other part of the larynx, sometimes one edge, sometimes both, are ulcerated and more rarely a V-shaped erosion forms at the tip and occasionally the epiglottis entirely disappears, leaving a small stump behind. As a rule, however, the ulceration is confined to the larynx or the upper portion of the trachea and to the under surface of the epiglottis, but sometimes it spreads to the upper surface of this structure, now and then invading the pharynx, on the posterior wall of which greyish ulcers of irregular forms are by no means rare; they are found occasionally also on the inner surface of the cheek.

In the larynx itself, the small ulcers are common on the thyroid cartilage, above and below the vocal cords, and look like minute pit-like depressions of the mucous membrane arranged more or less longitudinally, or, if extensive, they give rise to the worm-eaten appearance before described. Owing to the ulceration, the cartilages of the larynx, especially the arytanoid, are found in a state of necrosis due to inflammation and death of the perichondrium, and they occasionally present a blackened appearance, from their being laid bare and exposed to the air, through ulcerative changes. Occasionally portions of them are extruded into the cavity of the larynx, and coughed up in the expectoration. In the most advanced cases of laryngeal phthisis, the larynx is ploughed up by large ulcers, and the individual parts can hardly be recognised, the ulcers being quite as extensive as those found in laryngeal syphilis. Not uncommonly, at the bottom of the large ulcers, miliary tubercles are to be found, and careful examinations of sections of these ulcers, small and great, show the presence of abundant tubercle bacilli.

Dr. Percy Kidd has drawn attention to the existence of tuberculous tumours in the larynx in some cases of phthisis, and furnishes an interesting report of cases from his own practice and that of Dr. J. N. Mackenzie and Professor Schnitzler of Vienna. They varied in size from a pea to a hazel-nut, were generally rounded, sometimes single, sometimes multiple,[1] and originated from various parts of the larynx; from the ventricles in three cases, from the whole upper larynx in one case, the inter-arytænoid fold in two cases, the ary-epiglottic fold in one case, and the trachea in two instances. Ulcerative changes did not always accompany the tumours, but were found in a certain number. The tumours were all tuberculous and often were composed of miliary tubercle, and contained tubercle bacilli.

The *prognosis* of laryngeal phthisis is most unfavourable, as recovery is exceedingly rare and almost unknown, and death takes place, either in the ordinary way in phthisis, viz., from exhaustion, or from apnœa, or from both ; or again, from inanition, through the great dysphagia. No form of phthisis presents more thoroughly the aspect of advanced phthisis than this one. Tracheotomy is sometimes performed to relieve the urgent dyspnœa, but such relief is only temporary. The treatment will be considered under the palliative treatment of consumption.

[1] *St. Bartholomew's Hospital Reports*, vol. xxi.

CHAPTER XX.

CHRONIC TUBERCULAR CONSUMPTION.

Hopeful Prognosis—Good results of continuous treatment—Three cases of arrest of disease in First Stage—Case of arrested Second Stage—Seventeen cases of Third Stage arrested or retarded—Patients surviving first symptoms from three to forty years—Five cases of double cavity included—Case of contraction of double cavities, arrest of tubercular disease sixteen years, and death from renal disease—Case of arrested Third Stage—Calcareous transformation of tubercle—Death by intestinal perforation—Two cases of arrest of disease by mountain treatment.

CHRONIC TUBERCULAR PHTHISIS is too well known to require a minute description. It differs generally from acute tuberculosis and acute phthisis, not in the nature of the pulmonary lesions but in their extent, and principally in the rate of their formation. For while the sudden eruption of miliary tubercle or scrofulous pneumonia will speedily overwhelm a patient, a more gradual invasion of the lungs, by tubercle and the destruction of their tissue by softening and excavation of the same, will afford time for the constitution to react against the process, and under the fortifying influence of suitable medicinal, dietetic, hygienic, and climatic measures, present a limiting bulwark, in the form of fibrosis or emphysema, against further invasion. In course of time the tubercular mass is thus more and more cut off from contact with the blood-vessels and lymphatics, and either (1) completely undergoes fibrosis, or (2) some portions undergo fibrosis, while other parts caseate, and some cretify, but it is most usual to find a combination of caseous and cretaceous masses enclosed in fibroid material of great toughness and density, and not rarely surrounded by localised emphysema.

The residual portion of the lung in these cases often presents some hypertrophy, and thus compensates for some of the patient's lost respiratory area, but in other and unfavourable cases, when the fibrous bulwark is not so complete, tuberculosis takes place in the hitherto healthy lung, due to re-infection, and it is by no means rare to find one lung undergoing the changes of arrest of the disease, with a cavity contracting and fibroid metamorphosis, while in the other lung there is fresh eruption of tubercle.

It is in chronic tubercular phthisis that the greatest triumphs of treatment have been achieved, when time, money, and perseverance are not lacking, and we can in most cases promise prolongation of life, and in others, complete arrest of the disease. Such examples, as follow, demonstrate how the system of the tubercular patient will wage battle with the invading bacilli, and also how this contest may be converted into victory by the steady perseverance in those measures which tend to the conservation of the vital force and energy.

In the first edition of this work a large number of examples of arrest of phthisis were included, but from the introduction of so much new material, which recent researches have rendered necessary, it has been found needful to restrict ourselves now to a few striking cases of chronic tubercular phthisis, including examples of arrest of the disease in its various stages, and even in double-cavity cases, which we subjoin. These cases were, it need hardly be added, mostly taken before the modern tests of the expectoration, but the history and physical signs, and in some cases the *post-mortem* examinations, abundantly prove their tuberculous character.

Owing to some of the patients having been lost sight of, it has been impossible to bring all their cases up to the present date, and under these circumstances the latest report has been given. The duration is in many of them, probably, longer than that mentioned, but it has been deemed best to state only the last reliable information.

CASE 66.—*Chronic Phthisis—first stage.—Cured by enforcing treatment.*— Mr. ——, aged 23. June 19, 1863.—Quite well till twelve months ago, after close application in an office in Edinburgh, coughed up a dessertspoonful of blood. Soon began to cough, and this continued all the winter; worse in April, with expectoration, and rapid loss of flesh and strength. Is now returning from the south coast, weak, pale, emaciated, and in a most desponding state. Says that he cannot take cod-liver oil.

Much dulness, with loud reedy bronchophony and deficient breath in upper half of right chest, front and back. Tubular sounds within and above left scapula and below left clavicle.

As the only chance of saving him, he was urged to take the oil in a mixture of phosphoric acid, strychnia and orange. The chest to be blistered with acetum cantharidis, and a morphia linctus for the cough to be used at night.

July 1.—Takes the oil very well, and is already better. Only occasional nausea. *More breathing audible in right lung.* An occasional dose of blue pill and colocynth to be taken.

February 12, 1866.—Dr. W. heard that the patient had continued the oil and tonic ever since, and soon became wonderfully better. Has now hardly any cough, is regularly attending to business in Edinburgh, and is going to be married.

May 1871.—Heard from another patient that Mr. —— was quite well, and fully engaged in business.

The condition of this poor wasted young man, with his unhappy mother, was truly pitiable at the first visit. At the second, the unexpected discovery that he could take the oil had filled him with hope, and his amendment went on rapidly from that time to a happy issue.

CASE 67.—*Chronic Phthisis—first stage.—Arrested.*—A young gentleman, aged 15. May 11, 1857.—Maternal aunt died of phthisis. In last Christmas holidays, after a chill, had pain in right side, and some cough ever since. Ten weeks ago had a severe chill after playing at football, and since has suffered much from cough and weakness, loss of appetite and breath. *Dulness, defective motion and breath-sound in whole right chest. Tubular sounds at and above scapula.*—To take oil, with nitric and hydrocyanic acids in orange infusion, and use acetum cantharidis.

June 16.—Much better. Cough nearly gone. Says breath is not short now. *Still dulness, and tubular sounds at and above right scapula.*

November 22, 1866.—Took oil at times for several years, and much improved in general health and strength, but always found himself rather short-winded on exertion, and sometimes rather wheezy. Still he has kept his terms regularly at Cambridge, and studied with such success that he came out Senior Wrangler last winter. Has continued well till six weeks ago, when he caught a cold, and has had slight cough since, with an occasional feeling of faintness. *Bronchophony above right scapula. Percussion and breath-sounds generally clear.*—To take oil, with iron tonic.

1868.—Continues well. Has now become Fellow of his college.

May 3, 1871.—Has continued well generally, but had cough several times last winter. None lately. Is pale and out of condition. *Chest-sounds good generally, but still tubular sounds in upper right.*—To resume oil and tonic.

1887.—Alive and well, and actively engaged in scientific pursuits, experimenting, writing, and lecturing—thirty years after his first symptoms.

CASE 68.—*Chronic Phthisis first stage.—Arrested.* Mr. ——, aged 30. October 18, 1861. A brother and a sister had died of consumption. Another brother had hæmoptysis and calcareous expectoration repeatedly. Has been twelve years in Syria, and three years ago, when on Mount Lebanon, was chilled, and since has had cough, sometimes spasmodic and sometimes hacking. Is now pale and thin. *Dulness; tubular sounds at and above right scapula. Bronchophony above left scapula.* Oil ordered with phosphoric acid, and calumbo, and a croton-oil liniment.

February 25, 1862.—Improved on treatment, but lately has been much worried, and has now more cough and chest irritation. Has not used counter-irritation.—Ordered acetum cantharidis liniment.

September 2, 1864.—Passed two years in Cyprus, but never free from cough, and several times had hæmoptysis, once to amount of half pint. Now resides at Tangiers, but has been home for the last three months. Has lost some flesh. *Dulness, large tubular sounds at and above right scapula. Small tubular sounds above left scapula.*

October 10, 1865.—Has been in better health during the last year at Tangiers, taking oil regularly. Within the last few days he has spit a little blood.—Oil to be taken with phosphoric acid and hypophosphite of soda.

September 17, 1866.—Pretty well, but always has some cough. Has twice expectorated calcareous matter. Sometimes has bronchitis, as at present, after voyage home.

July 1868.— Has suffered more during the last two winters, from exposure at night at Tangiers. Last November had scarlatina, and expectoration has diminished since. Took oil and hypophosphite steadily, continuing them in the hottest weather, and has gained 10 lb. *Still dulness, tubular sounds at and above right scapula. Some tubular sounds in upper left chest.*

1887. – Still resides at Tangiers enjoying good health, with active habits. Twenty-nine years have elapsed since the first symptoms of disease.

CASE 69.—*Phthisis—second stage. Cured.*—Mr. ——, aged 38. May 5, 1864. Brother said to have lost one lung nine years ago, still living. Three years ago had perineal abscess, which healed in a week; but ever since has had morning expectoration. Not otherwise ailing till twelve months ago, when a cough came on, which has increased much in the last three months, and for two months he has been hoarse. *Dulness in upper half of right lung, with coarse crepitus. Tubular sounds above both scapulæ.* To take oil with phosphoric acid and calumbo. To blister with acetum cantharidis and inhale hot water with creosote, hemlock, and chloroform.

November 19.— Hoarseness gone and cough much better, with less expectoration. Breath still very short. Has inhaled constantly. *Signs not much altered.*

May 18, 1865.—Has wintered at Torquay, and been out daily. Cough and expectoration moderate. Breath better. Flesh and strength good.

Dulness, obstruction in right upper chest, with croak on deep breath, mostly in front, large tubular sounds above left scapula.

May 14, 1868.—Heard that he had, as advised, gone to Australia, where he had been quite well for two years, often riding fifty miles a day.

CASE 70.—*Phthisis—third stage. Arrested twenty years.*—A veterinary surgeon, aged 30. August 6, 1846. Six months' suffering from cough, copious expectoration, rapid loss of flesh, strength, and breath. Now sweats and diarrhœa. *Dulness in upper parts of chest on both sides; most right, where cavernous sounds are heard, with obstructed breath around.*

U

Ol. jecoris aselli, bis die : pil. cupr. sulph. c. morph. omn. nocte : liniment. aceti cantharid.

October 18.—Was blistered freely, and has taken oil regularly, increased to six drachms three times a day. Very much improved in every respect. Has gained 9 lb., can walk five miles, and can even run. Bowels now costive. *Still cavernous sounds in upper right, but no crepitus around, and more breath.*

March 6, 1848.—Was quite well, except breath rather short, and left off the oil twelve months. Fourteen days ago attacked with ' severe pleurisy ' of right side, after exposure to cold. Has been blistered, and is better, but very weak. *Still large tubular sounds above and below right clavicle, but percussion good. Loud sonorous rhonchus on deep breath. Slight crepitus at left base.*

July 29, 1850.—Took oil, &c., and soon recovered. Now walks twenty miles, and is stouter than ever ; but breath still short, and suffers from cold winds. *Tubular sounds above both clavicles, but the stroke and breath-sound clear in all other parts and expiration long.*

September 18, 1854.—Quite well, except occasional lumbago, and in last few weeks cough. No oil for nine months. *Some dulness, obstruction sound and croak in upper right.*

July 8, 1859.—Generally well, except occasional attacks of cold and cough, for which he blisters and takes oil, and is soon well, gaining 12 lb. or 15 lb. after illness. *Still some dulness and tubular sounds in upper right back.*

1868.- Heard continues well.

July 20, 1870.—Has been well, and attending actively to business as inn-keeper ; weight 11 st. In May got wet, and since has had cough, and lost flesh, but says his breath is good still. *Dulness and dry cavernous sounds in upper right, front and back. Good breath heard below.* Has taken oil only irregularly.

June 13, 1871.—Resumed oil and tonic steadily, and recovered flesh and lost cough ; but it returned in winter, with expectoration. He was shut up, and lost appetite and strength. Not so regular with oil as formerly, and has again lost much flesh. *Collapse below right clavicle, and harsh vesicular breath below second rib. Above and to mid-scapula loud dry cavernous sound. Obscure breath with subcrepitus below.*

August 7.—Gained 2 lb. Still cough and opaque expectoration, and cavernous sounds in right back.

This case records arrest of the disease for twenty years, with full enjoy-ment of life ; then renewal of disease in the same parts, and again ameliora-tion, twenty-five years after first attack.

CASE 71.- *Chronic Phthisis. Double cavity. Arrest of disease. Duration twenty-seven years.*—Miss — -, aged 21. April 27, 1849. Always delicate, but with much precocity and energy of character. Five years ago suffered for several months with cough and loss of voice. Well after till three months ago ; cough and short breath, and a glandular swelling above left clavicle. Not lost flesh. Has never worn flannel. *Slight dulness and tubular expiration above left scapula.* Iodid. potass. et sarza.; flannel.

May 30.—Glandular swelling diminished ; but still coughs, and has lost much flesh and strength. *Signs continue.* Oil, with nitric and hydrocyanic acid mixture ; capsicum liniment for the chest.

February 19, 1851.— Took the oil for six months, and lost cough and was quite well till last November, when suffered long from sore throat, and in the last month the lump has reappeared above left clavicle, with much depression of strength and spirits. This swelling afterwards threatened to suppurate, but subsided without discharging, and its subsidence was followed by a fever, called 'bilious.'

In 1854 the throat again became troublesome ; and then followed a cough, with opaque expectoration, which continued through a great part of the next year. Then were found *signs of disease in the left lung*, but the notes are not preserved. The oil and tonics were given as before, with the effect of removing cough and other symptoms.

In 1857, after much vocal exertion and exposure to damp, had an inflammatory affection of the left lung, with dulness and crepitus in the upper lobe, which lasted several months, but was at length removed under the use of oil and free counter-irritation ; but tubular sounds remained above both scapulæ ; and more or less cough and chest irritation continued during that and the following year.

In 1861 was so much improved in health and strength that she undertook the superintendence of a charitable institution, involving a great amount of mental and bodily exertion.

In 1862, in the midst of this work, she left off all stimulants, and soon had a bad carbuncle on the leg, which for several months caused much discomfort and weakness.

In 1863 began to suffer again in throat and chest, *with return of crepitus in left apex, which in the course of that year showed signs of excavation.* Remedies were then more steadily persevered with ; and two winters were passed at Torquay, with very beneficial results as to general health ; but the *left lung continued much obstructed, chiefly in its upper half, where there remained the large croaky sounds of a contracting cavity. The lower portion was imperfectly pervious, with emphysematous stroke-sound and crepitus ; and tubular sounds were heard at and above the right scapula.*

In the summer of 1867, feeling so much better, this lady could not be restrained from her beneficent work, and with a view to its continuance, remained in London during the winter. At the commencement of this season the trial of a coke-stove in her room brought on a severe attack of suffocative bronchitis, which very nearly proved fatal. Orthopnœa, with livid lips and nails, quick pulse and hot skin, a tight wheezy cough, and only scanty viscid expectoration—*Loud prolonged sonorous rhonchi in the chest, eclipsing all other sounds*, continued for several days ; yielding at length to ether, squill, and antimony, as expectorants, aided by blisters. The attack was of the nature of croupy bronchitis, and relief came with copious purulent and curdy expectoration ; and *then were heard moist crepitus in most parts of the lung, coarse and even cavernous in both scapular regions, chiefly the left ; but the right now for the first time showed evidence of active disease.* Although the suffocative symptoms were relieved, the reduction of flesh and strength was fearful, and a lapse into rapid consumption seemed inevitable. Happily, however, the appetite returned, enabling the patient to take food, and to resume the oil with the strychnia tonic which had served her so well during the previous years. Slowly and with interruptions during that winter, more decidedly and steadily during the following summer, improvement took place in flesh and strength, and imperfectly in breath, for only a portion of the upper and lower lobes of the right lung remained effective for respiration. *Between the second and fourth ribs in front, and at the upper half of the scapula behind was heard loud cavernous rhonchus, assuming a stridulous or grating character at times, when the breathing was more oppressed. The left upper region also presented cavernous sounds, croaking or dry at different times ; and in the lower portions the only sound was a short crackle on deep breath. Yet the stroke-sound was clear, indicating the obstruction to be more from emphysema than from consolidation.*

This interesting patient had other ailments to distress her, although it is probable that, while adding to her sufferings, they may have contributed to avert the worst results. A succession of boils—some large and painful—broke out in various parts of the body, ending in pretty complete suppuration. These were eventually checked under the use of sulphite of soda (gr. x. or xv.) two or three times daily, which for convenience was combined with a morphia cough mixture. Another plague was an eruption of eczema, which appeared in the limbs and bends of joints, as the chest symptoms became mitigated. This was relieved by a carbolic acid liniment, and disappeared at length after several weeks' use of arsenic. The oil and strychnia tonics once a day have been steadily continued whenever possible.

After the improvement in the summer, several bad attacks took place during the winters 1868–69 and 1869–70, reproducing the croupy suffocative symptoms of the former winter, and rendering recovery very uncertain. But she did recover, and during the summer 1870 made considerable advances on her former improvement, enjoying a residence in the country of several months.

The winter of 1870–71 was a greatly improved one. Although shut up entirely for seven months, she gained flesh and strength, had no severe attack, and has been able to do much writing and other official work in connection with charitable institutions, of which she has long been a mainspring. And this lady, so limited in breath and strength, yet still so useful to humanity, had cavities in both lungs, with the greater part of one in an emphysematous, imperfectly pervious state: a condition balancing between consumption and asthma. Happily, the vigorous mind has not been overcome by the infirmity of the body ; and its resolute good sense, in strictly following the advice given through all trials and difficulties, was mainly instrumental in prolonging a valuable life.

February 11, 1876. Died at Bournemouth, of bronchitis, about twenty-seven years after the date of attack.

Case 72.—*Phthisis - - third stage. Arrested twenty years.* - Miss —— , aged 16. November 4, 1856. Last April had scarlatina, followed by cough and irregularities of bowels and catamenia. Cough ceased during summer, but returned a month ago, with clear expectoration and feverish symptoms. Has been losing flesh all the time. *Dulness and tubular sounds at right scapula. Croak above, and loose crepitus below clavicle.*

Oil, with nitric and hydrocyanic acids and hop. Morphia linctus. Blistering with acetum cantharidis.

April 21, 1857.—Wintered at Ventnor, and improved very much till January, when a violent cough came on, like whooping-cough (which she had formerly), often ending in vomiting, and with much yellow expectoration. The oil was stopped for a month, and she lost much flesh. It has been resumed, and she is regaining, and the cough is much diminished.

Collapse of right front chest wall. Dulness of whole right side of chest, most behind, where, below and above scapulæ, are heard loud tubular sounds. More vesicular breath-sound in front, and crepitus on deep breath below clavicle.—To continue oil, with phosphoric acid and sulphate of iron.

April 20, 1858.- Wintered in Shropshire, going out regularly, except in an east wind. Has wonderfully improved in flesh and strength, and had little cough. Catamenia occurred only once (in November). *Less dulness and more breath-sound, but still loud tubular sounds in dorsal regions.* Aloes and iron pill every night. Continues oil.

July 16, 1863.--Has wintered at Torquay, and not confined to the house, but during winter there is generally some cough and expectoration after exertion, sometimes with calcareous concretions in it. Her flesh and general health have been good, and catamenia have been regular for two years, but breath is always short on exertion, and there is occasional pain of right chest. Has taken no oil since the winter. *Dulness and large bronchophony at and above right scapula. Weak breath-sounds in lower part of left lung, and tubular sounds above left scapula.* To resume oil and tonic.

December 8, 1865.—Has remained in Norfolk during the last two years, comparatively well, and sometimes more active than was prudent. There is always some cough and expectoration. *Loud tubular sounds above both scapulæ most on the right side, where they extend lower also, with dulness; but the breath is rough below left clavicle.* -The oil to be continued, with the addition of hypophosphite of soda to the tonic. Caution against exertion, &c.

May 16, 1867.—Improved very much, and had hardly any cough till last March, when one came on with catarrhal symptoms, and although then moved to Torquay, has continued to suffer with weakness and sickness. Being so much better, the oil and hypophosphite were not taken regularly through the winter. *Dulness and dry cavernous sounds in upper half of right back. Some vesicular breath in front. Tubular above left scapula.*—To take oil with simple strychnia mixture.

May 20, 1869.—After three months of suffering from sickness and renewed attacks of bronchial and pulmonary inflammation, was able to take the oil and tonic again, and rallied till end of winter, when the attacks recurred again, followed by improvement in summer. Has been generally in Norfolk, going for a month to Torquay in spring. *Physical signs little changed.* Flesh and strength not much diminished.

June 15, 1871.—Has been since to Norfolk, taking oil and tonic regularly, and has been generally free from attacks, but always has some cough and expectoration. Was out almost daily through the winter, but in the last three months has had inflammatory colds, and still suffers from their effects in pain on both sides of the chest, coupled with opaque expectoration, sometimes offensive, and some loss of flesh and strength. Catamenia regular. Lips and gums of good colour. *Moderate dulness with crepitus, and some breath in right front. The whole right dorsal region dull, with loud cavernous sounds in upper two-thirds, and coarse crepitus below. Tubular sounds with partial crepitus at left apex.*

July 7.—Much better.

In 1876 she was seen by Dr. Theodore Williams, who found the cavity increased in area, and the other lung more involved. Dyspnœa increased, and she died in the summer of 1876, about 20 years after the commencement of her illness.

CASE 73.—*Chronic Phthisis—third stage—Improved.*—Mr. ——, aged 35. May 4, 1865. Brother and two maternal uncles died of phthisis.

Two years ago, after a long bathe in the sea, became hoarse, with a cough and pain in the chest. Expectoration sometimes bloody. Passed winter of 1863 at a hydropathic establishment in Scotland, and lost much flesh. Last winter in Egypt, was better, and regained some flesh and strength.

Dulness and deficient breath in right chest, most upper front. Tubular sounds above right scapula.—Oil prescribed with nitric acid mixture, and tincture of iodine to chest wall.

May 15, 1867.—Has been to Australia and America since. Has not continued the oil, but flesh pretty good and no cough, only expectoration, sometimes bloody. Has had a fistula for last eighteen months. *Dulness and bronchophony below right clavicle; less behind.*

December 17.—In Scotland; cough returned with expectoration, and has lost 14 lb. weight. Fistula healed. *More dulness and crepitus above right scapula.* Has neglected the prescriptions; admonished accordingly.

May 15, 1868.—Wintered at Nice, taking oil, &c., regularly, and is much better, having gained 7 lb. and lost the cough, but there is still expectoration and short breath. *Less dulness, but still tubular sounds in upper right. Loud bronchophony below clavicle.*

May 25, 1869.—Gained 7 lb. more during summer in Scotland. Wintered again at Nice, keeping well, but in spring at Paris lived too freely, and took cold, and has had much cough and opaque expectoration. *Dulness and moist cavernous sounds above right scapula.* Has taken no oil for more than a month.—To resume oil with phosphoric acid and hypophosphite of soda.

May 26, 1870. Again improved during the summer. From October to December was yachting to Malta and Alexandria from Nice, and cough increased, with more expectoration, in which portions of lung-tissue were found. Has lately been more careful, and taking the prescribed remedies, and is much better. *Dulness and large tubular or dry cavernous sounds at and above right scapula. Tubular also above left scapula.*

1871.—This patient is still continuing his yachting, which he prefers to making the care of his health the first consideration, and, although at a risk, he is able to do this with much enjoyment.

CASE 74.—*Phthisis after Fever. Large Cavity. Great Improvement. Death from Pneumonia.*—Mr. M., aged 46. First seen by Dr. Williams, with the late Dr. Pye Smith, October 4, 1861. After twenty years in China, was long ill with fever, and has been very weak ever since. Cough for more than a year. In low state last summer at Dresden, and afterwards at Walton, where he coughed up half a pint of blood, and was much reduced, with profuse sweats, and expectoration has since become opaque, yellow, and copious. *Extreme dulness, and loud cavernous voice in left front as low as heart. Same signs audible, to less extent, behind, with some wheezy breathing. Dulness and tubular sounds at and above right scapula.* Oil ordered in nitric acid, calumbo and orange mixture; cantharides liniment; and an effervescing henbane draught at night.

April 28, 1862.—Improved greatly on treatment and wintered at Worthing.

Was out daily, and gained much flesh and strength. *Physical signs the same.*

July 26.—Has lived lately inland, taking oil only once daily, and cough and expectoration are increased.

November 16, 1862.—Omitted to state that had hæmoptysis ʒij. in May, and has had more cough and expectoration since cold weather, but also more strength and easier breath. Has lately taken Dugong oil. *Contraction, dulness, with small cavernous sounds in upper left chest; good breathing below.*

Continued in this improved state till the winter of 1864, when he was carried off by an attack of pneumonia.

CASE 75.—*Phthisis—third stage. Repeatedly checked by Oil.*—Mr. ——, aged 32. January 14, 1852. Well till two months ago he caught cold, and since has had severe cough, with copious greenish expectoration. Has been taking quinine in last three weeks, and cough worse, with much pain in chest, especially on the left side. Urine very thick. *Dulness and tubular sounds above right scapula ; moist crepitus below. The same on left side, but to less extent.*—Ordered cod oil in mixture of iodide and carbonate of potass, hydrocyanic acid, and orange-peel ; and liniment of acetum cantharidis.

February 11.—Taken medicines well, and is very much better in all respects. Has been at Hastings. *Crepitus gone. Breath-sound still weak in right chest.*

July 13.—Lost cough since May, and is stronger and more active than ever. In May had discharge from right ear, which is still deaf. *Still tubular sounds above the right scapula. Breath pretty clear below.*

December 6, 1859.—Continued well and active till last June, when, after much excitement and exertion at an agricultural meeting, he had an attack of bronchitis, with much expectoration, sometimes fœtid, and was much reduced. Has since improved, but still coughs. Lately very bilious, and taken no oil. *Dulness and tubular sounds in upper right. Small cavernous voice and croak above the right scapula.*—To take oil in mixture of calumbo and nitric acid ; and use cantharides liniment.

July 2, 1862.—Continued the remedies without difficulty, and regained flesh, strength, and activity. Gained 6 lb. and often walks twenty miles. Was out as usual last winter. Has still some cough and expectoration. Taken no oil for two years. *Tubular sounds at and above both scapulæ, left as much as right ; but no crepitus or rhonchus.*

February 21, 1867.—In the last five years has taken little oil, as it made him bilious. Cough has increased ; breath become shorter ; and now can walk only two miles. Nails convex. *Dulness, collapse, and moist cavernous sounds below the right clavicle. Breath much obstructed all down the right back. Tubular sounds above the left scapula.*—To resume oil in mixture of nitric acid, strychnia, &c. Upwards of *fifteen years* have elapsed since this patient's first symptoms. His latter deterioration came on after long suspension of treatment.

CASE 76.—*Phthisis—third stage. Arrested. Twenty six years' duration.*—An officer in the army, aged 35, consulted Dr. Williams July 26, 1861. He had campaigned in the Crimea, and was afterwards ordered to the West Indies,

and then to India, where he distinguished himself in the Sepoy mutiny; and
was quite well till his return to England last summer, when camped at Shorn-
cliffe, and found the place extremely cold. In March caught severe cold, and had
sore throat and cough, with hæmoptysis ʒi., and was much reduced. Improved
since taking oil and iron; but is still thin, and has short breath. *Much dulness
and tubular sounds upper half right chest; tubular sounds above left scapula.*
—Oil ordered with phosphoric acid and quinine, the application of acetum cantharidis, and a morphia linctus.

June 19, 1862.—Improved much, and went to Madeira in November, where
he took oil regularly, and rode on horseback. Lost cough, and gained much
flesh and strength. *Dulness; dry cavernous sounds upper right third of chest,
chiefly in back.*

February 6, 1865.—Recovered so much that he was on active duty in New
Zealand during the war of 1864, endured all kinds of hardships, wet, cold, and
starvation. In April had dysentery; in May spat blood, and was laid up with
inflammation of the left lung three weeks, which much reduced him, but he
recovered in all respects, except his breath being short. Improved much since
his return, taking oil regularly. *Moderate dulness: tubular sounds high up in
left lung: good breathing below. Tubular sounds above right scapula.*

May 1866.—Wintered at Mentone, without cough; gained flesh and
strength. *Physical signs improved.* The following winter left the army,
and has since resided in Madeira; where he was well, and active in spring
1871.

1887.—Returned to England, from Madeira, 10 years ago, and has since
resided at Arundel, enjoying excellent health and freedom from cough. His
chest was examined by Dr. Theodore Williams about 4 years ago, and there was
found *some tubular sound at the left scapula, with marked flattening of the left
front chest. Emphysematous crackle audible in lower portions of left lung.*
—Nearly 26 years have elapsed since commencement of the disease.

CASE 77.—*Chronic Phthisis. Cavity. Duration* 18 *years.* — Mr. M. F., aged 15,
was seen by Dr. Williams, October 30, 1853. For three years he had been subject
to occasional cough, which had become constant since June. Has been well
and active till within the last few weeks, when cough has become worse. No
expectoration at present, but a mucous rattle sound in the tubes. *Decidedly
tubular sounds in both scapular regions.* Ordered oil in nitric acid and
calumbo.

March 11, 1854.—Very much improved, but always has a loose cough.
*Large tubular sounds above scapula and within left scapula. Dulness at right
scapula, and some mucous rhonchus.*

November 16, 1867.— Grew strong and tolerably well; but breath was always
short, and often had attacks of cough, which have been worse during the last
few years. A year ago he had congestion of both lungs, relieved by blistering
and other treatment. He took oil; and went to Nice and Italy; and there
caught fresh cold; and since has had constant cough, with expectoration and
short breath. Then tried several German baths, but received no benefit; and
is now much reduced in flesh and strength. Nails convex. *Cavernous sounds
in upper right chest; obstructed breath-sounds below. Whistling tubular
sound at left scapula.*— Oil was ordered in phosphoric acid and strychnia mixture, and acetum cantharidis liniment for the chest. To winter at Cannes.

Improved much under this treatment, gaining both flesh and strength for twelve months.

August 1870.—Appears much broken in health, and is in great financial difficulties. Very feverish, and cough troublesome. Now quite an invalid, *with extensive cavernous sounds on right side and crepitation throughout left lung.*

Again improved on oil and tonics, and is living (May 1871), eighteen years since his first visit.

CASE 78.—*Phthisis, with Cavity. First Acute, but Arrested. Recovery.* —A delicate boy, aged 8, whose father had died of phthisis, was first seen June 28, 1853. Had cough, since influenza six weeks ago. Soon after was seen by Dr. West, who found *dulness and coarse crepitus in upper left chest.* Had lost flesh, and had profuse perspirations; but has taken oil for a fortnight, and much improved. *Dulness in upper half of left chest, with loud tracheal note on percussion at the top. Large tubular breath, and voice-sounds also, with coarse crackle in parts. Large tubular sounds above right scapula.*

The oil to be continued with acid tonic, and acetum cantharidis is to be applied from time to time on left chest.

1861.—For several years continued delicate, requiring constant care, and always taking the oil with varied tonics; residing generally at Tunbridge Wells, or on the south coast. *The signs of cavity became very distinct, but in two or three years they diminished* as the general health improved; and now there are only *dulness, tubular sounds below the clavicle and above the scapula and weak vesicular signs below.* Flesh and strength now pretty good for a lad of 16.

July 21, 1863.—Continued well and active till three months ago, when he had a severe cold, with signs of bronchitis affecting the right as well as the left lung; *and the large tubular sounds and coarse crepitus reappeared there.* After repeated blistering and continued use of the oil, he gradually recovered, and has had only moderate cough and rather short breath; but his strength is good, and he walks ten miles *Still coarse crepitus in upper left front. Dry cavernous sounds above left scapula. Large tubular sounds above right scapula, but breath quite good in other parts.*

1871.—Not seen since; but is reported to have recovered his health completely.

CASE 79.—*Phthisis, with excavations contracted probably 16 years before death. Death from renal disease 32 years after first attack. Autopsy.*—A gentleman, aged 22, first consulted Dr. Williams on October 4, 1847. Six months ago he had an eruption, followed by boils, and three months later cough and expectoration, which continued up to the present date. *Dulness and deficient breath in upper portion of the left side of the chest.*—Was ordered a mixture of nitric and hydrocyanic acids, tincture of hops, and henbane in decoction o Iceland moss; and counter-irritation with acetum cantharidis.

August 12, 1848.—Lost cough, and improved much in two months; but caught fresh cold a month later, and cough returned, with wasting and sickness, the latter having been caused by trying impure cod-liver oil.—*Dulness and dry cavernous sounds in both mammary regions; large tubular in upper back.*

December 5.—Has taken pure oil in a mixture of nitric and hydrocyanic acids, with tincture of orange-peel regularly at Ventnor, with increase of strength and appetite, and of weight amounting to 14 lb.; cough worse at Ventnor. *Physical signs the same, but croaky cavernous sound under left clavicle.*

May 5, 1849.—Generally better, and cough moderate. The oil occasionally sickens when exercise is not taken. *Dulness on left side; muffled cavernous sounds below left clavicle; cavernous sounds above scapula; tubular sounds above right scapula.*

September 25.—Continued well till a week ago, when he had hæmoptysis, six ounces.

May 4, 1850.—Had bilious fever some months ago, which much reduced him. Omitted oil for one month, but has since taken it regularly, and quite recovered flesh and strength; cough moderate; *physical signs the same.*

May 9, 1851.—Well since, and out all the winter at Ventnor, with only slight cough, and no expectoration. *No croaky or cavernous sounds in left lung; obscure bronchophony in scapular region, and breath weak above and harsh below. In right lung, respiration tolerably vesicular; expiration long, but not tubular.*

October 1852.—Embarked for New Zealand, but was wrecked in the Channel, and much exposed to cold and wet, without permanent injury. Wintered well at Torquay, and went to New Zealand in September, 1853.

October 28, 1863.—After arrival in Auckland, had more or less cough the first two or three years, but gradually lost it, and for the last six years has been quite well. Held a Government office in New Zealand till his return this year. Has been married several years. *Tubular breath in upper part of both sides of chest, most in the left, but good vesicular below.*

1867.—Has lived in England four years, wintering at Falmouth, and enjoys general good health.

1871.—Continues well.

1879.—He remained fairly well till March of this year, when œdema of the ankles came on, and the urine (s. g. 1010) was found to contain albumen. The renal symptoms became worse, and after a journey to Mentone, which much exhausted him, the albumen increased, vomiting and diarrhœa came on, followed by œdema of the lungs, and he died on October 27.

On *post-mortem* examination by Dr. Siordet and the late Dr. Sparkes, both lungs were found adherent in the upper portions; the right lung at its posterior apex contained an irregular honey-combed cavity, with fibrous bands traversing it in all directions. About one-third of the upper lobe was the seat of chronic induration, the remainder was emphysematous, the two lower lobes were œdematous. The left lung contained at its apex an irregular-shaped thin-walled cavity, the size of a walnut, and near it, imbedded in fibrous tissue, was a caseous nodule the size of a hazel-nut. The tissue round was indurated for a short distance, and the remainder of the upper lobe was emphysematous, the lower lobe œdematous. Neither cavity contained any secretion, nor were there any bronchi opening into it. No miliary tubercle was detected in the lungs.

Both kidneys were enlarged, and were typical instances of the smooth large white kidney of advanced Bright's disease.

The heart was healthy, but the pericardium, in common with the body generally, contained a thick layer of fat.

Liver fatty. Spleen and the rest of the organs normal. No lardaceous disease of organs was detected.

As **Dr.** Sparkes remarked, the patient *died* of renal disease, the old phthisical lesions having little, if any, direct influence in the fatal termination.

We may fairly conclude that contraction of the cavities occurred through fibrosis at least sixteen years before his death, and that emphysema was developed round the old lesions, and the patient enjoyed good health until attacked with renal disease apparently unconnected with the pulmonary. Thirty-two years elapsed between his first visit to Dr. Williams and his death.

CASE 80.—*Chronic Phthisis. Persistent Cavity* 20 *years.*—A lady, aged 40, who had lost a sister from consumption, first consulted Dr. Williams June 20, 1849. There had been slight cough and expectoration for a year and a half, for which she had been long treated with steel without benefit. *Amphoric stroke and loud cavernous sounds above right clavicle and scapula: the same signs below clavicle, though less marked.*—Ordered cod-liver oil.

July 13.—Heard that she was suffering from increasing weakness, diarrhœa and hoarseness, and copious expectoration. *Oil had not been taken.*—Ordered a tonic of nitric and hydrocyanic acids, and tincture of orange, to be combined with oil at a later date.

November 20.—Has taken oil regularly, and much improved in last three months, having lost her cough for the first time in two years, and grown fat and strong.

Dulness, and loud cavernous sounds above right scapula, and to a less extent below right clavicle.

1856.—General health good, no cough, but *physical signs the same.*

1868.—Alive and well 20 years after first symptoms. In this case the disease was limited to the very apex of the right lung, ending in a cavity which remained dry long after the patient's restoration to health.

CASE 81.—*Chronic Phthisis. Third stage. Arrested* 21 *years.*—A single lady, aged 46, consulted Dr. Williams December 2, 1858. Three years ago she had severe cold and cough and became very thin. She took oil and regained her flesh, but never lost her cough, and a year ago had hæmoptysis to the amount of ℥iij. Her breath is always very short. *Extreme dulness, deficiency of breath, and coarse moist crepitation throughout left lung. Large tubular sounds above left scapula.*

Oil was ordered in a tonic of nitric acid and calumbo and orange-peel. A saline opiate at night, and counter-irritation with acetum cantharidis on the chest.

July 27, 1859.—Has been very ill in the winter, suffering from cough, frequent hæmoptysis, and much loss of flesh; but after much blistering and taking oil and tonic, gradually regained flesh and strength. Now has fistula in ano. *Same as at last visit, except dry cavernous or large tubular sounds in upper left chest.*

September 24, 1860.—Has wonderfully recovered, and is fat and pretty strong, but breath short, and has lately fulness above the right clavicle from emphysematous protrusion. *Crepitation still throughout left lung, but with more breath.*

October 11, 1862.—Is wonderfully well as regards flesh and strength, and free from cough. Fistula discharges only occasionally. *Physical signs much as before.*

May 17, 1864.—Now has only occasional cough, and is out walking all day long.

Dulness and cavernous croak in upper left chest, with obstruction sounds below. Tubular sounds above right scapula.

July 16, 1868.—Has continued well, taking oil in the winter. Has lately suffered from weak eyes. Now has an attack of severe nephralgia. Was relieved by effervescing saline with opium (oxalate crystals in urine). Has no cough. *Collapse, with much dulness and croaky cavernous sounds in left front: croaky crepitation mixed with breath-sound in left back.*

January 1871.—Heard of her continuing well. Walks out on fine days. Died in 1876, aged 64, twenty-one years after her first symptoms of phthisis.

CASE 82.—*Chronic Phthisis, with Cavity, Arrested. Living and well twenty-five years after.*—A clergyman, aged 32, was seen by Dr. Williams in consultation with Dr. Hamilton Roe and Mr. Young, for the first time in March 25, 1846. He had lost four sisters by consumption. Three years ago, after great exertion of voice and close application to work, he became hoarse, and had remained so ever since. Cough came on five months ago, with expectoration and shortness of breath, but no wasting. Wintered at Ventnor. *Dulness and cavernous sounds in upper part of right side of chest. Breath rough below left clavicle. Sputa opaque and heavy.*—Was ordered a combination of nitric and hydrocyanic acids, iodide of potassium, and sarsaparilla, and to use iodine ointment. Lost cough at Ventnor, and went to Bridgewater, which did not agree with him.

August 21.—Has been at Minehead, taking cod-liver oil regularly, using salt friction three times a day, and keeping a blister wound open. Has improved, except in breath, and expectoration is still puriform and opaque. *Slight and irregular dulness in upper part of right chest; tubular sounds above and below right clavicle and scapula.*

June 10, 1848.—Wintered first at Malta, which he found too irritating to his chest; next at Pisa, where he grew weaker, but was improved by the voyage. *Still slight dulness; loud bronchophony and irregular breath-sound below right clavicle; but much good vesicular breath- and stroke-sound. Breath rather irregular below left clavicle.*—Ordered iron in combination with the oil, on which he gradually improved, wintering generally in Devonshire during the following years.

January 1868.—Quite well and active; conducting a large school, which he has done for the last twelve years. Can walk, preach, and bear exposure to any extent; and has no cough. *Still dulness and tubular sounds in upper right chest, most above scapula. Marked tubular sounds above left scapula.* Although the physical signs have not entirely disappeared, they are greatly diminished since the first examination twenty-two years ago, and may be supposed to indicate contraction and obliteration of the cavity. In fact, they may be considered signs of the vestiges of disease, rather than of actual disease, as the patient has enjoyed excellent health for the last twelve years. Alive and well in 1871. *Upwards of twenty-five years have elapsed since the first symptoms appeared.*

CASE 83.—*Chronic Phthisis. Cavities. Arrest of disease. Living eighteen years after.* A solicitor, aged 32, first consulted Dr. Williams November 27, 1856. His brother died of phthisis. Was attacked with influenza two years ago, and ever since had cough and grey expectoration, increasing in winter and diminishing in summer. In last three months sputum has become pink, and patient has lost flesh, strength, and breath. *Some dulness and decided tubular breath at and above both scapulæ, mostly left.*—Ordered oil in a tonic of nitric and hydrocyanic acids with tincture of orange.

June 1857.—Wintered at Pau, taking oil, &c., till April. Lost cough and expectoration, and gained flesh and strength. Then went to Eaux Bonnes, and left off the oil. Lost flesh, but otherwise remained well, walking three miles and riding twenty, daily. Cough returned in last fortnight. *Physical signs the same.* To resume the oil.

May 1858.—Wintered well in South of France and Italy, generally taking oil. Is stronger and stouter.

March 5, 1859. Well till middle of January; since then cough and opaque expectoration. *Dulness and tubular sounds in upper part of both lungs, especially in left, where there is some moist rhonchus.*

June 21, 1861.—Well, and taking oil till February, when he became bilious, and omitted oil for two months. After fresh cold, cough came on, accompanied lately by expectoration and pain in front of left chest. *Dulness, deficient motion and breath in upper part of left chest, front and back. Loud tubular sounds at and within scapula.* Urine scanty and high-coloured.— To continue oil in a tonic of strychnia and tincture of orange. Also to take an effervescing saline at night for a short period, and to use counter-irritation with tincture of iodine.

May 16, 1862.—Went to Scarborough, and gained three pounds. Physical signs also improved in autumn. Wintered at Pau, taking oil regularly, and out of doors a great deal. Cough slight, and strength good. *Dry tubular sounds at and above both scapulæ; mostly left, where there is dulness.*

October 15, 1862.—Well, and in London at his office the whole summer.

October 30, 1863.—Worked in London all the winter, and tolerably well till summer, when cough increased after exertion, and the expectoration became tinged with blood. Has taken the oil regularly. *Flattening and obscure breathing in left front; dry cavernous sounds above left scapula; loud tubular above right.*

May 16, 1865.—Continued well and in business, with only occasional cough till lately, when only took oil once a day (instead of twice). Patient at present weak and exhausted, with irregular pulse.—Ordered oil twice a day in tonic of hypophosphite of iron and strychnia. Soon improved, and has lost cough.

1867.— Has continued well, and at his business ever since. *Dry tubular sounds at and above both scapulæ.*

1874.—Well and active, except cough and occasional shortness of breath. *Physical signs the same.* Eighteen years since first visit.

CASE 84.—*Chronic Phthisis arrested.*—A gentleman, aged 15. Several paternal aunts died of phthisis. September 24, 1847.—Nine years ago, after gastric fever, was very weak, and cervical glands enlarged, and discharged several times on both sides for a year and more. A year ago had measles, followed by whooping-cough, and ever since has had cough and short breath, and been thin and weak.

Superficial dulness, tubular sounds, and moist crepitus below left clavicle. Loud tubular sounds and small spot of dulness within right scapula.

Prescribed cod-liver oil, and a vesicating liniment, occasionally.

May 1848.—Wintered in Madeira, but did not continue the oil. Cough and breath better. *Less dulness and crepitation in left front. Still tubular sounds over inter-scapular region.*

June 1860.—Another winter was passed in Madeira, and two in Egypt and Italy. Taking oil irregularly. Cough generally better, but not much improvement in breath or strength. Dr. W. then recommended long sea voyages for two years. Accordingly he went to Australia and India, and returned much improved in flesh and strength, and quite lost cough and expectoration. Has since wintered chiefly in Egypt, which agrees well. *Chest and respiratory power increased. Stroke clear (emphysema) and breath-sound rather coarse and rough in left front. Tubular sounds at right scapula. Still loud tubular sounds within left scapula.*

1865.—He enjoyed more uniform and vigorous health, and wintered three years in Scotland and England. Now in Parliament, and attends pretty regularly.

December 1866.—After skating in Scotland, attacked with pain in chest, dyspnœa, and fever, with obstructed breathing and crepitus in both lungs, most left. This soon subsided under treatment, and chest returned to former state; but, after wintering in South Hants, stomach and bowels were disordered more or less, till the end of the summer, when he passed six weeks at Kissingen under the goat's milk cure, and was much improved.

July 19, 1869.—Following winter at Cannes, which disagreed with him, causing much gastric and bronchial irritation. On return to England in spring improved, and regained flesh and strength. But stomach often out of order, and has occasional attacks of bronchitis, which now assume a wheezy character. Has lately had one; and also inflammation of the ear and eyelids, ending in abscess. Still cough and opaque expectoration. Takes oil once a day.—*Breath-sound superseded by crumpling crepitus in whole left front, and less in back. Large tubular sounds at and above left scapula and at right scapula.*

September 26, 1870.—Last winter at Ventnor, with more care, and less suffering from attacks. Flesh and strength fair. Still had recurrences of gathering in the ear and boils on the body. These have been much better since taking sulphurous acid mixture with the oil. When free from bronchial attacks, there are always the *loud tubular sounds in both scapular regions, and coarse breath-sound with a few large clicks in left front. A fresh cold always blocks this up with crepitus, sometimes with bronchial wheezing at the roots and summit of the lung.*

June 1871.—Another winter at Ventnor, and with improved health and more freedom from attacks. Flesh and strength fair. Attends to his parliamentary duties.—*Less crepitus in the lung, but, as before, the harsh breathing and tubular sounds of old induration and emphysema in portions of the lung.*—Continues oil and tonic.

1887. · Alive forty years after the first symptoms.

CASE 85.—*Chronic Phthisis. Double Cavity arrested twenty-one years.*—A gentleman, aged 26, first consulted Dr. W., June 15, 1848. He had long been subject to occasional cough, which had become constant during last two months, and was accompanied by expectoration, loss of strength and breath, though not to any extent of flesh. Had improved on iodide of iron and bark, and counter-irritation with tartarated antimony. *Dulness, crepitation, and loud tubular sounds above and below left clavicle; loud tubular expiration above right scapula.*—Ordered oil with nitric acid and tincture of orange, and counter-irritation with acetum cantharidis.

August 23, 1848.—Has taken oil, and is much improved in flesh, strength, and breath. Cough stopped till last few days, when he caught fresh cold. *Physical signs same.*

September 25, 1851.—Continued well, taking 'gallons of oil,' but cough has increased in last three months, otherwise strong and stout. *Dulness and cavernous sounds above left scapula.*

September 21, 1857.—Well since, taking oil occasionally. Has had hardly any cough till the last month. *Dulness and tubular sounds at and above both scapulæ, most left.*

February 17, 1862.—Continued pretty well, but always short breathed on exertion, and more so lately with pain in left side. Has had no oil for five years, but has been living well and taken beer freely. *Dry cavernous sounds in upper left chest; large tubular sounds above right scapula. Breath good in front.* Ordered oil in above tonic, with tincture of calumbo.

May 3, 1863. Out all the winter, but lately distressed about his wife who is phthisical, and suffering from mental depression. Has now more cough and expectoration. *Dulness and dry cavernous sounds above both scapulæ.*

March 21, 1866.—Looks stout and well, but has always cough and expectoration, which have increased in last three months. Also has piles. Lost his wife a year ago. *Cavernous sounds at and above left scapula, obstruction and crepitus in front. Tubular and bronchial rhonchus above right scapula.*

. October 27, 1866.—Has continued to take oil in various tonics, quinine and sulphuric acid, afterwards phosphoric acid, hypophosphite of soda, and quassia. When the urine was thick, occasionally an effervescing saline. Lately breath short, and occasionally blood in the expectoration.—Ordered oil in nitric acid and tincture of nux vomica. Living at Bognor.

October 23, 1867.—Has been tolerably well, except occasional slight hæmoptysis and piles. In last few days cough has increased. *Cavernulous sounds in upper left chest, obstruction and subcrepitant sounds in lower part, puerile breath in right, except at and above scapula, where sounds are tubular.*

In this case cavities formed in both lungs, but in the right cicatrisation probably took place. In the left the cavity contracted, and other parts became emphysematous. The puerile breathing in the sound part of the right lung testified to the extent to which its powers were taxed. This patient died suddenly of hæmoptysis, July 1869, on his way to the railway station; but up to that time presented a stout and ruddy appearance, and was able to enjoy life, but his breath was always short. He lived upwards of twenty-one years after the commencement of his illness.

CASE 86.—*Phthisis, with Cavity. Acute Rheumatism and Endocardial Murmur. Recovery.*—A married lady, aged 30, consulted Dr. Williams, October 13, 1860. Lost her mother, two sisters, and one brother from consumption. During the last four years she had had occasional cough, and hæmoptysis, which on one occasion amounted to three ounces. Did lose much flesh; but after taking oil, and applying it externally also, became fat and still remains so. Lately has been taking glycerine. Cough has increased in last two months, and now she has a bad cold. *Extensive dulness, cavernous sounds and surrounding crepitus in upper right chest. Tubular sounds above left scapula.* —Oil ordered with phosphoric acid, calumbo, and orange; a morphia linctus at night; cantharides liniment.

December 24.—At the end of October, severe rheumatic pains came on with

tenderness of wrists and ankles ; was soon relieved by opiate salines ; but sh
has been confined to bed for two months.   Looks well, but still has a cough.
In addition to former signs there is a loud diastolic murmur at mid-sternum.
February 5, 1861.—Much better in every respect.   Little cough.   *Physical
signs same, only a trace of crepitation above scapula.*
September 18, 1861.—Looks quite well, and has only slight cough and short
breath, with occasional pain in right chest.   Catamenia irregular in time
and quantity.   *Less dulness and more breath in upper right; no cavernous
but tubular sounds.*
November 10, 1862.—Weathered last winter fairly in Ireland, being free
from cough.   From May to September took a cold sponge bath every morning.
A scaly eruption appeared on arms and legs, which Dr. Neligan cured by
arsenic.   During last fortnight she has had a cold and cough.   *Still dulness and
tubular sounds in upper right chest, with slight crepitation.   Cardiac murmur not audible.*—Effervescing saline with opiate at night, and continue oil
and tonic.
May 30, 1865.—Living at Woolwich, and wonderfully well.   Only occasionally suffers from cough and rheumatic pains.   Has taken a great deal of
oil, but is less stout, though in good condition.   *Still dulness and tubular
sounds through upper third of right lung.*   Loud diastolic murmur audible to
right and along upper portion of sternum.
November 2.—Had hæmoptysis to the amount of 1 oz. a month ago, and
since then has had violent cough, with mucous expectoration.   Now she is
taking oil with sulphuric acid.   *Dulness, tubular sounds and coarse crepitation
in upper half of right lung.*   The patient was seen again in the summer of
1867 ; she was then stout, but her breath was short with some cough.
March 13, 1868.—Has lost cough, is ruddy and stout, but occasionally has
palpitation, and has lately had a vesicular eruption on the hands, which was
soon cured by lotion.   No catamenia for twelve months.   *Physical signs of
heart and lungs much improved.   Only slight dulness and loud tubular
sounds above the right scapula.   The second sound of the heart is clangorous,
but without murmur.*
1871.—Has been heard of as in good health, fifteen years after the first
attack.

CASE 87.   A brewer's clerk, age 25.   First seen September 13, 1847.   A
year and a half ago had syphilis, and was treated by mercury for three
months.   Eight months ago an eruption appeared on the skin, followed by
cough, slight at first, but afterwards becoming violent, with purulent expectoration, and lately accompanied by sweats, and loss of flesh, strength, and
breath.   *Marked dulness below left clavicle, extending to mammary region,
gurgling and crepitation around.   Loud tubular sounds within and above right
scapula.*   Ordered oil ; and counter-irritation with iodine liniment.
May 24, 1848.   Has taken oil regularly all the winter, and has quite recovered flesh and strength.   Cough slight, with scanty opaque morning expectoration.   *Still dulness in upper left chest, with dry cavernous sounds above
and below clavicle ; but no crepitation ; vesicular sound faintly heard in
lower chest ; less tubular sounds in right lung.*
September 1850.—Quite well, and taking active exercise.   *Dulness much
diminishing ; some tubular sounds in upper left chest, front and back.   No
cavernous sounds, but breathing harsh.*

December 26, 1856. Heard from Dr. Carhill that this patient had just died suddenly of peritonitis from intestinal perforation. He had been apparently well, and actively engaged in business till ten days before his death ; subject only to occasional attacks of headache and costiveness. After walking a mile he was seized with sudden and severe pain in the abdomen, with collapse and other symptoms of perforation, and died in two days. A *post-mortem* examination was made by Dr. (now Sir W.) Jenner. The abdominal walls were found to be covered with fat an inch thick. The ileum just above the cæcum was perforated by ulcers of tuberculous character, of which there were several.

Both lungs were strongly adherent at their apices, especially the left ; and in both cretaceous matter was found ; in the right in tubercles, varying in size from a pin's head to a pea ; but in the left lung there was a large mass of the same material, which quite filled an ancient cavity at the summit of the lung.

The patient had been free from chest symptoms for eight years.

This case is less remarkable for the duration of life after recovery from the third stage of consumption, than for the completeness of the cure of the chest symptoms and for the demonstration it afforded after death (from another cause) of the arrest of the tuberculous disease which had made rapid strides eight years before.

The subjoined are two good examples of the effect of mountain climates on chronic tubercular phthisis.

Case 88.—Mrs. D ——, aged 30, consulted Dr. Theodore Williams May 12, 1879. Two maternal aunts, two brothers, and one sister, had died of phthisis, the three latter of rather rapid forms of the disease. Her father had induration of the left lung of many years' standing. She had been failing in health and strength for about a year, and during the last six months had been steadily losing weight, the total loss being 1 stone $6\frac{1}{2}$ lbs. Cough and night-sweats appeared six weeks ago, the latter symptom having lately ceased, the former being slight and unattended by expectoration. At present. Tongue clean, appetite poor, catamenia regular, temperature 98° Fahr., pulse and respiration normal, weight 8 stone 7lb. *Dulness, scattered crepitation audible over the upper third of the left chest posteriorly.* Ordered cod-liver oil with hypophosphite of soda, dilute phosphoric acid and strychnia, and an iodine liniment to the left back ; also a dietary in which milk, sugar, and starch preponderated.

June 16.—Oil and tonic have been taken regularly. Appetite better, cough less. Has gained $1\frac{1}{2}$ lb. Weight 8 stone 9 lb. Physical signs worse. Crepitation is now audible below the angle of the scapula and in front just above the mamma. It is still audible in the upper left back, and the crepitation in the interscapular region has become coarser. Right chest measures just above the mamma $16\frac{1}{2}$ inches, left chest $15\frac{1}{2}$ inches. Ordered to Davos until the following spring.

After three months' stay at Davos I heard from Dr. Ruedi that she had considerably improved and gained 7 lb. in weight. He writes : 'Mrs. D—— has little or no cough, eats and drinks well, and ascends the Schatz Alp without fatigue or shortness of breath. Temperature 98° Fahr. Pulse 70, and respiration 16.'

In October, dreading the *ennui* of the approaching winter at Davos, she descended to Milan for a change, and was there attacked with violent diarrhœa followed by fever. Dr. Freeman, of San Remo, who happened to be at Milan, examined her, and advised a speedy return to Davos. After being laid up three days at Bellagio in consequence of the persistent diarrhœa, she safely reached Davos, and Dr. Ruedi found, in addition to the former physical signs, ' crepitation under the left clavicle.' The diarrhœa subsided, but returned a week later after the exertion of a long walk, and was accompanied by high temperature. She was confined to bed for six weeks with a fever, apparently of the intermittent type, accompanied by great sleeplessness, and eventually yielding to quinine given by day and morphia at night. The diarrhœa ceased, and in January 1880 she had recovered sufficiently to ascend the Schatz Alp, to sledge and to ' toboggan ' down the sides of the Davos-Thal. In February Dr. Ruedi wrote to me that the physical signs were nil.

She remained at Davos till April, and then after spending about a month at Berne and Paris reached England.

June 29, 1880.—Appears in robust health, and can walk ten miles at a stretch. Weight 8 stone 11½ lb. Has quite lost her cough ; breath not short in ascents. Pulse 70 strong, respiration 16. Chest measurements just above mamma. Right side, 17½ inches ; left, 16½, being an increase of 1 inch for each side of the chest, or a total of 2 inches in the whole circumference. The expansion appears to have taken place in both antero-posterior and lateral directions. The thorax appears full, but the movements are free. Several veins are conspicuous in the upper portions. *The whole chest is hyper-resonant, this being most marked in the posterior portion of the left side. The breathing is harsh on the right side, and at the left posterior apex, especially where the resonance is greatest, emphysematous crackle is heard on deep breath. Below this the breath is harsh anteriorly and posteriorly. No dulness, crepitation, or bronchophony can be detected anywhere. The heart is not displaced.*

January 24, 1881.—Last summer yachting, and has spent part of this winter in Cornwall and at Ryde. Has gained 6½ lb. more, weighing 9 stone 4 lb. Continued quite well till Christmas, when in the severe weather cough returned, and since then expectoration has been twice streaked with blood. Careful examination could detect no fresh mischief anywhere, though in the presence of so much emphysema it might well be masked. Heard since that the cough had subsided.

January 1887.—Has maintained excellent health ever since, and when I examined her last year (1886) I could detect nothing amiss in her lungs.

Remarks.—The result of mountain climate in this case is the more remarkable on account of certain decidedly unfavourable elements in the prognosis. *First.* Strong family predisposition. *Secondly.* The distinct increase of local disease previous to reaching Davos. *Thirdly.* The occurrence of intermittent fever and its consequent debilitating influence on the constitution. Nevertheless the arrest of the disease was complete and probably permanent, as upwards of six years have elapsed without any return of the symptoms.

Case 89.—Miss C., aged 25, consulted Dr. Theodore Williams, September 30, 1878. Mother died of phthisis, being consumptive at the time of Miss C's birth; one sister and one brother phthisical. Cough and expectoration with wasting came on about 3 years ago, when consolidation of the right lung was detected by her medical adviser. She had wintered at Pau and Nice twice without any marked benefit. Her symptoms had continued up to the present time, September 30, 1878. *Dulness on the right side from clavicle to the third rib and above the scapula. Cavernous sounds audible in the first interspace.* Weight 7 st. 11½ lb. Ordered to spend the winter at Davos.

December 21, 1878. Saw her at Davos in consultation with Dr. Ruedi. She had gained appetite and strength and could walk for several miles. Breath fair but short on rapid exertion. Pulse 80, cough slight; expectoration mucous. Measurements show no shrinking of the right chest which measures ½ inch larger than the left. *Dulness diminished over first interspace and chest fairly resonant below. Cavernous sounds as before. Tubular sounds audible from the second to the fourth rib. Dulness with râle (on cough) above the scapula.*

June 25, 1879. Quitted Davos April 9th, having had hæmoptysis to the amount of 1 oz. in March (at the commencement of the snow melting), and descended to Coire and there had slight hæmoptysis—then passed five weeks at Baden-Baden where she ascended the hills without difficulty. Has gained flesh (2½ lb.). Cough and expectoration slight. Pulse 80.

Resonance on percussion over the whole right front chest, except over the outer third of the first interspace, which is dull; dry tubular sounds (not cavernous) are heard in the first interspace. Marked resonance above the scapula.

	Right	Left
Measurements at the level of third rib	14¾ in.	15½ in.
„ „ „ mamma	14 „	14½ „
„ „ ensiform cartilage	13 „	12 „

June 2, 1880. Passed a second winter at Davos with great benefit. Is bronzed and strong. Still has slight cough but no expectoration. The chest is now of most peculiar shape and shows enormous flattening and contraction of the right front chest wall, with some bulging of the lateral and posterior regions.

	Right	Left
The measurements are at the level of the third rib	15 in.	14½ in.
„ „ „ mamma	14¼ „	13¼ in.
„ „ ensiform cartilage	12 „	11 „

The whole right side is now completely resonant. Tubular sound audible in the first interspace. On the left side breath is harsh and resonance very marked.

October 27, 1880. After residing for some time in a damp part of Somersetshire, and over-exerting herself, profuse hæmoptysis occurred August 26th, and persisted to a large extent for 5 days, the amounts varying from 2 to 7 oz. a day. After this the expectoration was streaked up to September 12th, when it became clear again. Dr. Brittain, of Clifton, reports the right lung to have become consolidated throughout after the hæmorrhage, but to have gradually cleared, and the patient recovered with breath much shortened. A fortnight later (Nov. 10) I saw her and found in addition to the old signs in the right lung dulness and crepitation over the lower third of the posterior base and crepitation above the left scapula.

1887. She returned to Davos and has since that period resided there and in the Engadine till last autumn, when she came to England to see her father, who had been ill. She is wonderfully well and active, still has slight cough and some expectoration which contains a few tubercle bacilli. She is able to take walks, but breath is short on ascent, and the patient looks thin though she weighs about the same as formerly (7 st. 12½ lb.). Examination of the chest shows dry tubular sounds over the upper quarter of the right lung, and dry tubular sounds in the first and second interspaces on the left side. The rest of the lungs are apparently in a condition of hypertrophy and there is great expansion of the lower portions of the chest.

Remarks.—The points of interest in this case were the great changes in the thorax, due partly to the contraction of the cavities and the emphysematous lung extension at a later date. There is little doubt that the patient was saved by her steady residence at high altitudes, and that this influence has prevented the rapid progress of hereditary phthisis. About 11 years have elapsed since her first symptoms.

CHAPTER XXI.

THE DURATION OF PULMONARY CONSUMPTION.[1]

Estimates of Portal, Laennec, Andral, Louis, and Bayle compared with those of the Brompton Hospital, Fuller, and Pollock—Differences explained by class of patients and mode of treatment—Author's Thousand Cases selected from wealthy Classes -Ground of selection explained—Method of tabulation of cases—Sex—Age of Attack—Family Predisposition—Origin and first Symptoms—Cases of inflammatory origin; their proportion and course of symptoms—Hæmoptysis—State of Lungs at first visit as evidenced by Physical Signs—Classification of Stages adopted with restrictions—Majority of Patients in First Stage, and consequent favourable prognosis—Mortality in each Stage—State of Lungs at last visit—Classification of 'Healthy,' 'Improved,' 'About the Same,' and 'Worse;' and Percentage of each—Relative liability of lungs to attack, excavation, and extension of Disease—Number of Deaths—Causes—Long Duration—Living patients more numerous and with higher average duration - Present State described as 'Well,' ' Tolerably well,' and ' Invalid '—Large Proportion of first two classes—Hopeful Prognosis— Causes of long Duration — Influence of Age and Sex on Duration—Among Females duration shorter, Age of Attack earlier, and Age at Death less advanced than among Males— Great age reached by some Patients—Relation of Age of Attack to Duration—Prolonging effect of inflammatory origin, pneumonic, pleuropneumonic and bronchitic—Duration of Pathological Varieties of Consumption difficult to determine.

WE need not dwell on the importance of the subject which we propose to treat of in this chapter. In a country where, according to the Registrar-General, one death in every eight is caused by phthisis, it is obvious that a true knowledge of the duration of the disease and the conditions which modify it, is of the greatest consequence to the community. Many estimates have been formed of the duration of phthisis in this and other countries, and these estimates will be found to vary to such a degree that a reader may well despair in attempting

[1] This chapter is an abstract of a paper on the 'Duration of Phthisis Pulmonalis, and on certain Conditions which Influence it,' contained in the LIV. volume of the *Medico-Chirurgical Transactions*.

to harmonise them. A due consideration, however, of the
conditions under which each estimate was made—*i.e.* of the
number and of the social class of the patients, of the form of
disease, of the mode of life and of the treatment pursued, will
serve to explain many of the variations. Portal's saying that
phthisis may last from 10 days to 40 years is undoubtedly
true, but far too indefinite for our present state of knowledge.
Laennec gives 24 months as the mean duration ; Andral the
same ; Louis and Bayle 23 months, founded on the examination
of 314 cases.

The first Brompton Hospital Report, in 215 fatal cases,
found that 40·8 per cent. died less than one year after attack,
45·3 per cent. between 1 and 4 years after, and 6·5 per cent.
had a duration of more than 4 years. Austin Flint, excluding
cases of acute tuberculosis, found from an analysis of 112 fatal
cases the average duration to be ' a fraction over thirty-three
months.'

Dr. Fuller,[1] in 118 cases investigated by himself at St.
George's Hospital, found that by far the greater number died
from 3 to 18 months after first attack, whereas in 46 cases from
his private practice he found the usual duration varied from
1½ to 7 years, and he remarks that this discrepancy cannot be
wholly explained by the social position of the sufferers and the
advantages the latter enjoyed with respect to medical treatment, change of air and proper regimen. He accounts for it
by the greater jealousy with which the upper and more
educated classes are wont to watch their health, and note the
earlier inroads of disease. Dr. Pollock,[2] in his valuable work,
which has contributed more than any other to our knowledge
of the prognostics of consumption, gives from 2¾ to 3 years as
the average duration of 129 cases ending in death. These
occurred among 3,566 hospital out-patients, the rest of whom,
at the end of 2½ years, were living and in a state of health
favourable to the expectation of life for a considerable term.

Louis' and Laennec's cases seem to have been chiefly of a
rapid kind, treated with depletion, antimony, starvation, &c.,
or else on the expectant method, with little or no medicine.

[1] *Diseases of Chest.*   [2] *Elements of Prognosis in Phthisis.*

It has been urged in connection with this that a more acute form of consumption prevails in France, but no facts have been hitherto adduced in proof of this; and, on the other hand, if anyone compares English hospital cases of forty years ago with those graphically described in the pages of Louis, he will find the greatest possible similarity in symptoms and duration, and in what, to my mind, affords some explanation, in their treatment.

In the days of bleeding, antimony, &c., the great majority of cases of phthisis were distressing tragedies, as those who can look back on a very long experience of consumption strongly testify, and at that time the prognosis of English physicians was as unfavourable as that of French, as far as the disease was concerned, though the different constitutions of the two races may have exercised some slight modifying influence on it. Moreover, from what I have seen of French consumptive patients, during my numerous visits to the South of France, there appeared to be no material difference in the nature of the cases, but a very great one in the hygienic and medicinal treatment. Climate was almost entirely relied on; cod-liver oil, tonics, and a special dietary, though recommended, were seldom persevered with.

The estimate of the first Brompton Hospital Report refers to deaths occurring among the in-patients; and those, owing to various causes, and chiefly to their having to wait so long before admission, are exceptionally bad cases. Some died within a week after admission.

Dr. Pollock's statistics are taken from the broader and very extensive set of cases which the out-patient department at Brompton furnishes. These may be said to embrace all classes below the wealthy one, and what is more to the purpose, all varieties and degrees of the disease; the fortnightly visit to Brompton, not as a rule, interfering with the necessities of occupations or home cares, and thus securing the attendance of a large number who could not afford to become in-patients; while, at the same time, information as to the state of those not able to attend, is given through a form of note supplied to the patients at the hospital, or else by a

letter from the relatives.   Dr. Pollock's statistics, when
viewed in relation to the few deaths and the expectation of
life for the survivors, give the most favourable results for the
lower classes ever published.

The cases on which our estimate of the duration of phthisis
is founded amount to 1,000, and have been selected from private practice, the patients, for the most part, belonging to the
upper and middle classes of society, and consequently enjoying many advantages over hospital patients in the
avoidance of those ills which arise from poverty, exposure to
cold, unhealthy atmospheres and occupations, and in the
opportunities of rest, change of climate, better living, and
exercise.   As statistical information of disease among the
upper classes is rare, we hope that these statistics may prove
acceptable, as affording some facts capable of comparison with
results of hospital experience, which have been well set forth
by some of the above-mentioned authorities.

The broad definition of pulmonary consumption, as stated
at the beginning of the book, includes all the cases we have
now to deal with, and also includes the most acute forms of
the disease, as acute tuberculosis and scrofulous pneumonia;
but some of the worst forms of these are excluded owing to
the grounds of selection, which will be shortly explained.

The 1,000 cases have been selected from the records of
patients who first consulted Dr. Williams between the years
1842 and 1864, a period of twenty-two years.   The chief
ground of this selection has been the time during which the
patients have been under observation.   Considering phthisis
to be, in most instances, a chronic disease, and that observations of its course and how it can be modified by treatment
can hardly be satisfactory, unless carried on for some length
of time, we have judged it advisable to select, out of a mass of
records, those cases which have been under treatment twelve
months and upwards.

A large majority of the patients who consult physicians
are seen once only, or two or three times within a short
period, and there may be no opportunity of learning their
subsequent history.   Such cases, although supplying useful

information as to the origin and varieties of the disease, are of no value in relation to its treatment, results, or duration.

Yet because these cease to attend is no proof that they derive no benefit. Many come only to ascertain the physician's opinion, and are unable, through scanty resources, or through distance from town, to repeat their visits. We must not conclude, however, that because they do not continue to attend, they are unfavourable cases, and likely to terminate within the year. On the contrary, all the evidence at our command points to a different conclusion. Patients frequently appear on the scene years later, having, after one or two visits, been lost sight of, who have been prevented, by various causes, from visiting the physician, but had been carrying out treatment steadily.

Were we, however, to include all the cases, our numbers would be enormously swollen, but the addition could only be a large quantity of indefinite and useless material, more likely to obscure the statistics than to render them lucid.

Still we must not lose sight of the fact that certain cases of phthisis prove fatal within twelve months; for instance, the forms known as acute tuberculosis and scrofulous pneumonia, though the latter does not always terminate rapidly, but is sometimes brought by treatment into a comparatively chronic state, and in this condition may last on some years. These early fatal cases, which have been excluded, form, among the mass of consumptive patients, a very small percentage; estimated variously at three or five per cent.

Against these we would balance the much larger number of patients reported as having much improved after a few months' treatment, and as affording promise of permanent recovery. As our limitation shuts out these, the few deaths may also be fairly excluded.[1]

---

[1] In order to form some estimate of the proportions which the cases fatal within the year bear to those more or less improved within the same period, I have carefully examined the records of every case of phthisis occurring during one year, the year being selected at hazard, as a sample, from the period of 22 years. Of 433 consumptive patients who consulted Dr. Williams for the first time in 1863, 245 were seen only once, and no more was heard of them; 84 were one year and upwards under observation, and were among

As the duration of phthisis is such an important subject, we must crave the indulgence of our readers, if we give a considerable amount of preliminary information about the 1,000 patients, before stating the results of the statistics.

The cases were extracted from the note books of Dr. Williams, and arranged in tables containing twenty-five each, under the headings of

Age.	Date of first Visit.
Sex.	State of Lungs, as evidenced by
Family Predisposition.	Physical Signs.
Date of first Symptoms.	Treatment by Medicine, Climate, &c.
Origin of the Disease.	Result.
Occurrence of Hæmoptysis.	Duration.

The obituary of the newspapers has been closely watched, and where the patients had been lost sight of for several years,

those, therefore, selected for our tables. Of 104 patients whose subsequent history was known for periods under one year, 8 died, 13 were at the last visit rather worse, 3 were about the same, 75 were more or less improved, and 5 were quite restored to health. Thus those improved and cured were ten times more numerous than the deaths. It can hardly be said then, that in taking the fact of the patient being at least one year under observation as the basis of our selection, we increase the balance of favourable results, but we thereby deal with facts more carefully observed, and more conclusive in relation to the real efficacy of treatment.

Note by Dr. C. J. B. WILLIAMS : - In determining in the first instance to select for analysis only those cases which had been under my care for a year and upwards, I was guided by the desire to obtain more sure and reliable results than could accrue from cases during shorter periods of observation. I wished to ascertain the power of nature, aided by art, to control or arrest the course of pulmonary consumption; and knowing the deep-seated and enduring nature of the disease, I distrusted all results not confirmed by time, and I rejected reports of temporary amendment or even cure, as unsatisfactory and inconclusive. Deaths were indeed conclusive, although not satisfactory ; but the few deaths which did occur within that period were the issue of that degree and form of the disease, over which treatment never had, or is likely to have, any control. I already knew such cases to be hopeless—too rapid and overwhelming to be stayed by human power— therefore I put them out of calculation. I am quite content if our accounts are debited with the three or five per cent. which such deaths may be supposed to amount to; for although, as my son argues, the ten times more numerous 'improved and cured' cases may be 'set off' against them, yet this is balancing a certain against an uncertain quantity, which brings no definite result. But it cannot be fairly said that our selection gives nothing but chronic cases, for many of the cases are acute at commencement, or in some part of their course, and are reduced to a chronic state by treatment.

without having been announced as dead, a correspondence was opened, either with themselves or their friends, to ascertain whether they were alive and in what state of health—a correspondence which, when addressed, as it often had to be, to the individual whose life was suspected, sometimes evoked ludicrous answers. Reference to the various lists, the 'Army and Navy,' 'University,' 'Clergy,' and 'Law,' to the 'Court Guide,' and to the 'Peerage,' has often afforded valuable information; and on this point private practice has great advantages over hospital practice, for in respect of the former, by some means or other, patients can be traced through a number of years, whereas in the latter they are generally lost sight of when they quit the hospital.

Nevertheless, a certain number of the tabulated cases could not be traced up to the present time; and of these the date when last heard of, with notice of their state, is registered.

SEX.—Of the 1,000 cases, 625 were males and 375 females, or 62·5 per cent. of the former, and 37·5 per cent. of the latter. The preponderance of males cannot be regarded merely as accidental, for it is closely in accordance with the evidence of the first report of the Brompton Hospital, where the percentage of males was 61, and that of females 39. Among Dr. Pollock's out-patients, 60·75 per cent. were males, and 39·25 females.

AGE.—The ages of the patients have been arranged in the following table. The table differs in one point from many

Age at Time of Attack of 1,000 Cases of Phthisis.

Age at Time of Attack	Males	Percentage	Females	Percentage	Total	Percentage
Under 10 years .	10	1·60	3	·80	13	1·3
10 to 20 . . .	86	13·79	96	25·60	182	18·2
20 to 30 . . .	245	39·20	173	46·13	418	41·8
30 to 40 . . .	183	29·28	66	17·60	249	24·9
40 to 50 . . .	70	11·20	24	6·40	94	9·4
50 to 60 . . .	22	3·52	8	2·33	30	3·0
60 and upwards .	9	1·44	5	1·33	14	1·4
	625		375		1,000	

Average age at time of attack, Males .   .   .   29·17 years.
        „   „   „ Females .   .   26·06 years.

similar records.   Instead of the age at first visit, the age at
first attack is tabulated ; and this is arrived at by subtracting
the history from the age at first visit.   The date thus obtained is of far more consequence in estimating the duration
of the disease, and the conditions which modify it, than the
age at first visit, which depends upon shifting circumstances ;
as, for instance, the feelings and opportunities of the patients,
who may come under the observation of the physician either
at the commencement of their disease, or many years after,
near its termination.   The record of their age at the time of
the first visit would therefore afford us but slight information
as to the time of attack or its duration.   It may be objected,
that it is difficult to arrive at accuracy as to the date of first
symptoms.   And undoubtedly this is true in the case of hospital patients, with whom it is necessary to pursue a system of
close cross-questioning, in order to evoke the necessary information.   Dr. Fuller[1] truly says on this point : 'The average
duration of the complaint is ordinarily, I believe, very much
understated, from the fact that the inferences respecting its
duration are drawn from the statements of hospital patients,
who pay little heed to the earlier, and, as they imagine, unim-
portant symptoms of the disease, and pertinaciously date their
malady from the occasion on which they first experienced pain
in the chest, or were frightened by the occurrence of hæmoptysis, or found themselves unequal to their daily work.'
Private patients, with whom we have now to deal, hardly err
on this side, for the upper classes generally remember and
narrate, almost too fully for the physician, every symptom,
early or late, of their illness.

   The results of this table accord with the commonly received
opinion as to the period of attack.   Taking the sexes collec-
tively, 41 per cent. were attacked between 20 and 30 ; about
25 per cent. between 30 and 40 ; 19·5 per cent. under 20 ; and
13 per cent. under 50.   When we examine the relative liability of the two sexes in the various decades, we find some
important differences to exist.   Between 20 and 30—the most
common period of attack for both sexes— about 7 per cent.

[1] *Op. cit.* p. 413.

more females were attacked than males; and again, between
10 and 20, 11·8 per cent. more.   On the other hand, after 30
the reverse was the case.   Between 30 and 40 the males attacked exceeded the females by 11·68 per cent., and above 40
by 6 per cent.   These results may be said nearly to agree
with those of the first Brompton Report.

The average age of attack was—for the males 29·47, and
for the females 26·06.

Family Predisposition.—The results under this head have
already been given in Chapter XV., to which the reader is
referred.   This feature is traced in 48·4 per cent. of the 1,000
cases.

Origin and First Symptoms.—In 385 cases the disease
came on without any antecedent illness, and was characterised
by the usual group of symptoms, more strongly marked in
some cases than others, and it pursued its course free from
complications, other than the ordinary ones of phthisis.   In
315 cases it followed closely after other diseases, as the subjoined table will show :—

Phthisis was preceded by Pleurisy and Pleuro-pneumonia in 149 Cases.

,,	,,	,,	Bronchitis	118	,,
,,	,,	,,	Asthma (spasmodic) . . .	7	,,
,,	,,	,,	Scrofulous Abscesses . . .	12	,,
,,	,,	,,	Fistula	5	,,
,,	,,	,,	Whooping Cough . . .	6	,,
,,	,,	,,	Croup.	1	,,
,,	,,	,,	Scarlatina	4	,,
,,	,,	,,	Measles	2	,,
,,	,,	,,	Continued Fevers . . .	3	,,
,,	,,	,,	Peritonitis	1	,,
,,	,,	,,	Malformation of the Chest . .	2	,,
,,	,,	,,	Injuries to the Chest and other		
			Organs	5	,,

315

The number complicated with pleuro-pneumonia and bronchitis is very large, reaching a total of 267, or more than one
quarter of the whole, and deserves attention as showing statis-
tically the influence of these diseases as predisposing the
patient to bacillar attack.   It is well-known to physicians
connected with hospitals for diseases of the chest, how often a

neglected case of pneumonia or bronchitis, under depressing causes, passes into one of consumption ; but statistics proving this frequency are rare, if not wanting. This number, 267, or 26·7 per cent., is high, considering that it is taken from a class which has opportunities of protecting itself from many depressing conditions ; but high as it is, it is probably much below a correct estimate for hospital patients, among whom the prevention or rapid cure of these diseases is much more difficult, and therefore less common than among their wealthier brethren.

In the cases of phthisis arising from pleuro-pneumonia or interstitial pneumonia, the course of events was generally as follows. After the attack, some portions of the lungs remained consolidated or compressed by dense pleuritic adhesions, or both these lesions existed, and tended to cripple the lungs for their respiratory work.   The breath remained short ; the patient seldom or never lost the cough, which a fresh cold or some disordering influence caused to increase, muco-purulent expectoration and sometimes hæmoptysis accompanying it. Signs of softening were detected in one or both lungs, followed by those of excavation, and the case assumed a consumptive aspect.

Of the cases of phthisis following bronchitis, which may be termed catarrhal phthisis, some followed acute attacks, others chronic.   These last patients generally lost their cough and other symptoms in the summer, or in warm weather, but were subject to a return of them every winter, or during inclement weather.   A longer, or more severe attack than usual, greatly prostrated them, and the cough now remained persistent, and was also accompanied by permanent feverishness, heat of skin, and wasting.   On examination of the chest, in addition to the ordinary bronchitic sounds, patches of consolidation were detected ; these did not clear up, and in some cases softening and excavation eventually took place, and it was evident that bacillar destruction had been proceeding.

Of the 149 cases following on pleuro-pneumonia, in 85 no family predisposition could be traced ; and this was also the case in 57 out of 118 instances complicated with bronchitis.

Hæmoptysis.—This symptom was recorded to have been present in various degrees, at some period of the patient's history, in 569 cases out of the 1,000 ; *i.e.* 57 per cent.—a percentage lower than that of the First Medical Report of the Hospital, which was 63 per cent., but nearly agreeing with that of Dr. Cotton's [1] 1,000 hospital cases, which was 53·6 per cent., and that of Dr. Pollock's [2] 1,200 hospital cases, which was 58·4.

State of the Lungs as evidenced by Physical Signs.—We shall now endeavour to describe, as briefly and succinctly as possible, the state of the lungs of these patients when they came first under observation, and afterwards to give some report of the changes which had taken place at the date of their last examination ; and the reason we do so is to give our readers some account of the local changes, whether for the worse or better, which took place in these patients, and thus enable them to form an opinion as to how far the improvement in the general health was accompanied by improvement in the state of the lungs. The relation, or in many cases the want of relation, between those two, must strike all physicians. How often does a patient gain flesh and strength and colour, and improve in breathing in a few months, and yet the physical signs show no perceptible improvement, but remain stubbornly at about the same ! The converse is more rare, though we have known instances of cavities contracting and the general health making no great progress.

The record of the physical signs has been perhaps more carefully carried out than any other point in these cases ; and in perusing it, a fair idea can be easily obtained of the amount of disease present in each case, with its subsequent progress ; but the selection of similar cases for the purposes of statistics, and their arrangement into as few classes as possible, has been attended with great difficulty. The classification of the conditions of the lung, consolidation, softening, and excavation, into first, second, and third stages, is open to objections, because such stages are not always well defined, it being difficult, sometimes impossible, to distinguish between the end

[1] *Op. cit.*   [2] *Op. cit.*

of the second and the beginning of the third, and again various parts of the same lung may be in different stages. What different amounts of consolidation, too, may not the first stage include! Sometimes only a small portion of the lung, like that underlying the supra-scapular or the inter-scapular, or the infra-clavicular region, is consolidated; in other cases two-thirds or more are involved. However, it has been found difficult to avoid some such classification for the purpose of statistics, and therefore that of stages has been adopted, with the understanding that the first stage embraces various amounts of consolidation, and the second and third are some-times only different degrees of the state of softening and excavation. In none of the present cases is the evidence of physical signs alone accepted; in all it has been amply confirmed by the clinical symptoms and the course of the disease. The results have been embodied in a table, divided into two parts, showing the 'state at first visit,' 'state at last.' From this it will be seen that 660 patients or two-thirds were in the first stage at the first visit; 181, or 18 per cent., in the second; 145, or 14·5 per cent. in the third ; and 14 patients presented the physical signs of other lung diseases, namely, bronchitis, pneumonia, pleurisy, and asthma, on which shortly afterwards supervened signs of consumption. Those in the second and third stages hardly constituted a third of the total, which shows how large a proportion came in the stage of consolidation, of which the prognosis was likely to be more favourable. As regards the relative liability of either lung to disease, of those in the first stage both lungs were affected in 282; the right alone in 241, and the left alone in 137. Of those in the second stage, 46 had the right alone affected, 59 the left alone; 76 had both lungs involved, and in many in-stances both in the second stage. Of the 145 in the third stage, 29 had the right lung alone affected ; 34 the left, and 35 both ; but in only 4 cavities were detected in both lungs. This indicates a greater liability of the right[1] lung to consoli-

---

[1] This agrees with Laennec's conclusions; but it is at variance with Louis' and Cotton's, both of whom found the left lung more frequently affected.

Stage	No	Percentage	State at First Visit	State at Last Visit						
				Dead	Healthy	Improved	About the same	Worse	Unknown	
1st	660	66·0	241 had the right lung alone affected / 137 had the left lung alone affected / 282 had both lungs affected	104	30	184	60	233	49	= 556
			660							
2nd	181	18·1	46 had the right lung alone affected / 20 had the right in the 2nd stage and the left in the 1st	20	1	17	6	21	1	= 46
			59 had the left lung alone affected / 45 had the left lung in the 2nd stage and the right in the 1st	25	1	34	7	32	5	= 79
			11 had both lungs in the 2nd stage . . .	3	—	2	—	3	3	= 8
			181	48	2	53	13	56	9	= 133
3rd	145	14·5	29 had the right lung alone affected / 5 had the right lung in the 3rd stage and the left in the 2nd / 37 had the right lung in the 3rd and the left in the 1st	22	—	24	12	7	6	= 49
			34 had the left lung alone affected / 1 had the left lung in the 3rd and the right in the 2nd / 35 had the left lung in the 3rd and the right in the 1st	22	1	17	12	17	1	= 48
			4 had both lungs in the 3rd stage . . .	—	1	2	1	—	—	= 4
			145	44	2	43	25	24	7	= 101
	14	1·4	presented physical signs of other diseases, but the signs of phthisis supervened after first visit / 4 had signs of bronchitis / 4 „ pleurisy / 3 „ pleuropneumonia / 1 had signs of asthma / 2 had doubtful physical signs	2	—	—	4	8	—	= 12
			14 Totals	198	34	280	102	321	65	= 802

dation, but of the left[1] to softening and excavation : a conclusion confirmed by the evidence of the second report of the
Brompton Hospital, and by other authorities.   Having briefly
considered the state of the patients at first visit, let us turn
our attention to their state at last report.   Of the 1,000
patients, 198, or nearly one-fifth, died ; the deaths being distributed as follows :—

Of those who came in the first stage, 104, or 15·75 per
cent. were ascertained to have died.

Of those who came in the second, 48, or 36·51 per cent.

Of those who came in the third, 44, or 30·34 per cent.

Thus we see that the percentage of mortality of those who
came in the second and third stages was very much higher
than those of the first ; the third showing actually a double
proportion of deaths ; and the fact must not be overlooked as
demonstrating that, although cavities may be tolerated for
years, yet the danger from blood infection, and from hæmoptysis through rupture of pulmonary aneurysms, after their
formation, is considerably increased.   In 80 out of the 152
in the first and second stages, cavities were ascertained to have
formed before death, but they probably were present in a
larger number.

The state at last visit of the *living* patients is arranged
under five headings : (1) *Healthy* ; where the physical signs of
disease had entirely disappeared, and could no longer be detected. (2) *Improved*. (3) *About the same*.   This last term
is used to include, not only the cases in which no change has
taken place, but also those which, after various fluctuations
towards better or worse, presented at the last about the same
amount of disease as at the first. (4) *Worse*.   This heading
is intended to signify extension of the disease, either in the
same lung or the opposite one, as well as progress in the way
of softening and excavation. (5) *Unknown*, signifying that
no recent physical examination had been made.

The table shows that among 802 living patients, the last

---

[1] Cotton, Walshe, and Pollock confirm this, but Dr. Ewart's figures show
that the difference of liability between the two sides is not very great.
Vide p. 51.

recorded state of the lungs was 'healthy' in 34 ; 'improved' in 280 ; 'stationary' in 102 ; 'worse in 321 ; and 'unknown' in 65. Excluding the unknown ones, the relative percentages are:—'healthy,' 4·6 per cent ; 'improved,' 38 per cent ; 'worse,' 43·53 per cent, and 'stationary,' 13·39 per cent. If we take the cases in stages, and compare the numbers under 'healthy' and 'improved' with those under 'worse,' we find that, whereas in the first stage the 'worse' somewhat outnumber the 'improved,' in the second they are nearly equal; and in the third the ratio is entirely changed, the number of the 'healthy' and 'improved' being nearly double that of the 'worse.'

Some further particulars about the changes that took place in the lungs may not be unacceptable. Where registered as 'healthy' or 'improved,' the improvement in the physical signs of patients in the first stage consisted of dulness diminishing, either in extent, or degree, or in both ; of the breath- and voice-sounds becoming less tubular, and more vesicular ; and, in some few instances, of the signs disappearing altogether, the percussion- and breath-sounds being normal. In those of the second stage the crepitation diminished, and was replaced by breathing generally having some roughness or tubular character, which, in some instances, eventually gave way to healthy sounds. The favourable change in the physical signs of the third stage was shown by the dulness decreasing, the moist cavernous sounds becoming croaking and drier, and pectoriloquy being less marked and audible over a smaller portion of the lungs, sometimes being replaced by the dry whiffing or crackling sounds of emphysema, but generally by tubular breathing and bronchophony. These last signs have, in some instances, disappeared, except above and within the scapula, where, with some remaining dulness, they generally could be detected after they had vanished from other parts of the chest.

The cases of restoration to complete health number 34, and include 30 recoveries from the first stage, 2 from the second, and 2 from the third stage, in one of which, wonderful to relate, were cavities in both lungs—but they were small,

and the long duration of the case, viz., 22 years, afforded
time for their contraction and obliteration. In 16 cases out
of the 1,000, calcareous expectoration is noted; in 20, contraction of cavities; in 2, contraction of the lung without the
formation of a cavity; and in 16, emphysema of the lungs
were recorded. So much for the 'improved' and 'healthy'
classes.

Under the heading of 'worse' we find that, in cases of
the first stage, in 77 or 15·18 per cent., cavities formed in one
lung; in 10 in both lungs; and that softening took place in
24 others. Of those in the second stage at first visit, cavities
are reported to have formed in 32, or 28·8 per cent.

In order to arrive at satisfactory data as regards extension
of the disease from one lung to the other, the results of the
deaths were included, and thus the whole number of cases was
brought into use. We find that, exclusive of 454 patients
who had disease of both lungs, 546 had one lung only attacked at first visit; among these a spread of disease to the
other lung was noted in 107 instances. The numbers indicated that, after a certain period, the disease has less tendency
to spread, but is rather apt to remain limited to one lung.
As regards the relative tendency of the two lungs, the right
seems rather more liable to extension than the left, and this
greater liability exists in whatever stage of disease the lung
may be.

The results of the changes in the lungs may be summed
up as follows:

A cure was effected in 4·6 per cent. of the cases; great improvement in
38 per cent.; the disease was stationary in 13·4 per cent.; but in 13·5 per cent.
there was more or less increase.

The right lung was attacked more frequently than the left; but the left,
when attacked, was more prone to softening and excavation.

Where the disease extended from one lung to the other, the right lung was
more liable than the left to such extension.

In former times it was hardly admitted that phthisical
disease of the lung was ever cured, though it might be sometimes arrested. The 34 cases, however, mentioned as cured,
were undoubted instances, as far as the disappearance of all

physical signs can attest the fact. Yet how few were they, contrasted with the whole number of phthisical cases, and especially with the 'improved' class, in which the various steps towards arrest of the disease were to be found.

We have now laid before our readers sufficient information to show the nature of these cases ; and we think that we are not far out in stating that they include and fairly represent all forms of phthisis, except the very acute cases, which are rare. The main questions of this chapter can now be considered. How long did these patients live ? What did they die of ?

Of the 1,000 patients, 198 are ascertained to have died ; and the greater part of these succumbed to the gradual waste and decay of phthisis ; 15 died of phthisical complications as seen below :—

4 died of hæmoptysis.	2 died of dropsy (from contraction of	
1 „ hæmoptysis and diarrhœa.	lung).	
2 „ diarrhœa.	3 „ pneumothorax.	
1 , diarrhœa and dropsy.	1 „ emphysema.	
	1 „ ulceration of the intestine.	

How long did these patients live ?—

8 lived 1 year and under 2	31 lived 10 years to	14 inclusive
22 „ 2 years „ 3	12 „ 15 „	19 „
18 „ 3 „ „ 4	9 „ 20 „	30 „
23 „ 4 „ „ 5	— —	
75 „ 5 to 9 inclusive	198	

Of 21 patients who survived their first attack from 15 to 28 years,—

2 lived 15 years.	1 lived 22 years.
2 „ 16 „	2 „ 24 „
6 „ 17 „	1 „ 26 „
1 „ 18 „	2 „ 28[1] „
1 „ 19 „	—
3 „ 21 „	21

The average duration of the disease in these 198 patients was 7 years 8·72 months, *the highest average duration among deaths from phthisis yet published.*

---

[1] Cases of longer duration are included in the abstracts of cases published, but in these death occurred after the completion of the statistics.

The chronicity of these cases is very remarkable; and it may be noted that 64 per cent. lived five years and upwards, while only 36 per cent. lived less than that period. In the above list the greatest number is included under ' 5 to 9 years;' the next under ' 10 to 14,' and the smallest number under ' 1 to 2 years.' Taking the duration of life by stages,—

In 106 of the first stage, the average duration was 7 years 11·8 months (nearly 8 years).

In 49 of the second stage the average duration was 8 years ·04 months.

In 43 of the third stage the average duration was 6 years 8·3 months.

What results do we obtain from the 802 patients who were alive when last heard of? The average duration of life in these has been 8 years 2·19 months, a somewhat higher duration than among the deaths (which were probably the worst cases), and one, which considering the still favourable state of many of the patients, bids fair to increase further.

The average was thus composed :

71	have lived	1 year and less than	2		124	from	10 years	to	15
97	,,	2 years	,,	3	54	,,	15	,,	20
96	,,	3 ,,	,,	4	65	,,	20	,,	30
68	,,	4 ,,	,,	5	3 have lived 30 years and upwards.			
221	from	5	to	10				

This table shows that 332 or 41·4 per cent. have lived from 1 to 5 years, and that 470 or 58·6 per cent., have already lived 5 years and upwards. The class of 10 to 30 years' duration is a large one, forming 30 per cent. of the whole, and affords remarkable evidence of the chronicity of the disease. Still more remarkable is the fact of as many as 68 patients having lived 20 years and upwards, and the distribution of these it is worth our while to note further.—

11	have lived 20 years.		3	have lived 28 years.	
7	,,	21 ,,	1	has lived 29	,,
13	,,	22 ,,	1	,,	33 ,,
12	,,	23 ,,	1	,,	36 ,,
10	,,	24 ,,	1	,,	47 ,,
3	,,	25 ,,	—		
2	,,	26 ,,	68		
3	,,	27 ,,			

The question naturally arises as to the state of the 802

living patients at last report? Were they complete invalids, lingering out a miserable existence? or was their health sufficiently good to permit of their returning to the duties, if not to the pleasures of life? Observation on this point leads us to divide the patients into three classes:—

Firstly, those who have apparently quite recovered their general health, and are able to follow their occupations without any recurrence of their former symptoms. These we describe as 'well.'

Secondly, those who are able to follow their occupations more or less actively, but, owing to their being subject to a return of their symptoms, are obliged to use precautions and to limit their exertions. These we designate 'tolerably well.'

Thirdly, those who are obliged to devote themselves entirely to the care of their health are described as 'invalid.'

The 'well' class numbered 285, or $35\frac{1}{2}$ per cent.; the 'tolerably well' 293, or $36\frac{1}{2}$, and the 'worse' 224, or 28 per cent. The two first classes, therefore, comprise 72 per cent. of the whole, and show a great preponderance over the 'invalid' class, which is only 28 per cent. This is remarkable, and proves what reparative power nature puts forth, if only the time is allowed for her to do so. In considering the patients, we must remember, that though their social status exempted them from absolute want, it by no means exempted them from exposure to other injurious influences. Among these patients were men of every profession, members of parliament, officers in the army and navy, clergymen, practitioners of law and medicine, men of business, &c., who were therefore liable to the trials consequent on each calling: as exposure to great varieties of temperature, from which military and naval men suffer; or close confinement in hot rooms and occasional pressure of work, the lot of many professional and business men; or again, the strain on the lungs which public speaking entails on members of parliament, clergymen, barristers, public lecturers, and the like. When we remember these facts, it must be considered highly satisfactory that so large a majority are found in the 'well' and 'tolerably well' classes. The greater part of the 'well' class could not be distinguished in ordinary life from healthy persons, and many are sufficiently strong to undertake exertion of an arduous kind, whether physical, like long walks and mountain ascents, or mental, like close application to study or business.

Numerous illustrations of the arrest and retardation of the disease will be found among the 'Abstracts of Cases.'

If we compare our results with those of the authorities given at the beginning of the chapter, we find Dr. Fuller's 46 private cases to be the only ones which resemble our own in duration ; which might be expected, as they are taken from the same class of society.   As regards the French authorities, the contrast is most striking ; the average duration of our 1,000 cases is four times greater than Louis' or Laennec's, and far exceeds any estimate yet formed.   We must not forget that the restrictions which we have adopted exclude the very acute cases; this very limitation, however, indicates a decidedly favourable inference, viz., that if the average duration of life, in consumptive patients who survive their first symptoms one year, is 7 or 8 years, and the possibility of longer life, extending to 10, 15, 20, 30, 40 years, and even to the natural term, is often realised ; then surely the time is come when we can hold out a fairly hopeful future to the consumptive patient.   We can tell him that if he is prepared to make certain sacrifices of time, of money, and of liberty for some years ; to rigidly carry out certain common-sense rules which long experience of the disease inculcates, he may, under favourable circumstances, live on for a long period, even to the ordinary span of life ; and, as he lives on, may gain sufficient strength to resume his former occupations and duties.

The long duration of these cases may be attributed—

Firstly, to the early detection of the disease, two-thirds of the patients being in the first stage when they came under observation.

Secondly, to the perseverance with which they carried out the various healing measures at their disposal, whether medicinal, hygienic, or climatic.

The average duration of pulmonary consumption having been ascertained, let us see how far it may be modified by certain varying conditions, as those of age, sex, origin, hæmoptysis, and family predisposition.

The influence of the last two elements on duration having been fully discussed in their respective chapters, need not be entered into here, but we will direct our attention to the other points.

Sex exercises an important influence on duration. Among females the disease lasts a shorter time than among males, as the following abstract from our tables demonstrates :—

Average duration of disease in 119 males (dead), 8 years 4·72 months.
„   „   „   „   79 females „   6   „   8·67 „

This shows a difference of 1½ years in favour of the former. When we call to mind that the age of attack with females was earlier than with males by an average of 3½ years, we see clearly that women succumb much more quickly to the fatal disease. This is borne out by an examination of the average age reached by the sexes before death. The females died, on an average, at 34½, the males, on an average, at 40—showing a difference of 5½ years between the expectation of life in the two sexes.

Age has always been held to exercise considerable influence on the duration ; and, according to our researches, the age of the patient at the time of attack exhibits this most strongly.

Our annexed table shows that this feature is more marked in the males than in the females :—

TABLE.—*Showing Influence of Age of Attack on Duration in 198 Deaths.*

Age when attacked	Males	Duration yrs.—mo.	Females	Duration yrs.—mo.	Total	Duration yrs.—mo.
Under 10 years .	2	16 –11	1	7—0	3	[1]13— 7
10 to 20 . .	15	6— 6	19	6— 6·47	34	6— 6·23
20 to 30 . .	40	8 – 9·12	35	6— 6·97	75	7— 4·28
30 to 40 . .	36	8 –11·13	15	6—10·06	51	8— 3·76
40 to 50 . .	15	8 – 2·20	3	6—10	18	[1]7 –11·5
50 to 60 . .	8	8— 0·12	2	6— 4	10	[1]7— 8·1
60 and upwards	3	2—11·66	4	8— 5·25	7	[1]6— 1·14
	119		79		198	

Of those attacked in the decade 10 to 20 the duration is the

[1] Numbers too small to yield a fair average.

same for both sexes; but of those attacked between 20 and 30 the duration for the males is 8 years 9 months, for the females 6 years 7 months—a difference of more than two years; and the result was much the same in the decade from 30 to 40.

TABLE.—*Showing Age at Death of* 198 *Patients.*

Age at Death	Males	Females	Total
Under 10 years .	0	0	0
10 to 20 . .	6	7	13
20 to 30 . .	19	32	51
30 to 40 . .	38	17	55
40 to 50 . .	28	14	42
50 to 60 . .	19	4	23
60 to 70 . .	8	1	9
70 to 80 . .	1	4	5
	119	79	198

The ages that some of our patients reached were remarkable; about 50 per cent of the males, and 29 per cent. of the females, survived 40; 9 males and 15 females lived over 60; and of these 1 male and 4 females lived beyond 70. The majority of the males died between 30 and 40, and of the females between 20 and 30. These are the results which our 198 deaths give; but we may add that they are amply confirmed by the evidence, as far as it goes, of the living cases. 21 per cent. of these were between 40 and 50, at last report; 13 per cent. between 50 and 60; 21 per cent. between 60 and 70; 5 per cent. between 70 and 80; and one patient was over 80. Thus, 60 per cent. have passed 40.

The results of the influence of age and sex on the duration of consumption may be thus summed up :—

1stly.—The duration is longer in proportion as the age of attack is later, the retarding influence of age being more conspicuous among males than among females.

2ndly.—Among the females the time of attack is, on an average, earlier than among males.

3rdly.—The duration of the disease is shorter.

4thly.—The age reached by consumptive females is less.

These conclusions naturally give rise to some speculations

as to the different operation of the causes, exciting or pre-disposing, of phthisis on the two sexes.

Although phthisis is more common among males than females, yet we see that the female frame, when it is subjected to the action of any decided cause of phthisis, offers less resistance to its attack. Why is this? Is it because the causes which particularly affect the female sex are more powerful and less likely to be resisted, or is it because females attain full development and growth at an earlier age than males?

If the former supposition were true, and disordered menstruation, pregnancy, and lactation were more potent causes than any which affect men, it follows that the disease would be more frequent among women. This not being the case, it seems reasonable to have recourse to the latter hypothesis, and to suppose that a ratio exists between the period of cessation of growth and development in each sex and the onset of the disease. The shorter duration of the malady in females may be explained by the stronger frame and better power of resistance possessed by the male, which enable him to battle with the disease for a longer time, and allow more chance for treatment, &c., to have effect.

Origin.—Does the mode in which a case of phthisis commences affect its duration? or is it immaterial, when the bacillar disease has once attacked the lungs, what the pre-disposing cause may have been, whether it be inflammation, or starvation, or the like?

There is little doubt that this is not the case with phthisis, for the mode of origin has great influence over the form and character of the disease and its duration. Compare a case of phthisis preceded by bronchitis, which is gradual and local in its development, and the general eruption of miliary tubercle following an attack of typhoid fever! Our statistics do not at present include a sufficiently large number of instances of the different modes of origin to estimate the effect of each on the duration, but we are able to do so in the case of inflammatory attacks. It will be remembered that in 149 of the cases the disease was preceded by attacks of pleurisy and pleuro-pneumonia, from which the patients recovered,

with lungs more or less crippled by adhesions, by consolidations, or by both.   Did these patients live a longer or shorter
time than the average ?   Among 29 who have died, the mean
duration was 9 years 6¾ months, and the 120 who still survive
have on an average also lived 9½ years, thus exhibiting an
extension of life beyond the ordinary, of nearly two years for
cases complicated by inflammation.   In 64 of these cases
hereditary taint was traced ; but it is not worth while to consider the duration of these separately, as the number of deaths
is small, and it has been already demonstrated that family
predisposition exercises no curtailing influence over the duration of the disease.

To further investigate the influence of the inflammatory
complication on the duration of consumption, at Dr. Burdon
Sanderson's suggestion, I selected a small number of cases
which exhibited the inflammatory connection most strongly,
and were entirely free from family predisposition.   Not only
did the disease follow directly after the pneumonic or pleuro-
pneumonic attack, but in every case lesions more or less
extensive, the result of such attack, remained behind, and
were easily detected by the physical signs.   The duration of
these cases confirms still more strongly the conclusion, that
inflammatory origin has a prolonging influence over the
duration of phthisis.   Among 10 patients who have died, the
average duration was 12 years 10 months : among 20 who
still survive, it is 11 years 8¼ months.

Bronchitis ushered in the disease in 118 patients —19 dead
and 99 living.   Here a different conclusion presents itself,
though we hesitate in accepting it on account of the small
number of deaths.   The average among these was less than
6 years ; among the living 99 it was slightly over 8½ years ; a
great contrast to the deaths, and one which rather invalidates
any conclusion arising from them.   We may assume, however,
that if the origin from bronchitis has any prolonging
influence on the duration of phthisis, it is not equal to that of
pleuro-pneumonia.

CHAPTER XXII.

TREATMENT—PROPHYLACTIC AND ANTIPHTHISICAL.

History of the treatment of Consumption—Dr. Williams' testimony as to the great improvement—Effect of introduction of cod-liver oil—Prophylactic and Antiphthisical measures suitable for childhood—For adults—Food —Importance of milk Koumiss—Kéfir, artificial koumiss—Meat—Transfusion of blood—Vegetables—Stimulants, their uses - Clothing, under- and over-clothing—Respirators—Habitation, site, soil, surroundings, and shelter — Cubic space and Ventilation—Importance of exposure to sunshine—Sleeping with open windows—Removal of dust—Destruction of tubercular sputum—Exercise—Its aims, in Consumption—Active exercises—Rowing— —Swinging Hodge's gymnast – Skating—Tobogganing—Walking—Mountain ascents - Passive exercise—Carriage—Sailing—Riding Bicycling and tricycling—Massage -- Its results—Aero-therapeutics—Effect of Compressed air baths—Rarefied air—Mountain climates—Baths—Twofold effect of douches–Sponging better than immersion Mineral waters—Sulphur - Arsenic—Bromo-iodine.

THE treatment of phthisis has undoubtedly undergone some modifications since the discovery that it belongs to the group of diseases, of which our knowledge is daily extending, due to the action of organisms ; and necessarily our measures must be directed to destroying and eliminating the tubercle bacillus from the body.

This, however, is not all, for our present knowledge goes to show that a soil suitable for the cultivation and multiplication of these organisms is as necessary as the organisms themselves; and while the conditions of daily life and intercourse render it probable that most persons are liable to receive the bacillus into their lungs or intestines, in only a comparatively small number does it settle and colonise. The fact of the lungs and air passages being usually the first and often the sole seats of attack, indicates that the organisms are generally conveyed through the air, and another fact, viz., that tubercular ulceration of the intestines is usually associated with excavations of the lungs, and often with laryngeal ulceration, points to the intestinal lesion being due to re-infection from swallowing tubercular sputum.

The fact of some individuals being more prone to the destructive action of these tubercle bacilli than others, has been explained by peculiarities in the structure of the epithelium of the trachea and bronchi, such, for instance, as the ciliæ being less abundant or absent in the former case, but we can hardly assign to the epithelium the sole resisting power of the lungs, and certainly complete protection is not given by the epithelium of the trachea and larger bronchi, as we know by Veragut's experiments, that the bacilli-laden air may penetrate into the alveoli.

We must not be too certain that because bacilli enter the alveoli, and by their irritating influence give rise to epithelial proliferation, that therefore they succeed in destroying the nuclei of cells and the cells themselves. Metschnikoff[1] of Odessa's interesting account of the Daphnia and its struggles with the spores of a fungus furnishes an illustration of the methods and powers of resistance employed by insects. The blood corpuscles of this insect, like those of other invertebrates, are colourless, and move about in lacunar spaces. When these spores gained entrance into the tissues of the daphnia, the corpuscles attacked them and digested the spores, but if a spore was too much for one cell, two, three, or even more, assisted and fused to form a giant cell which repelled the invaders. It thus became a mortal combat between the daphnia and the spores. If the cells conquered the spores, the daphnia lived, but if the invaders conquered, then the daphnia, overrun with spores, died. Similarly we may conclude, that in animals the resistance of the individual to such destructive parasites as the tubercle bacillus is shown in, first, cellular proliferation around the invading bacillus; secondly, in the formation of giant cells. The conversion of these at a later date into fibrosis is always fatal to the life of the bacillus, and thus it should be our aim in every way to promote the development of this tissue.

[1] Quoted by Mr. J. Bland Sutton in *Proc. Roy. Med. Chir. Soc.*, December 1884.

The treatment of phthisis must always include a large share of constitutional measures which, if applied before the onset of the disease, act in a prophylactic sense, and prevent the occupation of areas of lung territory by the bacillus, or, if carried out vigorously in the early stages of the disease, do much to aid the system powerfully to react against the invader now in occupation, to repel further advance, and to limit its destroying power.

By constitutional measures we thus fortify an individual against the bacillar attack, and if we look back on the history of the treatment of phthisis, we shall find that as long as it consisted solely of measures directed to reducing the local manifestations of the malady, the disease was unchecked in its fatal career, but that when constitutional measures, such as tonics, cod-liver oil, generous diet, and appropriate climates were brought to bear on the consumptive, the duration of life was gradually prolonged. Perhaps the best example of this was given by Dr. C. J. B. Williams in the first edition of this work, where he stated that, during the first 10 years of his practice, the beneficial effect of treatment was limited to incipient cases, and specially to those who were able at an early stage to take long voyages, such as those to Australia and to India. Dr. Williams says further : ' My general recollection of the histories of the developed disease at that time is that of distressing tragedies, in which no means used seemed to have any power to arrest the malady ; and life was rarely prolonged beyond the limit of 2 years, assigned by Laennec and Louis as the average duration of the life of the consumptive.

' In the next period of 10 years (from 1840 to 1850) a marked improvement took place in the results of treatment, apparently in connection with the allowance of a more liberal diet, and the habitual use of mild alterative tonics, as they might be termed, particularly iodide of potassium with sarsaparilla, or other vegetable tonics. These were first given in conjunction with liquor potassæ, or an alkaline carbonate ; but the lowering effect of the alkali led to the substitution of a mineral acid, generally the nitric, and a combination of this description (iodide of potassium 2 grains, dilute nitric acid 15

drops, tincture of hops and compound fluid extract of sarsaparilla of each one drachm, with an ounce of water or infusion
of orange-peel) became the favourite prescription, until it was
superseded by something which was much more efficacious.
Several of the early cases recorded were treated in this way
and with improved results, in respect of the general health of
the patients and diminution of the cough and expectoration.

‘ It was in the latter half of this period that chemists began
to produce cod-liver oil of sufficient purity and freshness to be
fit for the human stomach ; and I have no hesitation in stating
my conviction, that this agent has done more for the consumptive than all other means put together.’

This would probably be endorsed by all practitioners of
large experience, and important proofs have been already
supplied in the preceding chapters, but perhaps one of the
most convincing was furnished by me in a comparison of the
duration of life in 250 consumptive patients who wintered at
various foreign health resorts.[1] The total duration of life
among them, from the date of the first symptoms, was, for
those who died, 8 years, for those still alive, nearly 9 years.
Of these, 40 took the oil irregularly, or not at all, of which
number 17 died, giving a duration of 4 years and $8\frac{1}{2}$ months,
little more than half the duration of life of the total number.
Yet at the commencement of treatment these cases were not
more unfavourable than the rest. Other important constitutional influences succeeded in arresting and curing pulmonary
consumption, long before the discovery of the tubercle bacillus
and its relation to the disease, and we must conclude that
they acted by improving the bioplasm, and increasing constitutional resistance, thereby enabling the various cells to become the destroyers and not the victims of the bacillus.
Such are, among others, the so-called mountain cure, of which
more will be said later, the effects of a generous dietary and
certain tonics, such as arsenic, quinine, and the hypophosphites
of lime and soda.

[1] ‘ Influence of Warm Climates in the treatment of Pulmonary Consumption,’ *Medico.-Chir. Transactions*, vol. lv. p. 233.

The inflammatory conditions of the lung accompanying the tubercular formations call for as much attention as the tubercular lesions themselves, and a great part of the treatment of phthisis, to be successful, must consist in active means to reduce the local inflammation. Very important also is the antiseptic, or what may be termed the bacillicide treatment of the disease, *i.e.* the measures to be directed against the organism itself, which will be fully considered later. First, we will deal with the prophylaxis of phthisis, and the treatment of the early stage.

Prophylactic and antiphthisical treatment.—We now propose to consider the measures best adapted (1) to prevent the development of phthisis in those predisposed by heredity or other weakening influences, and (2) to enable the constitution to react against the disease in its early stage, when the amount of lesion is limited in extent, and unaccompanied by pyrexia.

As the same measures are for the most part suited to the requirements of both classes of individuals, they may for convenience be considered together.

Our prophylactic measures should commence in infancy, and, in the case of the child of a consumptive mother, should first consist of the substitution of a healthy wet nurse for the consumptive mother's milk for twelve months. If this is not available, the use of asses' or goats' milk, or cows' milk diluted with water, in as large quantities as the child can digest, is advisable. After a year, beef-tea or meat may be gradually introduced once a day, and the whole-meal bread or good country seconds (now, alas, so seldom to be got) should during childhood be supplied instead of ordinary white bread. In delicate children with whimsical appetites raw pounded meat mixed with sugar often ensures steady persistence with animal food when it would otherwise be omitted, and thus bridges over a temporary difficulty.

The great point is, of course, to secure an abundant supply of the various kinds of food, albuminous, starchy, saccharine and saline, and especially a supply of phosphates in a form easily assimilated by the child, and the diet in each case must be carefully regulated with due regard to the digestive powers.

Attention should be paid to the state of the lymphatic system, and care be taken to remove all sources of irritation, such as decayed teeth, impetigo capitis, cutaneous sores or ulcers in the head and neck leading to enlargement of the cervical or submaxillary glands, which may become a centre of bacillar growth and caseation. If glands caseate or suppurate, they had best be not simply incised, but scraped out, which, if skilfully done, leaves no more scar than an incision, and effectually removes a source of future danger to the lungs and system. Above all, let the child be brought up in the country, or at the sea-side, far from town atmospheres tainted with the exhalations of closely-packed thousands, and swarming with pathogenic and other organisms, or else polluted by chemical and other impurities and sadly lacking the life-giving oxygen.

Let clothing be carried out on common-sense principles, and no naked arms or legs be visible in cold weather, but let the extremities as well as the body be cased in woollen material varying in thickness according to the season, and precluding the necessity of too much muffling with overcoats, which would interfere with the free play of muscle and limb, which it is most desirable to encourage, as tending to the promotion of physiological activity and development of the frame. A child too should sleep in flannels, and the nightdress be so arranged as to prevent the risk of catching cold if the bedclothes are kicked off.

The daily bath, the temperature being regulated according to the season, and occasionally rendered more stimulating by the use of sea water or Tidman's sea salt, must be part of the invigorating process, and is far better for very young children than forcing them into the sea, at risk of chills and giving them distaste for an element which later on may be highly beneficial.

All exercises which tend to the development of the body and expansion of the thorax are valuable, such as are afforded by the gymnasium and various athletic games, when not too violent, by running, tennis, cricket, &c. Residence in a mountainous country is often desirable for reasons we will give further on, but the main point is life in the open air, and in a dry, bracing climate.

When the time for serious education arrives, we must remember that the children of tuberculous parents are, as a rule, precocious, and only too ready to acquire knowledge; therefore, anything in the way of forcing or over-pressure in education is sadly out of place, and the curb is more requisite than the spur. The type of child I allude to is only too intelligent, and if the faculties are stimulated may indeed become an infant prodigy, only to sink into phthisis before reaching manhood or womanhood.

Let the minds of these children be at first trained chiefly by the study of nature in her various forms, which can be accomplished through the senses, in the open air, and without confinement, or bending over desks, in close rooms, and cramming up books the contents of which will be forgotten when the purposes of reading, such as examinations, have been attained.

Catarrhs and coryzas, when contracted, should be rapidly cured, and every care should be taken that the diseases of childhood, such as whooping-cough and measles, should leave no troublesome sequelæ in the form of chronic cough or enlarged glands.

Passing from childhood let us consider many of the above measures in some detail in reference to adult life, bearing in mind that the age most prone to tubercular attack in both sexes is from 20 to 30. What we desire to do is to raise the standard of assimilation and nutrition, and thus produce better lymph, chyle, and blood, and lead to a fuller life and more perfect development of organs.

Food.—Much can be done in phthisis by systematic feeding alone, and, as has been shown by Sir Risdon Bennett, Weir-Mitchell, Debove, and others, the system will tolerate and even welcome food when the appetite is capricious, or altogether lacking. When the appetite is good and a fair amount of open-air exercise can be taken, a dietary like the subjoined generally results in gain of flesh.

Breakfast, 8 to 9 A.M. Bread (whole-meal if possible) and milk (¾ pt.), or porridge [1] and milk, rendered a little more digestible by the addition of a little ground malt, or A.B.C. cereals, white wheat, or Durber's wheat or hominy with milk; fried bacon, egg, or fish, or poultry; a cup of coffee or cocoa and bread and butter.

1 to 2 P.M. Luncheon or early dinner. Plenty of tenderly cooked meat, with potatoes and fresh vegetables; light farinaceous puddings, a little ripe or stewed fruit, and a glass of sound sherry, or a larger one of claret, or ½ pt. of bitter ale.

4 to 5 P.M. ½ pt. of milk with a rusk or biscuit.

7 to 8 P.M. Dinner, or supper. Plainly dressed white fish, to be followed by meat, mutton or beef, alternated, to give variety, with poultry and game; vegetables; sweets as at luncheon, and a glass of sherry with water, or a larger one of claret, hock, or chablis, followed by a cup of hot coffee or a glass of hot water, if needed, to assist digestion.

In cases where the appetite is very capricious, as in weakly women, less food can be taken at a time and more frequent feeding is requisite, so that a dietary something like the following is to be preferred.

7 A.M. ½ pt. of warm milk with a dessert-spoonful of brandy or rum.

9 A.M. Breakfast: milk with cocoa or coffee, bread and butter, bacon, fish or poultry.

11 30 A.M. Egg flip, i.e. one egg beaten up with a dessert-spoonful of brandy; or ½ pt. of milk, or a glass of koumiss or kèfir.

1 to 2 P.M. Luncheon as in first dietary.

4. Same as at 11 30, or, if desired, a cup of tea with milk and biscuit, or slice of bread and butter.

7. Beef-tea with toast and a glass of wine.

10. Some farinaceous food, such as milk gruel, arrowroot, &c.

These forms may, and should be, endlessly varied according to the digestive capacity of the individual, provided always that the relative proportions of the food-stuffs are

[1] Porridge and the two following items correct constipation.

maintained, and the stimulants introduced in amount and kind to assist digestion and prevent waste.

A few words on some of the leading varieties of food may not be out of place when discussing the diet of the consumptive.

Milk is of such vital importance to the consumptive that some form of it must always be insisted on. But two cautions are necessary—one, that it must never exceed one quart a day, and that if cod-liver oil be taken, the quantity shall be limited to a pint; another caution is that milk should not be taken as a drink with meat meals, inasmuch as instead of stimulating the appetite it satisfies it, and thus prevents the desired amount of meat being consumed. Whether the milk is to be taken warm from the cow, or is to be allowed to cool first, or to be boiled, are points on which our Continental brethren lay much stress, but which may, I think, be settled by individual experiment; but if it be not tolerated in the ordinary form, it will probably be so, if diluted with barley water, or soda water, or with the addition of lime water (a tea-spoonful to a tumbler); or again it can be rendered more digestible by being peptonised,[1] according to Sir William Roberts' method. For some, who appear to experience insuperable difficulties in assimilating cows' milk, asses' milk (which closely resembles human in composition) or goats' milk, the peculiar flavour of which may be entirely covered by the addition of a few drops of orange flower water, is clearly indicated. In some cases also whey may be tried. Koumiss and kèfir and similar preparations are also substitutes largely used on the Continent, being varieties of fermented mares' or cows' milk. Koumiss is the fermented milk of unworked mares grazing at large on the Kirghis Steppe of Southern Russia, and has for centuries been used as an article of food by the Kirghis themselves, whose reported exemption from consumption forms the basis of the koumiss cure for that disease. The fermentation results in the production of alcohol (1·7 to 2 per cent.), carbonic and lactic acid, and in the elimination of a large proportion of casein. The amount of carbonic acid varies from ·9

---

[1] Peptonised milk is supplied by the Aylesbury Dairy Company.

to 1·1 per cent., and that of lactic acid from ·475 to ·831, both increasing up to the 10th day after the commencement of the process.

Dr. Carrick, of St. Petersburg, a former clinical assistant of the Brompton Hospital, exhibited at the National Health Exhibition of 1884, a group of Kirghis mares, with their picturesque attendants, and supplied fresh koumiss daily to visitors ; and Dr. Jagielski had some years previously introduced into London koumiss made from the cow, which is now largely supplied by the Aylesbury Dairy Company.

Koumiss has an acidulous alcoholic taste, which, combined with the effervescence of the carbonic acid, forms a drink grateful to an irritable gastric membrane and pleasant in hot weather to a thirsty palate. Ssamara, the region where the best koumiss is produced, according to Dr. Carrick, has a dry, warm climate in May, June, and July, inviting thirst, and it is at this season that the koumiss is most perfect. Moreover Dr. Carrick states that the koumiss treatment cannot be effectually carried out save in Tartary, under the special climatic and dietetic conditions of the Russian Establishment. Biel's[1] researches go to show that the region inhabited by the mares is of little importance, but the life of liberty, combined with absence of work, is the best explanation of the milk having a special character which closely approximates it to human milk, and that, when these conditions are reversed, this character is lost.

If koumiss be prepared from cows' milk, the cows must be granted similar advantages and pastured free and not shut up in stables. This koumiss is now prepared in all the large capitals of Europe, and closely resembles the mares', but is slightly richer in alcohol and lactic acid, and contains less carbonic acid. At Ssamara the quantity taken is considerable—a glass or two before breakfast, three or four in the forenoon, and the same number in the afternoon—and constitutes the patient's sole beverage, no alcohol or sweets being allowed, and the dietary consisting of mutton, poultry, eggs, butter and bread.

---

[1] Jaccoud : *Treatment of Pulmonary Phthisis,* loc. cit.

The cure, as it is called, occupies two or three months, spent mostly in the open air, riding or walking.

Biel states, that when koumiss is the sole food, the daily excretion of urea, phosphoric and sulphuric acids is increased, and that of uric acid diminished. He claims a reduction of the patient's temperature and an improvement in the general and local symptoms.

Many years ago koumiss was tried at the Brompton Hospital on a limited number of consumptive patients with negative results, the great difficulty being one often since experienced, viz., to induce patients in moist cool England to imbibe such large quantities of acidulous fluid as is required, the element of excessive thirst being wanting. My own experience of koumiss, and I have now tried it largely, is that it has no specific influence on the tubercular disease, but that it is a most valuable article of dietary and acts as a substitute for fresh milk. In cases of phthisis accompanied by persistent vomiting not due to hard cough, such as is seen in young females, and where there is no reason for suspecting intestinal or gastric ulceration, where various kinds of medicine have been tried in vain, I have often discontinued all food and medicine, and have placed the patient on koumiss, three or four glasses a-day, combining a few tea-spoonfuls of Brand's essence of meat; and the result has been to entirely check the vomiting, to restore the appetite, and after a few days the patient has been able to return to ordinary diet.

We may, therefore, regard koumiss as a valuable adjunct to the milk group, and though it is not indicated in all stages of the disease, it may prove useful in the prophylactic and first-stage periods. It should be given at the rate of a bottle a day, and, like milk, between meals and not with them.

Dr. Edmonston Charles, of Cannes, gives the following receipt for making koumiss, which I subjoin as useful for travelling patients.[1] 'Nearly fill a quart bottle with fresh milk, leaving enough room to shake it easily; add a spoonful of crushed lump sugar and a bit of German yeast the size of two ordinary 5-grain pills; cork and tie down with wire and

[1] *Diseases of the Lungs.* Dr. D. Powell, p. 422.

string. Keep in a cool place and shake twice a day. The koumiss will be ready on the 6th day, but earlier in hot and later in cool seasons.'

Kèfir or Kef [1] (to which my attention has been drawn by Dr. Lauder Brunton) is another palatable form of fermented cows' milk used in the Caucasus, resembling koumiss in composition and taste, and can be made from either fresh or boiled milk. It is used in the same way as koumiss, and claims to closely resemble the variety made from mares' milk.

Meat.—Whenever it is possible, fresh meat plainly cooked and unalloyed with rich sauces or gravy or stuffing should be taken two or three times a day, and the diet varied by the occasional introduction of poultry, fish, and game, always retaining a preponderance of butcher's meat. If meat cannot be taken cooked it should be eaten raw, finely pounded and mingled with sugar or salt to render it more palatable, measures being directed to prevent possible entozoic disease arising from this source, except in the case of mutton, which recent researches have shown to be absolutely free from entozoa. Of late years the drinking of lamb's blood has been practised in France to promote more complete sanguification and to correct anæmia, and with the same object, in Germany, the transfusion of either human or lamb's blood has been carried out by Hasse and others. Hasse tried it in upwards of a hundred cases, and Dr. Redtel, a consumptive patient, wrote a most interesting and graphic account [2] of the process as performed by Hasse on himself, and especially of his own sensations and symptoms. He was suffering from tubercular consolidation of the left lung and laryngeal ulceration, with dysphagia. About $3\frac{1}{2}$ fluid ounces were transfused from the artery of a lamb into the median basilic vein of the patient for 95 seconds. The first sensations were warmth in the arm, formication, redness of the face, followed by dyspnœa so intense that the operation had to be stopped. Violent

---

[1] It can be obtained of Messrs. Spring & Co., 21 Blenheim Road, St. John's Wood, N.W.

[2] By Dr. Frank's of Cannes desire I translated it and communicated it to the *Obstetrical Journal*, December 1874.

pains in the loins succeeded, lasting some hours, and assumed a pulsatile character, being synchronous with each arterial beat, and were assigned by the author to pressure of the distended inferior vena cava and abdominal aorta on the lumbar sympathetic. The operation was followed by a rigor, slight cyanosis, and eventually profuse perspiration, the pulse rising to 140 and the respiration to 32. On the second day the urine contained albumen, and on the fifth a rash of urticaria appeared, lasting two days. The result of the operation seems to have been that no improvement took place in the symptoms except a lessening of the dysphagia, and at the end of three weeks from the operation Dr. Redtel was gradually becoming worse. We cannot, therefore, advise transfusion of blood for consumptive patients.

Fresh vegetables, plainly dressed, and a small amount of ripe fruits, are to be included in the dietary; but rich pastry, cheese, savoury dishes, salads containing much vinegar, pickles of every kind, and the like, are to be carefully avoided, both as tending to upset digestion, and as interfering with the prolonged use of cod-liver oil. With regard to butter, cream, suet, and various other oily or greasy matters, we must bear in mind that the stomach and liver are already somewhat tried by the regular administration of cod-liver oil, one of the most easily assimilated members of this group, but still occasionally giving rise to symptoms of biliousness and gastric disturbance, and it is, therefore, highly desirable not to tax these organs further by the introduction of large quantities of fatty or oily material. Great moderation should be observed in the use of these articles, and, as has been mentioned before, the quantity of milk should be limited. At the same time it must be remembered that there are individuals who can assimilate almost any quantity of fatty matter, and to these the above recommendations do not apply; indeed, such can and do sometimes take cream, in addition to the oil, with benefit. Moreover, where much exercise is taken more fat can be tolerated.

Stimulants.—Though we cannot go as far as Dr. Flint in the importance to be attached to the use of stimulants in consumption, yet we highly commend them when taken *with food*, and not alone at odd times between meals, as is done by many persons—a custom more sociable than wholesome, and specially injurious to the stomach, for the gastric juice is thus stimulated to secretion, and having no food to digest, acts on the walls of the viscus, giving rise to flatulence and loss of appetite.   When the meal-time comes, the food is not thoroughly relished, and, on account of the waste of the gastric juice, imperfectly digested.

The principal uses of stimulants in consumption are, firstly, to increase appetite and promote digestion ; secondly, to check the waste caused by tissue combustion ; thirdly, to stimulate the heart's action, and thus obviate the tendency to death by syncope.

For the last purpose it is only required in the very advanced stages of the disease ; but in the first and second lies its principal utility, the only drawback being that stimulants are apt to increase the cough and local irritation, though they are less likely to do so, if mixed with water. As regards the choice of different kinds, much must depend on the state of the organs of digestion and circulation. If the patient be not of a bilious habit and the cough be not troublesome, malt liquor— in the form of bitter ale, table beer, or even stout—is a capital appetiser ; but in case of liver disturbance, sherry mixed with water, or hock or chablis, answers the purpose better. If the cough is at all troublesome, the amount of stimulant should be diminished, and sometimes its use discontinued altogether, but the least irritating to the lungs appears to be good sound claret, St. Julien, or Medoc, or Hungarian Carlowitz ; Burgundy and port are rather too fiery for this purpose. Champagne is only to be employed in cases of extreme weakness, and then but for a limited period. Brandy, gin, rum, and whisky are most useful in the last stages of the disease ; but they are best tolerated when combined with nourishment in the form of brandy and arrowroot, egg-flip, rum and milk, and other numerous combinations which the physician and nurse have to employ to ensure a proper amount of food and stimulant being taken by the patient. The custom of taking a cup of rum and milk in the morning before dressing is very beneficial to weak subjects. The use of tea, coffee, and cocoa is recommended in moderation, as tending to check the waste of the system, and often assisting by their warmth free expectoration, but in no sense are they substitutes for alcohol, and when taken with meat meals rather reduce than increase appetite, as may be proved by comparing the different quantities consumed at a late dinner and a ' meat tea.' If alcohol be not taken with lunch, dinner, or supper, let the meal be accompanied by some effervescing table water such as St. Galmier, Apollinaris, Salutaris, and the like, or even by ginger-ale, ginger-beer, or lemonade, which rather promote the appetite than otherwise, and do not excite the cough.

Clothing.—Consumptive patients, who are more susceptible than others to the process of ' catching cold ' and in whom it often sets up intercurrent pneumonia and bronchitis, should clothe warmly, though not to such an extent as to produce excessive perspiration, or to preclude a fair amount of exercise.

The most important point to be attended to is the under-clothing, which for at least eight months in the year should be of flannel, lamb's-wool, knitted shetland, or some other woollen material, and should not only cover the chest completely, but also encase the whole body and lower extremities. The double-breasted lamb's wool jersey answers the purpose well, and should be worn with drawers of the same material, or of flannel, and with woollen socks or stockings. In the summer months a thinner clothing of merino may be substituted, but the change must be carried out with great caution. Ladies should dress in ' combinations ' of lamb's-wool in winter, and of merino in summer, and should so far imitate the stronger sex in making warmth, not weight, the object of clothing. Overcoats and wraps are also of consequence, more especially when the patient is driving out in a carriage ; and in this particular he can hardly be too careful, for with a weak circulation, and but little means of exciting it, he must prevent the chilling effect of radiation from his body by wrapping up warmly in furs and rugs, and if this should be insufficient, by supplying extra heat by a hot flask to his feet.

The great advantages of woollen underclothing are (1) that being porous it admits of transpiration through from the surface of the skin, which is so desirable, and (2) that profuse perspiration is absorbed by the woollen fibres and eventually evaporates from the outer surface of the clothes, which is not the case when the underclothing is of linen, calico, or even silk, and this may be proved by the time such material remains damp after profuse sweating. This is the reason why flannel underclothing is largely used in India and other hot countries.

The much advertised 'Jäger' clothing is right, though by no means new, in principle, but the texture is closer than is desirable, and according to my experience it is inferior in quality and wear to lamb's-wool, merino, or flannel of British production and manufacture.

The wearing of leather vests, even when perforated, though they prevent radiation, also obstructs the healthy process of transpiration, and can only be sanctioned in very cold weather or for very long journeys. A better friend is the long-sleeved wool waistcoat or cardigan, which can be worn over or under an ordinary waistcoat, and is permeable. The use of chest protectors made of flannel and leather cannot be based on any scientific principle, for in addition to their being generally impermeable to moisture, they only add to the warmth of the thorax, while the extremities are left unprotected, and consequently become cold, and it is not uncommon to find under these circumstances a perspiring thorax and cold, almost livid, hands and feet.

The question of respirators is not always an easy one to settle, for whilst undoubtedly these appendages protect the lungs from damp, fog, and cold, owing to the warm expired air heating the cold incoming air, they often, unless the intervals between the wire meshes are of fair width, impede the act of respiration. Where there is great tendency to contract catarrh, they may be of use, but, as a rule, one or two layers of woollen scarf passed over the nose and mouth, as coachmen and omnibus drivers often do, protect the neck and throat and act as a sufficient respirator, without obstructing free breathing. For ladies a good Shetland veil or a fleecy 'cloud' drawn across the mouth will answer the same purpose.

Habitation.—The residence of the consumptive patient should be situated on a dry soil of sand or gravel, free from admixture with clay or other material likely to collect or retain moisture. It should stand on slightly elevated and sloping ground, so as to ensure thorough surface drainage. Vegetation should be present in order to keep up the supply of oxygen, and it should not be rank, succulent, or in excess, but short, herbaceous, heathery, and flower-bespangled, like the grass fields of a dry and open country, or the downy herbage of a hill-side or elevated common. Shrubs, such as brown furze, thorn, and thin copse-wood, are welcome, and a few trees scattered or arranged with a view to shelter; but dense woods and large deciduous trees, the chief ornament of park scenery, do not add to the healthiness of the air, but greatly increase its humidity. There is no objection to the neighbourhood of pines, especially on the north side of the dwelling, their dry shade and fragrant odour being very pleasant, and forming the great attraction of certain localities, as Bournemouth and Arcachon; indeed, the turpentine inhalations from this source are held in great repute in many parts of the Continent, and form one of the numerous 'cures' so congenial to the German mind.

The vicinity of marshy or swampy ground is of course to be avoided, as likely to cause malaria, and even that of low lands and valleys with a clay soil or subsoil, for moisture is generally excessive in such districts.

Peat bogs, on account of some antiseptic property, do not engender malaria, but are not exempt from the imputation of dampness; and any one who has witnessed the swarms of midges on a Scotch morass can testify to their capacity of breeding one kind of plague at least. The close proximity of the house to lakes or ponds of fresh water, or of slowly running streams little below the level of the ground, is not desirable, as the amount of moisture in the air is thereby increased; but this does not apply to the margin of the sea, for there is certainly something corrective, if not antiphthisical, in salt water, and the vapour arising from it, which renders sea damp less injurious than inland damp.

The house should be protected from northerly and easterly winds, and well open to the south and west. The walls should be thick and the windows large, so as to allow, if necessary, of thorough ventilation ; the cubic space should be from 1,500 to 2,000 cubic feet per head ; the rooms should be lofty and airy, and the temperature kept as near 62° F. as circumstances will admit of. Whilst means are taken to maintain this degree of warmth, others should not be omitted, to ensure a frequent and abundant supply of fresh air, either through the top of the window or through inlets opening from outside, and also for the removal of the impure air. A good bright fire, in an open grate, with Boyle's talc ventilators inserted in the chimney near the ceiling, will act as an admirable extractor of used-up air, and change the atmosphere of a room (according to my experiments) $2\frac{1}{2}$ times in the hour, and an ample supply of fresh air may be secured, when the external temperature precludes opening windows, by the insertion of Tobin's tubes in the outer wall, arranging that the delivery of air takes place at a level of at least five feet from the floor, thus preventing any chance of disagreeable draughts.

Both bed and sitting-room should be exposed to the sun's rays, whose warmth and purifying influence continue long after sunset, and cannot be supplied by artificial means, as the contrast between the atmosphere of a north and of a south room will show. These vivifying influences must also be brought to bear on the person of the patient, and much is to be said for Dr. Hermann Weber's [1] recommendation of balconies or large verandahs, where even the febrile cases can be exposed to the air and sunshine, while protected from the wind and rain. This is commonly done in continental hospitals with great benefit, and the fresh air encircling the patient stimulates appetite, improves digestion, and promotes sound sleep.

The question of sleeping with open windows at all seasons of the year is often raised, and must depend to a great extent

[1] *Croonian Lectures*, p. 64.

on the climate of the locality and on the individual susceptibility to cold of the patient. Certain it is that invalids and
very delicate ones sleep with open windows at Davos and St.
Moritz, when the external temperature is as low as 4° F., and
suffer no harm, whilst much stronger invalids attempting the
same in moist England, when the temperature is as high as
45° F., awake with symptoms of catarrh or coryza more or less
severe, the difference of effect being probably due to the
relative amounts of atmospheric moisture present. If the
windows are to be kept open at night, care must be taken to
direct the incoming current upward by a Hinckes-Bird wedge
or similar arrangement.

In the rooms dust should be avoided as much as possible,
as the inhalation of such, whether it consist of inert matter or
of various kinds of organisms, is likely to irritate the already
wounded alveolar and bronchial membranes, and lead to
further inflammatory or septic processes. The curtains and
carpets should be moveable and shaken outside, while, if the
patient is confined to one room, dusting the furniture may be
avoided by wiping with a damp cloth, which in some cases
may be moistened with an antiseptic solution. The patient
should sleep alone on a hair mattrass or a spring curtainless
bed, not overloaded with bedclothes. No gas should be
allowed in either sitting or bed rooms unless there be some
arrangement for the immediate escape of the products of
combustion, as in the Benham lamp. Electric lights, such
as the Swan-Edison, are admirable, and candles or lamps,
when used sparingly, are admissible.

The sputum of the patient should be received, not on a
handkerchief but into a vessel presenting the smallest open
evaporating surface and containing a strong disinfectant, and
this should be emptied and scalded at least once a day, it
having been proved by Schill and Fischer's[1] experiments
that dried tubercular sputum may be kept for periods of
from 95 to 120 days and yet, when inoculated into guinea-pigs,
may produce tuberculosis. Decomposition did not seem to
affect the virus, which resisted the action of a great many

[1] *Mittheilungen aus dem K. Gesundheitsamte.*

disinfectants, but succumbed to admixture for ten hours with absolute alcohol (1 part sputum to 5 of alcohol), carbolic acid (5 per cent), salicylic acid, caustic ammonia, and a saturated aniline solution (sputum 1 part, aniline 10 parts). Steam was found a most effective bacillicide, fifteen minutes' exposure being sufficient. At the Brompton Hospital, after various experiments, the following has been found the most effective method of destruction. The sputum is received into the ordinary earthenware covered cups, containing sanitas fluid, which are emptied two or three times a day and carried from the wards and galleries in galvanised iron buckets to the furnace used for the various heating and ventilating apparatus of the hospital, where it is well mixed with small coal and burnt in the furnace. In this way a dangerous element is got rid of, with some saving in the amount of coal consumption.

The stools of consumptive patients with intestinal ulceration are in the same way removed and destroyed.

Exercise.—It may be safely stated that, in all cases of phthisis, exercise in some form or other is beneficial, and the good derived where the patients are able to avail themselves of it, is very evident, as seen by the increase of appetite, by the quickened circulation, and the sounder sleep which so often follow when exercise is taken by the patient. Whether it should be of the active or passive kind, and what varieties of each are admissible, depend on the stage and type of the disease, and also on the strength of the patient. In the early stages, where the symptoms are not active, where there has been no recent blood-spitting, and where the cough is not hard or frequent, those varieties of active exercise are of most advantage which most effectually expand the upper portions of the chest, thereby bringing into play the upper lobes of the lungs, so generally, from want of use, the seat of tubercular lesions ; and by causing the blood to circulate freely through the pulmonary tissue, they prevent local congestions and fresh exudations, and aid materially in the absorption of old ones.

Important as chest expansion is in the first stage of

phthisis, it is still more so as a prophylactic against the disease, specially in the case of the ill-developed pyramidal chests of many young men and women.

What are the varieties of exercise which best accomplish this end? Those in which the upper extremities are raised, and the muscles connecting them with the thorax brought into activity. When the arm is raised, the numerous muscles which arise from the ribs and are inserted into the bones of the upper extremity, *e.g.* the pectoralis major and minor, the sub-clavius, the serratus magnus, &c., in contracting, raise the upper ribs, and thus increase the size of the chest cavity. This necessitates the inspiration of a larger amount of air. Dr. Silvester has called attention to this important principle, and on it has founded his excellent system of restoring respiration in cases of drowning, narcotism, &c. He has also recommended a modification of it in the incipient stages of phthisis. The forms of exercise which best carry out this principle are : rowing, particularly the pull and backward movement; the use of the alpenstock in mountain ascents; swinging by the arms from a horizontal bar, or from a trapeze; climbing ladders or trees. Dumb-bells, as commonly used, are calculated to develop the arms more than the chest, and rather tend to depress the latter by their weight. Various special gymnastic exercises, of which there is a great choice now-a-days in good gymnasia, may more or less answer the purpose; but there is one form which is particularly applicable to the object above mentioned, viz. the *gymnast,*[1] invented by the late Mr. Hodges. To make this instrument answer the purpose of a chest elevator or expander, it should be fixed, not, as it is sometimes done, at the height of the operator, but considerably above his head, in or near the ceiling, with the handles reaching to the level of his shoulders : then, by holding the handles and walking a few paces forwards and backwards, the arms are brought into a species of action which, while it exercises the whole body, especially tends to expand and elevate the upper part of the chest. Skating and toboggan-

[1] To be had of Matthews & Son, Charing Cross.

ing are good exercises, the benefit of the latter not depending on the swift and exciting descent of the snow slope, but on the slow and gradual ascent of the hill in returning; the slipperiness of the path, and the slight check which the drag of a toboggan causes to progress, precluding over-hurry or over-exertion.

Walking exercise, as a rule, does not work the upper extremities or raise the upper ribs, but acts generally on the system by drawing the blood to the extremities and quickening the circulation through the lungs. In mountain ascents and in fast walking the quickening of the circulation brings the whole lungs into play, and in this way the upper lobes come into full use. If the alpenstock be used in mountain climbing, the beneficial local effects of raising the upper ribs may be combined with the general advantages of walking. Walking exercise can be taken in most stages of phthisis, provided there be no active symptoms present. Even where cavities are formed, if there be no recent inflammation, a limited amount, and performed on level ground, is beneficial, but great care must be taken not to overtax the patient's strength.

Passive exercise may be used by the weak and delicate, even in advanced stages of phthisis, or when it is of the inflammatory type. Open carriage exercise, sailing, or being rowed in a boat or carried in a hammock, are instances in all of which little muscular exercise is involved, and they may be considered as a means of supplying a constant change of air, with least fatigue, while their effect in improving the circulation and appetite, and in promoting sleep, is often very apparent. But even these make some demand on muscular and nervous power, and must not be carried to the extent of producing exhaustion in weak subjects.

Riding exercise, from the time of Sydenham, has been generally acknowledged to be peculiarly beneficial to consumptive patients who are strong enough to bear it; and it is difficult to find a form of exercise which so admirably answers the purpose of giving plenty of fresh air and thoroughly warming both body and extremities with so small an amount of fatigue. Bicycling and tricycling combine the advantage of fresh air in abundance and a great amount of exercise chiefly of the legs, but there is no doubt that the whole muscular system is brought into play, though often too strongly. In cases of weakness and great loss of appetite and flesh, when at the same time the tubercular disease is not very active, 'Massage' of the muscles of the body after the Weir-Mitchell method can be practised with considerable benefit and with decided gain of weight. Under its influence appetite and digestion improve, and the circulation becomes more vigorous, colour returns, and the quantity of food consumed is sometimes astonishing.

Aero-therapeutics.—The use of compressed and rarefied air has been much in vogue lately in the treatment of phthisis. For the former we make use of (1) one of the numerous forms of transportable apparatus invented by Waldenburg, Hauke, Schnitzler, Cube, and Oertel, by which the patient breathes from a mask, tight-fitting to the face, air compressed by placing weights on the top of a hollow cylinder standing in water, or air rarefied by withdrawing of some of the water from the same cylinder, and by diminishing the weights placed on it;[1] or (2) the pneumatic chamber, into which extra air is pumped up to the pressure of $\frac{1}{2}$ to $1\frac{1}{2}$ atmosphere. Simonoff[2] claims that compressed-air baths cause absorption of inflammatory exudations in the lungs of phthisical patients, and Oertel[3] considers these baths of more importance, specially in the early stage of phthisis, than climatic influences, and it is urged that compressed air may open up alveoli blocked by tubercle and inflammatory exudations.[4]

I have made extensive observations on the effect of compressed-air baths in phthisis at the Brompton Hospital, where an excellent apparatus has been at work for more than three years, and the result of my experience is that:

1. They have no power to open up alveoli or lobules once

[1] See an excellent address on the use of these apparatus by Professor Gamgee, F.R.S. *British Medical Journal*, December 18, 1886.

[2] Oertel: *Respiratory Therapeutics.* [3] *Op. cit.*

[4] For further information the reader is referred to my Lectures on the 'Use of the Compressed-Air Bath in the Treatment of Disease.' 1885.

A A 2

invaded or blocked by tubercular masses, nor do they promote to any appreciable extent the absorption of inflammatory exudations, except by improving the general health.

2. That they, nevertheless, as in health, cause dilatation, and hypertrophy of the healthy portions of the lung tissue, as proved by increased cyrtometric and other measurements, and by the spirometric results, and that they are therefore valuable as a prophylactic against the disease.

3. That their physiological and chemical influence is excellent, showing itself in improved nutrition, increased oxygenation, gain of strength, colour, and weight.

4. That they are contra-indicated in pyrexia and hæmoptysis, and that they are liable to give rise to the last symptoms even in cases of phthisis hitherto exempt from them. Cases where cavities exist, as there is a possibility of vessels lying exposed in their walls, should be prohibited from using these baths, great hæmorrhage having occasionally arisen in the bath itself.

The use of natural rarefied air, such as we obtain from the diminished barometric pressure at high-altitude stations, yields wonderful results in phthisis. I allude, not to the general influence on the consumptive patient showing itself in increased vigour and improved appearance, which will be spoken of later (see chapter on Climate), but to the local effects on the thorax and its contents, which have been proved by the observations of Walshe, Jourdanet, Hermann Weber, Kellet, McCall Anderson, Denison, Ruedi, Holland, and myself,[1] to take place in all mountain regions above 5,000 feet, and to be the following:

1. Hypertrophy, or more complete development of the healthy lung tissue, shown by the physical signs and increased respiratory power.

2. Emphysema of the portions of lung in the neighbourhood of the cavities and tubercular masses, inducing isolation of the tubercular septic centres, emptying of the

[1] 'Treatment of Phthisis by Residence at High Altitudes.' *International Medical Congress Trans.*, 1881.

vessels and capillaries, caseation and cretification of the tubercle, and arrest of the disease.

3. In consequence of the above changes, there is expansion of the thorax, which increases in circumference at various levels from one to three inches, such increase being independent of any augmentation of fat or muscle, as it takes place in patients who are losing weight, and occasionally in bedridden ones.

4. The above thoracic expansion is always accompanied by diminution in the number of respirations, which become deeper, and by a slowing of the pulse.

5. The rate at which expansion takes place varies in different cases, but generally requires some months, and is dependent on the amount of tubercular disease, on the presence or absence of adhesions, on the yielding or non-yielding character of the thoracic wall, and on the amount of exercise taken by the patient. It is not always permanent on returning to low levels.

The great local advantage of respiring rarefied air is the necessity for taking deep inspirations, thus bringing into play the whole pulmonary area, and preventing the local congestions and blocking of the smaller bronchi and alveoli with inspissated secretions. Hence the great value of mountain climates, both as a prophylactic for phthisis and as a means of arrest in the first stage of the disease.

Baths.—The use of courses of bathing, whether in salt water or the many varieties of mineral waters, must be advised with considerable caution in phthisis. For prophylaxis, undoubtedly a cold shower-bath or douches well directed against the thoracic wall are beneficial, and even a few jugs or cans of cold water poured rapidly over the person, with free sponging, followed by vigorous dry rubbing, have an excellent effect, not only from their fortifying influence on the system, thus enabling it to resist chill more easily, but also, as Brehmer has well shown, from the shock to the chest wall, causing deeper and fuller inspirations to be taken. Salt water is preferable to fresh, on account of its stimulating effect on the skin, and in the absence of the genuine article, the addition of Tidman's sea salt to the water is a tolerable substitute. Careful drying and friction with rough towels must follow, and in case of weak circulation the temperature of the water must be raised to the proper degree to suit the sensibility of the patient.

Immersion in baths, or in the sea, is not so good for the purposes of promoting skin reaction as the application of douches or sponging, as above described, and long continuance in any bath is generally followed by great weakness and prostration, and therefore should be avoided.

Mineral Waters.—The French physicians have long recommended the use of the sulphur waters of Eaux Bonnes, Eaux Chaudes, Bagnères de Bigorre, Bagnères de Luchon, and others, for the early stage of phthisis, and it would seem that whatever benefit arises from their use is due, according to Peter,[1] to the catarrh-healing qualities possessed by the sulphurous acid.

My own experience is not favourable to the treatment of phthisis by sulphur waters,[2] nor can I quite understand its principle, unless M. Peter's explanation be accepted, but mineral waters containing arsenic, such as Mont Dore, La Bourboule and Royat, have certainly benefited phthisical patients, and to a less degree those containing iodine and bromine, such as Wildegg, Soden, Saxon and Kreutznach, which are highly praised by Jaccoud and others for the treatment of the scrofulous form of phthisis.

But we are inclined to agree with Jaccoud that the climate is the first consideration, and the mineral waters to be obtained in the locality the second, as more depends on the former than on the latter. At the same time, knowing as we do the complex character of many natural mineral springs, and the wonderful tolerance of their action exhibited by most patients, we must not set them aside as inert or useless agencies, but give them a fair trial when opportunities occur, only preferring to them well-known curative agents, such as cod-liver oil, and certain climates.

[1] *Clinique Médicale.*
[2] Sulphur waters, as those of Eaux Bonnes, are used in the Bergeon treatment. *See* Chapter XXVI.

CHAPTER XXIII.

CLIMATIC TREATMENT OF CONSUMPTION.

Treatment based not on pure meteorology but on results of climate in similar cases—Aseptic atmosphere not necessarily found in mountains nor peculiar to them—Principles of classification of climates—Table of climates— 1. *Moist temperate of British Channel—Biarritz and Arcachon.* 2. *Moist and warm Atlantic*—Madeira and Teneriffe—West Indies—Blue Mountains of Jamaica. 3. *Warm Pacific climates*—Santa Barbara and Los Angelos. 4. *Sea Voyages* to Australia—to Cape of Good Hope—to North and South America—Their relative advantages—Sedative and tonic influences—Time for starting—Clippers and steamers—Dangers of Red Sea —Drawbacks of sea voyages. *Dry climates of the Mediterranean Basin*— Malaga—the Riviera—Islands of Corsica, Sicily, Malta—Algiers and Tangier. *Calm temperate inland climates*—Cold and moist—Pau and Amélie les Bains. *Calm and cold climates*—St. Paul's, Minnesota and Canada. *Very dry and warm inland climates*—Egypt—Cape of Good Hope and Natal—Australian highlands. *Mountain climates with atmosphere more or less rarefied*—Swiss Alpine stations—Rocky Mountain sanitaria—Andean sanitaria—Himalayan and other hill stations—South African highlands—Selection of climate for each class of consumption.

WE do not propose in this chapter to attempt to discuss in detail the characteristics of the various climates available for invalids during winter and spring, the seasons when absence from an English home is often advisable, or the causes of their relative meteorology, which we have described elsewhere,[1] nor even the statistical returns of the consumptive patients under observation for periods of varying duration, as this has been done by us already ;[2] but after a brief survey of the leading groups of climates, we will treat the subject practically by sketching out the line of climatic treatment suitable for each class of consumptive patients, and we must ask our readers to

[1] *Climate of the South of France*, 2nd edition. Longmans.
[2] *The Influence of Climate in the Prevention and Treatment of Pulmonary Consumption.* Smith & Elder. 'Influence of Warm Climates in Consumption,' *Medico-Chir. Transactions*, vol. lv. 'Treatment of Phthisis by Residence at High Altitudes,' *Trans., Internat. Med. Congress*, 1881.

excuse what may occasionally appear to be somewhat dogmatic statements, but which are in reality conclusions drawn from many hundreds of carefully recorded cases, extending over a long series of years, and not merely speculative opinions.

The amount of information to be obtained as to the meteorology of various climates is very large, and information about the journey to and means of communication with a climatic station are easily attainable, but how few practitioners ever contribute facts as to the effects of these climates on the various forms of disease, and yet it is only on these data that we can ever frame a proper system of climatic treatment. The 'immunity' ground is often urged in behalf of a climate, which means that, because there is a great rarity, or absence of, certain diseases in a particular locality, that locality should be recommended for that disease. This is not reliable, because it overlooks all differences of race and food : for Europeans become attacked with intermittent fever, where negroes or natives of the countries are entirely exempt. Until our 251 cases of consumptive patients making trial of foreign climates and our 243 of patients wintering in English ones were published, there existed no mass of information to draw conclusions from, and it is on these, supplemented by later experience, especially with reference to high altitudes and sea voyages, that the deductions of this chapter are founded; and if we have not largely referred to other writers, it is because our opinions and decisions have been mainly based on our own experience.

Much has been urged of late years in favour of the use of an aseptic atmosphere in phthisis, and certain climates, viz., the mountain ones, have been recommended on this ground alone, the reasoning apparently being that if such an atmosphere presents conditions unfavourable to the growth and multiplication of other germs, this will apply also to the growth and development of the tubercle bacillus. We should, therefore, expect a diminution in the number of these organisms in the expectoration of consumptives at Davos and St. Moritz, as an early indication of improvement. I have made careful inquiries on this point, comparing the results of Dr. Ruedi's bacillar observations at Davos, and Dr. Holland's at St.

Moritz, on the same patients, whose sputum I had frequently examined in London. We all work with the same microscopic powers, and estimate bacilli in the same way. The result is that there is no perceptible difference in the relative numbers of tubercle bacilli in London and the Grisons. Where, with great improvement, the amount of expectoration is markedly diminished, as may happen anywhere, the number of bacilli also diminishes, but where the amount of discharge, as from a chronic cavity, remains stationary, the number of bacilli also remains stationary, as is the case in London. Fresh excavations and tuberculisation are, as here, heralded in by an increase in the number of bacilli.

We may therefore conclude that the aseptic atmosphere of the mountains and the ocean acts generally on the constitution by strengthening it to withstand the bacillar attack, rather than by actually destroying the invaders.

Any classification of climates on the basis of one element alone, such as altitude, must necessarily lead to error, as it would involve the inclusion within the same category of a number of localities differing vastly in latitude, hygrometry, and their relation to the sea. Nor again, if latitude or dryness alone be selected, should we be better off, as we should be compelled to class together places of bracing and of relaxing character, simply because of their lying within the same parallels. The principle I prefer is that of groups, each with its characteristic features, which are generally a combination of several meteorological elements, and not based on one climatic factor.

The subjoined list is by no means exhaustive and includes all the types of climate likely to prove useful, according to our present knowledge, in the treatment of phthisis, and some that are rapidly falling into disuse. We have omitted all mention of Rome, Pisa, Venice, Naples and other large cities, because neither their climatic qualities or sanitary condition render them suitable for invalids, and several others, which are suitable, are omitted on the ground of their closely resembling those already on the list, or because they are devoid of accommodation for invalids.

TABLE OF CLIMATES.

MARINE, CHARACTERISED BY SALINE ATMOSPHERE.

1. MOIST TEMPERATE:

(a) British Channel and other stations under influence of the Atlantic warm current and the accompanying winds. Winter mean temperature ranging from 39° F. to 44° F. Annual rainfall from 23 to 41 inches, and number of rainy days from 132 to 178. Climate equable, stimulating but moist.—Penzance, Torquay, Bournemouth, Ventnor, Worthing, Brighton, Hastings, Channel Islands, Tenby, Ilfracombe, Llandudno, and Cove of Cork.

(b) Biarritz, similar to above but warmer.

(c) Arcachon, somewhat dryer.

2. MOIST AND WARM ATLANTIC:

(a) Madeira. Winter mean 63° F. Rainy days 88. Rainfall 30 inches. Climate warm, moist, and sedative.—Teneriffe. Warmer and somewhat dryer.

(b) West Indies. Tropical and moist. Climate sedative.

3. WARM PACIFIC:

Santa Barbara. Winter mean 54° F. Rainfall 16 inches. Rainy days 17. Climate dry and equable.—Los Angeles, San Diego and San José.

4. Sea Voyages to Australia, New Zealand, Cape of Good Hope, and America. Mean temperature varying from 50° F. to 80° F. Equability of temperature. Climate very moist, but partly sedative, partly stimulating.

5. DRY CLIMATES OF THE MEDITERRANEAN BASIN:

(a) Malaga. Winter mean 56° F. Rainfall 16½ inches. Rainy days 40. Well sheltered. Climate dry, warm, and stimulating.

(b) The Riviera. Enjoying a winter climate warmer than England by at least 3° F., much less moist, and far more stimulating. Winter mean 47° F. to 51° F. Rainfall 25 inches. Rainy days 45 to 80. Well sheltered from cold winds. Climate stimulating.—Hyères, Costabelle, St. Raphael, Cannes, Nice and Cimiez, Monte Carlo, Mentone, Bordighera, Ospedaletti, San Remo, Alassio, Pegli, Nervi, Rapallo, Sta. Margherita, and Spezzia. (c) The Isles of Corsica, Sicily, and Malta. Ajaccio. Winter mean 53° F. Large number of rainy days.—Palermo. Winter mean 53° F. Warmer and moister than the Riviera.—Catania and Aci Reale (Sicily). With winter mean of 52° F. and a climate rather less moist than that of the Palermo.—Malta. Winter mean 57·46° F. Drier and hotter than Palermo. All the above less stimulating than the Riviera. (d) Algiers. Winter mean 56° F. Rainfall 32 inches. Rainy days 87. Moister and warmer than Riviera.—Tangier. Climate probably intermediate between Algiers and Madeira.

INLAND. WARM OR COLD.

1. CALM TEMPERATE (MOIST):
   Pau. Winter mean 42·8° F. Rainfall 43 inches. Number of rainy days 119. Climate cold and sedative.—Amélie les Bains.

2. CALM AND COLD CLIMATE:
   St. Paul's Minnesota—Canada. Climate very cold and dry.

3. VERY DRY AND WARM:
   (a) Egypt: Cairo. Mean temperature 58·52° F. Rainfall 1·339 inches (Marcet), and the number of rainy days 12. Climate of Upper Egypt still drier.
   (b) Cape of Good Hope and Natal.
   (c) Hill districts of Australia at some distance from sea.

MOUNTAIN. ATMOSPHERE MORE OR LESS RAREFIED.

1. CLIMATES OF THE SWISS ALPINE SANITARIA:
   Altitude 4,771 feet to 6,000 feet.— Davos, 5,105 feet. Winter (October to March) mean 28·1° F. Number of rainy or snowy days 52.
   St. Moritz, 6,000 feet; Maloja, 6,000 feet; Wiesen, 4,771 feet. Climate very cold, dry, and remarkable for diathermancy and absence of wind. Very bracing.

2. CLIMATE OF ROCKY MOUNTAIN SANITARIA (COLORADO):
   Altitudes varying from 5,200 feet to 9,000 feet. Climate resembling that of the Alpine, but warmer and with less snow.— Denver, 5,290 feet; Colorado Springs, 6,000 feet; Manitou Springs, 6,370 feet; Poucha Springs and Waggon Wheel Gap, 9,000 feet; Elkhorn, 7,500 feet; Pagosa, 7,000 feet; Santa Fè (New Mexico), 7,000 feet (warmer).

3. CLIMATES OF THE ANDEAN SANITARIA:
   Altitudes varying from 8,000 feet to 13,500 feet, and, owing to more southern latitude, enjoying warmer and more equable climate than the above. Often a mean temperature of 60° (perpetual spring) at all seasons.—Santa Fè di Bogota (New Granada), 9,000 feet; Quito, 10,000 feet; Arquipa, 9,000 feet; Jauja and Tarma, 10,000 feet (circa); Huancayo (in Peru) 10,718 feet; La Paz (Bolivia), 13,500 feet (much colder). Climate generally dry, warm, and bracing.

4. CLIMATES OF HIMALAYAN AND OTHER INDIAN HILL STATIONS:
   Altitudes varying from 3,500 feet to 8,000 feet.—Himalayan, from 4,000 feet to 8,000 feet. Winter, heavy rainfall, 70 inches to 132 inches. Winter mean temperature from 35° F. to 87° F.—Darjeeling, 8,000 feet; Simla, 8,000 feet; Landour, 7,300 feet; Nynee Tal, 6,200 feet; Dugshai, 6,000 feet; Subathoo, 4,000 feet; Nilgiri Sanitaria, with an altitude of from 5,000 feet to 7,000 feet. Mean temperature varies from 54° F. to 70° F., and rainfall from 50 inches to 60 inches.

5. CLIMATES OF SOUTH AFRICAN HIGHLANDS:
   Cape Colony (N.), Orange Free State and Transvaal. Altitude 5,000 feet (circa). Moderately warm, very dry, and bracing climate.

1. *British Channel Climates.*—These owe their mild climate to the influence of the warm equatorial current and to the winds it generates, which, while causing a remarkable equability of temperature, largely increases the rainfall and the number of rainy days. This is well shown by Dr. Tripe, in a comparison with the climate of inland places.[1]

My statistics[2] of 243 consumptives who wintered at one of the British Channel health stations demonstrated that the easterly, such as Hastings and Ventnor, acted more beneficially than the westerly ones of Torquay and Penzance—*i.e.*, that the full influence of the warm current, so advantageous to Devonshire vegetation, is less so to consumptives than when modified and rendered more bracing, as in the climate of the Sussex Downs. The advantages of the British marine health stations are—that the climate is devoid of great extremes, that they are generally moderate in prices and easily accessible, and are well provided with good food and invalid comforts, and moreover can be lived in all the year round; but their drawbacks are the want of sunshine, the high percentage of humidity and the large number of rainy days, and consequently of days of confinement to the house for invalids.

The *Atlantic stations* of Arcachon and Biarritz resemble the British health resorts in climate, except being somewhat warmer and enjoying more sunshine. Arcachon resembles Bournemouth with its pines and sandy soil, and, being placed on the shores of a land-locked sea basin, is protected from the Atlantic blasts.

The *warm Atlantic* group includes the well-known and far-famed station of Madeira, the great Peak of Teneriffe, so vividly described by Professor Piazzi Smythe and Dr. Marcet, where the recently erected Grand Hotel of Orotava and Mrs. Turnbull's boarding-house will supply the wished-for accommodation for visitors, so that it is now likely that this benign climate, which appears to be dryer than that of Madeira, will be turned to good account.[3] Warmer than these are

---

[1] 'On the Winter Climates of English Seaside Resorts.' *Quarterly Journal of Royal Meteorological Society*, April 1878.

[2] *Lettsomian Lectures*, 1875.

[3] *See* some graphic and interesting letters by Mr. Ernest Hart on 'A

the West India Islands, Jamaica, Barbadoes and Trinidad, with a moist and somewhat relaxing atmosphere, beneficial more on account of the outward voyage than for the exposure to tropical climate, which residence in them involves. The Blue Mountains of Jamaica, rising 4,000 feet above the sea, are an exception, and, from their altitude and moderate warmth, suit the requirements of many invalids, and a fine invigorating breeze is always enjoyed there.

The beautiful climates of the Californian sea coast, such as Los Angelos and Santa Barbara, San Diego and San José, from their dryness and equability of temperature, seem likely to rival many of the European ones, and to become very attractive to Americans, though their distance from England precludes their being recommended largely to English people. I have known several cases of phthisis with cavities do very well at Santa Barbara.

Sea voyages have long held a prominent place in the methods of climatic treatment for phthisis, and our statistics confirmed this by showing improvement in no less than 89 per cent. of the patients who undertook them.

The good influence seems to consist in the abundant supply of fresh pure air, free from organic or inorganic dust, and from germs, rich also in ozone and available to invalids without fatigue, and from the marked equability of temperature during a well-timed voyage, without fear of chill or exposure to extremes of temperature. There is a large amount of moisture too, for, according to Capt. Toynbee, the difference between the dry and wet bulb varies only from 2° to 5° F., a great contrast to most of the land climates, where there may be registered even a difference of 10° to 24°. The drawbacks are the difficulty of obtaining a proper amount of exercise, and the impossibility in the case of long voyages, such as that to Australia and New Zealand, of turning back when a start has once been made. Again, the accidents of a sea-trip may tell unfavourably on patients, such as running short of provisions owing to an unexpected length of voyage, and in stormy weather the confinement to the cabin, a con-

Winter Trip to Fortunate Islands.' *British Medical Journal*, April and May 1887.

dition by no means favourable to health. This last is often a serious matter, as the small ill-ventilated cupboard in which he may be doomed to live for days, or even weeks, is certainly not the best atmosphere for a phthisical patient, and more likely to be injurious than the sea-sickness, from which recovery is usually rapid.

Of the various sea voyages the one to Australia or New Zealand round the Cape of Good Hope is the most beneficial, and observations show that the temperature on board a clipper ship does not vary greatly, whatever season of the year this voyage is undertaken, but ranges from 50° to 80° F. But the season at which Melbourne, Sydney and Auckland are reached may greatly influence the consumptive invalid, and it is therefore advisable to arrive in Australia during their summer, hot though it be, and thus secure with the English one, two, or three summers in succession. For this purpose a start should be made in October or November, by one of the fine clipper vessels of Messrs. Green & Co., Devitt & Moore, Anderson & Co., Houlder Bros. & Co., Shaw Savill & Co., &c., to Australia or New Zealand, and after a voyage of about three months in a sailing vessel or thirty-seven to forty-five days in a steamer, Australia is reached. Shaw Savill and the Albion Company ply to New Zealand as well. A stay there of two months spent in the hills is highly advisable, and the return voyage should, if possible, be by the Cape of Good Hope. In this way all extremes of temperature are avoided, and the patient is fairly submitted to continuous saline influence under the best auspices. According to Dr. Maclaren, the first influence of the voyage before the South-West trades are encountered is sedative, showing itself in reduction of cough; and during the after part the gradually cooling atmosphere and strong fresh breezes form a tonic element, and improvement of appetite with gain of weight takes place. When the return voyage follows, as it too often does, the Cape Horn route, the patient is liable to a great fall of temperature, and the thermometer may vary from a few degrees above the freezing point to a few degrees below it, owing to nearing the Antarctic Circle, and therefore great precautions in the way of warm clothing are necessary. The Australian voyage is undoubtedly the most satisfactory experiment, but the shorter one to the Cape of Good Hope in the Donald Currie or the Union Companies' steamers is often beneficial. Where it is desirable for a patient to pass two winter months out of England, the three weeks' voyage, with a stay of fourteen days at Wynberg, and the homeward trip, is a very good method of getting a spell of undoubtedly warm weather for a short period.

Another winter trip is that offered by the Royal Mail Steam Packet Company's steamers to the West Indies, which, as the vessels touch at most of the islands and several of the ports of Central America, occupies from six weeks to four months, spent chiefly in warm regions. Another trip for the winter is the Brazilian one, occupying two to three months, when the vessel touches at Lisbon, Teneriffe, Pernambuco, Bahia, Rio de Janeiro, Monte Video and Buenos Ayres, keeping the passengers for a considerable period in South American waters.

The trip to Australia or to India and Japan, *via* the Suez Canal and Red Sea, lacks the equability of temperature, the even rise in approaching the Equator and the gradual fall on leaving it, which characterises the long voyage round the Cape. The increased heat in the Suez Canal, and the still greater rise in the Red Sea, where the thermometer marks about 98° F., is most unsuitable for invalids, who often suffer greatly from perspiration and anorexia in consequence. On the return voyage, too, the passing from the hot atmosphere of the Red Sea into the cooler Mediterranean climate, especially if the tramontana be blowing, checks perspiration, and, as a very intelligent Peninsular and Oriental Company's surgeon informed me, induces temporary albuminuria, probably from renal congestion.

This gentleman found albumen in the urine of every passenger on board one vessel after entering the Mediterranean from the Red Sea. For further information on the subject of sea voyages the reader is referred to Mr. Wilson's admirable work,[1] which is full of useful details. Voyages for the summer months offer plenty of choice, and include the trip to

[1] *The Ocean as a Health Resort.* Second Edition. J. & A. Churchill, 1881.

America and back, and numerous yachting expeditions, which often do great good to consumptives.

The *dry climates of the Mediterranean* contrast greatly with the last group, and are, as a rule, characterised by dryness and stimulating qualities, combined with a fair degree of warmth. These features are well typified in the climate of Malaga. The Riviera climate is too well known to need description, but we must bear in mind that the effects of nocturnal radiation show themselves very markedly along the whole coast, and the differences between individual health resorts depend partly on their degree of shelter from the cold winds, and partly on their proximity to, or distance from, the sea. This region now abounds in excellent hotels, provisions, invalid comforts, and in good doctors ; and, if the patients were placed a little more under medical direction and control, would furnish better results.

The Mediterranean Islands of Corsica, Sicily and Malta are warmer and moister than the Riviera, but are consequently less stimulating. Algiers differs largely from the north coast of the Mediterranean in its greater warmth and increased moisture, and Tangier stands as an intermediate between the Atlantic and Mediterranean climates.

In the first division of the *Inland Climates* stands Pau, once so largely recommended for the treatment of consumption, but which has been shown by my statistics[1] not to have deserved its patronage as a winter climate, but to prove a useful adjunct to treatment, as a spring resort, when the mistral and tramontana prevail elsewhere, its leading characteristic being calm stillness of atmosphere. The winter mean temperature is below that of Torquay, and its influence is not stimulating but sedative.

The *cold dry climates* of Minnesota and Canada were largely used before the high-altitude stations came to be recommended, and certainly exercise a bracing and invigorating influence, allowing of abundant exercise and out-door life.

The *very dry and warm climates* include Egypt and its

[1] *Warm Climates in the Treatment of Consumption. Medico-Chirurgical Trans.,* lv.

desert atmosphere,[1] which has yielded remarkably good results
in the treatment of consumption. The winter climate is warm
and occasionally intensely hot, but remarkable for its dryness
and small number of rainy days. The winter rainfall does
not exceed 1·339 inches (Marcet). The difference between the
dry and wet bulbs amounts on the Nile sometimes to 24° F.
The radiation after sunset is the most remarkable, amounting
according to Dr. Marcet[2] to 17° and 18° F. The nights are
consequently cool and refreshing, but require precautions to
be taken accordingly. The weather seems to be uniformly
fine, and the desert air is entirely free from carbonic acid
and strongly antiseptic (Zagiell), as has been shown by meat,
after three weeks' exposure, drying up without decomposition.
Alexandria is a moist climate, and unfit for invalids' residence,
but Cairo has an equable winter temperature, and abounds in
all requisites except good drainage. The best climate is to be
obtained in Upper Egypt, by ascending the Nile in a steamer,
or, better still, in a dahabeah to Assouan, and living as much
as possible in the desert atmosphere. Luxor is a good halting-
place, with a fair hotel, and a good dragoman will provide the
dahabeah with every comfort and luxury. Helwân, sixteen
miles south of Cairo, and three from the east bank of the Nile,
where sulphur baths exist, is also well spoken of, on account of
its good hotel accommodation, and fine desert air.

Some of the Australian hill stations, in proportion as they
approach the desert in the interior, present a remarkably dry,
clear, warm climate, and parts of the Cape Colony and Natal
may be embraced in the same category.

The *Mountain Climate* group is one of growing importance,
and my statistics when complete, as I hope they will be
shortly, will, I doubt not, furnish most successful results.
The group, which includes localities differing greatly in temperature, hygrometry, and shelter from wind, is bound together by the following leading characteristics :

[1] For good information on the subject of Egypt and its climates see Marcet's
Southern and Swiss Health Resorts, p. 205, and W. H. Flower's *Notes of
Experiences in Egypt*.

[2] *Quarterly Journal of Meteorological Society*, October 1885.

1. Diminished barometric pressure, and consequent rare-faction of atmosphere.

2. Diathermancy of the air, or the increased facility by which the sun's rays are transmitted through the attenuated air. This tends to an increase in the difference between the sun and shade temperatures, according to the elevation, in the proportion of 1° F. for every rise of 235 feet.[1] The consequence of this appears in the remarkably high temperatures in the sunshine, exceeding those recorded in the plains of the same latitude, and the very low night minima ; and thus we have, as is to be seen at Davos, a record of the solar radiation thermometer at 166° F. with a night minimum as low as −16° F., a range probably unequalled except under similar conditions of climate, and requiring great care to guard against its injurious effects.

3. Absorption of atmospheric oxygen by the blood takes place more readily, while at the same time the carbonic acid formed within the body passes outward through the pulmonary tissue into the air, which is breathed with a greater degree of facility than at lower altitudes (Marcet[2]).

4. Aseptic quality of the air and comparative freedom from organic germs.

The Swiss Alpine sanitaria and those of the Rocky Mountains combine with the above qualities dryness of atmosphere and a certain degree of winter cold. The Andean ones are dry but warmer, even at 10,000 feet presenting the climate of perpetual spring. This great advantage enables a patient to use them throughout the entire year, and at the same time gives a wide choice of altitudes; and we must ever remember that our first knowledge of the 'mountain cure' was derived from Archibald Smith, whose experience was based entirely on clinical observations in the Peruvian Andes, as I have shown elsewhere.[3] The Himalayan, with very varying mean temperatures, have a climate of great humidity, and probably less diathermancy on that account. It is possible that the less favourable view

[1] Denison, *Rocky Mountains Health Resorts.*
[2] *Southern and Swiss Health Resorts*, p. 334.
[3] *Influence of Climate in Consumption*, p. 21.

held by Indian medical men of the value of mountain sanitaria is due to the amount of humidity noted at the Himalayan hill stations. The Nilgiri stations of Ootacamund, Kotagherry, Wellington, and Coonoor, while they present a fair altitude (5,000 to 7,361 feet) and good mean temperature, have the advantage of fewer rainy days.

The South African highlands (4,000 to 5,000 feet), the advantages of which have been urged by Dr. Symes Thompson,[1] offer splendid sites for sanitaria in a magnificent dry climate, and have already furnished me with several admirable examples of arrest of consumptive disease. Aliwal North, Cradock in the Cape Colony, the Orange Free State and the Transvaal, are places of resort for invalids, and both Cradock and Aliwal North, where the hotels are good, are connected by rail with the ports of Port Elizabeth and East London respectively. At Bloemfontein, to which a railway will in time be made from Kimberley, are to be found all the necessary arrangements for an invalid. In this district the healthiest mode of life is to charter an ox-waggon and go ' on the trek ' for weeks together, shooting and riding, sleeping in the waggon nearly in the open air, and inhaling a pure atmosphere, and developing appetite to consume, and skill to supply, a larder full of varieties. Several of my patients have completely recovered in this way, and one (Mr. Nixon) has written a sensible book on the inhabitants and mode of life in this region.[2]

The effect of mountain climates on the human body is a bracing and stimulating one to the various vital processes. The influence of greater diathermancy is seen in the tanning of the skin of patients, even in winter, the appetite and digestion are decidedly improved, and gain of weight generally, though not invariably, takes place. The nervous system sometimes becomes excited, though not so much as we see in the stimulating Riviera climate, and in summer, especially, patients complain of loss of sleep. The temperature of the healthy body does not seem much affected, but the pyrexia of phthisis is often intensified, instead of reduced, by the moun-

[1] *Medico-Chirurgical Transactions*, vol. lvi.   [2] *The Boers.*

tain climate, whereas the soft moist atmosphere of the British Channel Islands tends to reduce it, when not due to active and extensive pulmonary lesions.   The gradual lowering of the pulse and respiration-rate, which have been demonstrated to coincide with the remarkable expansion of the thorax, are the specific influences of high altitudes, and have been proved to occur in all mountain sanitaria of 5,000 feet altitude and upwards, wherever situated, whether in the Rocky Mountains, Andes, Alps, or Himalayas.   This widening of the thorax appears to be due to the pressure of the hypertrophied lung structure, and to the localised emphysema developed around the old pulmonary lesions, on the thoracic parietes, as I have pointed out elsewhere.[1] When it is urged that development of the unaffected lung may proceed to the utmost possible limits at no higher level than the galleries of the Brompton Hospital, and can be often noted in chronic phthisis in process of arrest, as compensatory to the contraction of the diseased lung, it may be answered that undoubtedly such cases do occur everywhere under favourable conditions, but that the enlargement in this case is compensatory, and depends on the extent of lung surface abolished by tubercle and fibrosis.   It may be also admitted that localised emphysema takes place round cicatrising cavities.   These pulmonary changes differ both in degree and effect from those noted after residence at high altitudes. Chest expansion, accompanied by hypertrophied lung, is present in all dwellers on heights, be they sick or sound, and is characteristic of mountain races, being especially well shown in the Indians of the Andes, who are remarkable for their large chests and powers of continuous walking.   The most prominent instances I have noted are the Chamounix guides, who have enormous thoraces in proportion to their heights, and whose pulse and respiration are both slow. This enlargement of the chest is accompanied by an increase in the vital capacity, as measured by the spirometer, by more vigorous climbing power, also by slowing of the pulse and

[1] 'Treatment of Phthisis by Residence at High Altitudes' (*Trans. International Medical Congress*, 1881).

respiration. The emphysema around the old lesions noted after a mountain residence is far more extensive than anything witnessed at low levels, and what strikes the medical examiner most, is the general hyper-resonance of the chest, not only in the region of the old lesion but in all regions of the thorax, extending to the lowest pulmonary limits.

To prove that these changes are not special to high altitudes it would be necessary to show that they were present not only in certain cases of contracting cavity, but that all the patients and nurses living in the same gallery of the Brompton Hospital underwent some chest expansion and lung hypertrophy. I may add that I have measured a number of patients before, and after, wintering at low level health stations and after sea voyages, and they do *not* show these thoracic changes.

We will now give a few indications for the selection of a climate in phthisis, bearing in mind that, while the meteorology of the place and the clinical features of the case are the scientific grounds for a decision, a great many other considerations, as the food of body and mind, the facilities for exercise, the journey to and fro, the accessibility of medical aid and home comforts, and the feelings of the patient, must be more or less considered.

There are a certain number of cases which may be put out of consideration, where the best of climates can avail nothing, such as cases of acute tuberculosis, tuberculo-pneumonic phthisis, laryngeal phthisis, acute phthisis, except in a few of those instances (see p. 256) where the intensity of the tubercular process has been reduced by extensive excavation and has passed into quiescence. We must also exclude all cases accompanied by continuous pyrexia, or in which the processes of tuberculisation or excavation are actively proceeding; also advanced phthisis accompanied by intestinal ulceration and albuminuria.

The high altitude stations are most beneficial for the following :—(1) Cases of strong hereditary predisposition, in which phthisis is either threatened, or in a state of early development. (2) Imperfect thoracic or pulmonary develop-

ment. (3) Hæmorrhagic phthisis. (4) Chronic pneumonia (without bronchiectasis), which does not resolve. (5)
Chronic pleurisy where the lung does not expand after removal
or absorption of the fluid. (6) Phthisis accompanied by
more or less pneumonic consolidation. (7) Chronic tubercular phthisis in its various stages, provided the lung surface
be not too largely involved to admit of proper respiratory
change taking place at the high altitude, and there be no
pyrexia. (8) Cases of anæmia. (9) Spasmodic asthma without emphysema.

Where mountain climates are contra-indicated (in addition to the cases mentioned in which no change of climate
is advisable) we may mention cases of (1) emphysema, and of
phthisis with emphysema. (2) Chronic bronchitis and bronchiectasis. (3) Diseases of the heart and great vessels. (4)
Affections of the brain and spinal cord and states of hyper-
sensibility of the nervous system. (5) Diseases of the kidney
and liver. (6) Diabetes. (7) Catarrhal phthisis. (8) Phthisis
with double cavities with or without pyrexia. (9) All cases
of phthisis in which the pulmonary area is already encroached
upon too largely to admit of the proper performance of the
respiratory functions. (10) Cavity cases accompanied by
profuse hæmoptysis, pointing to the probable existence of a
pulmonary aneurysm. (11) Cases of phthisis in which there
is great irritability of the nervous system ; and (12) Of people
of advanced years, or those too feeble to take exercise. The
cases where the mountain treatment is of greatest benefit
are those of ordinary first-stage phthisis, where the physical
signs do not amount to more than a few crepitations audible
at one apex with slight dulness and bronchophony, where the
cough is not troublesome and the expectoration is scanty,
though containing a few tubercle-bacilli, and where loss of flesh
and night-sweats have continued for a few months. Many such
patients have returned to me after six months, or even sometimes only three months, of Davos or St. Moritz life, having
lost all signs and symptoms and with every appearance of
robust health, and have after a lapse of five or six years
still remained well and strong. Even where both apices were

affected, I have known complete recovery in one winter. In one case it took place in three months, and the patient three years later passed through a severe attack of pneumonia without any recurrence of the phthisis. Where the consolidations are more extensive, and if a cavity be present, a longer stay is desirable, and often two or three seasons may be required to complete the closure of the cavity and the full development of the healthy lung. Of twenty-two cases of phthisis which I published five years ago, ten recovered sufficiently to return to their occupations. I have now notes of upwards of 120 consumptives under mountain treatment, a large proportion of whom have recovered their health and strength, and I propose shortly to offer the particulars to the profession.

The plan to be pursued by a consumptive who proposes to try the Grisons resorts in winter is to arrive at Davos or St. Moritz, or the Maloja, in September or early in October, and become accustomed to the climate before the snow falls and the true winter season commences, but in all cases the consumptive must be settled at his station before the snow falls. I dwell on this because many contract catarrh by travelling from England during severe weather, and lose much valuable time by subsequent and necessary confinement to their bedrooms after arrival. The question of a gradual ascent to the high altitudes is not so important (*pace* Jaccoud) as the gradual descent from them, which in seasons of rapid snow-melting must be made about the end of March or the beginning of April, and is often an inconvenient process.

During the winter the consumptive passes a carefully regulated life under strict medical supervision, and well it is that the Alpine stations staff include able and experienced medical men, who are able to regulate not only the medicine to be taken, but also the clothing and the kind and amount of exercise. Indeed this last requires the greatest attention and good management, and the system consists of a graduated scale of bodily exertion, beginning with skating and walking on level ground, and then, after gentle and limited ascents of the neighbouring hills, proceeding to tobogganing, followed by short walking tours and even mountain ascents. It is this combined system of hardening and development which does so much for the consumptive of either sex, especially for the young, and enables them to resist later on the frequent changes of our variable weather.

The patient at the close of the first winter should have attained a vigorous state of health, with a widened chest and increased vital capacity. He is free from cough and expectoration, and is equal to considerable exertion. In order to keep him in this condition, and to make it a stepping-stone to yet better things, the descent must be cautiously managed. With this aim in view he must avoid, first, all damp and relaxing localities, such as the shores of lakes, which used to be favourite intermediate stations of German physicians; secondly, warm Italian valleys and places where hot springs exist, like Baden-Baden, the moist atmosphere of which may relax his system and increase his expectoration. Even a return from the mountains *viâ* the Riviera in spring is not in my experience beneficial, but rather the reverse.

When the snow begins to melt, generally about the beginning of April, the roads and paths become slushy and the atmosphere very damp, and therefore it is desirable that patients should leave for a season, and descend to a lower level, where this process has already been completed. A return can be made to the heights in May or June, if continuous residence at high altitudes is thought desirable.

For patients returning to England, stations of intermediate altitude are specially requisite, as the rapid descent to lower levels is fraught with danger, and is sometimes accompanied by hæmoptysis. For those wintering at Davos a halt should be made at Thusis (2,448 feet), or Gais (2,820 feet), or Weissbad (2,680 feet) in Appenzell, for a week or 10 days, and next at Basle or Berne (1 week); or after a stay at Thusis (10 days) another halt can be made at Badenweiler (1,380 feet) in the Black Forest, for 10 or 14 days, and then England can be reached in the first week of May. Those returning from St. Moritz and the Maloja can be halted at Wiesen (4,771 feet) for 10 days, which place is exceedingly well adapted for the purpose, or at Mühlen (4,793 feet), and again for 10 days at Thusis or Ragatz (1,709 feet), all of these places having fairly comfortable accommodation and resources; and after a few days spent at some low level town, away from lakes, the return can be safely made to England in May. The high altitude invalids do not as a rule, like the Riviera patients, find the English spring trying. They fear not the east wind, and only object to the relaxing influence of moist warm weather, and I have seen some patients return to London even at the end of March without harm : though this proceeding is not as a rule advisable.

In some seasons the snow-melting is so gradual that it is hardly necessary on that account to quit the high levels. The Rocky Mountains and Andean stations do not much suffer from this drawback, and patients reside there all the year round.

For American and Canadian patients Colorado [1] with its wide plateau, or the National Yellowstone Park, give a splendid choice of sites and elevations for residences, where camping out also can be carried on during the summer. More than one young consumptive patient of mine, in whom the disease has not quenched the spirit of enterprise, has embarked in a ranche in this region, and has recovered his health and filled his pockets after one or two years' residence. Colorado Springs (6,000 feet) and Elkhorn (7,500 feet) have both been the scenes of remarkable recoveries [2] among my patients from comparatively advanced disease, and are both supplied with good boarding-houses and doctors.

The Andean health stations will some day, when Peru is a country of settled government and Bolivia ceases to quarrel with its neighbour Chili, and revolution is not so rife— also when the Andean railway is complete in its branches as well as in its main lines—attract numbers of European as well as of American consumptives, who may fairly expect the same wonderful results as are furnished by Archibald Smith's and Guilbert's [3] successful cases. Some of the stations on this

---

[1] Full information is given about these places in Dr. Denison's *Rocky Mountain Health Resorts*, 2nd edit.

[2] *Clinical Transactions*, vol. lvi.

[3] *De l'Influence du Climat des Andes sur la Phthisie.*

line are above 10,000 feet, and one (Chosin) is 12,200 feet, above the sea, the climate being remarkably dry.

Of Inland Climates the still ones of Pau and Amélie les Bains were formerly in great vogue for consumptives, and were said to be specially indicated in erethic phthisis, as it is termed by the French, where the disease is accompanied by great irritability of the nervous system and of the gastric mucous membrane. This is one of the varieties of phthisis where mountain climates are contra-indicated.

My statistics assigned very unfavourable results to Pau as a winter climate, but from its calm atmosphere it may be recommended as a spring climate with some advantages, especially to consumptives returning from the Riviera.

Egypt and the Cape of Good Hope and the very dry climates are most useful in cases of large secreting cavities, or in phthisis with bronchiectasis, where fresh catarrh may prove dangerous by reducing the breathing space; also in phthisis with emphysema; for in the dry aseptic climate of the desert the secretion becomes lessened in quantity, and the patient respires more easily in this warm, exhilarating region, and especially so in the air of Upper Egypt, so easily reached by steamers or dahabeahs.

The Mediterranean climates are specially indicated in all case of phthisis complicated with, and determined by, inflammatory attacks of the lung, whether pneumonic or broncho-pneumonic, or even pleuro-pneumonic. Such cases are generally benefited by the combination of warmth and sunshine, the absence of fog and damp, and the presence of the stimulating saline element of the Mediterranean Sea. Ofttimes a patient who has not quitted his room for months during a winter in England, if transported in a day or two to one of the Riviera resorts, suns himself all day, discarding his comforter and respirator, and is soon able to take plenty of exercise, this bearing fruit in the improved appetite and muscular power.

We must remember that the bracing part of this climate, caused by the cold winds and dryness, benefits the patients quite as much as the warm element, originating in the bright

sunshine and the stimulating and equalising influence of the
Mediterranean Sea, and accordingly we often find the cold
winters of the South produce as much, if not more, improvement than the warm ones, as long as the patient is able to be
out sufficiently in the open air.

Chronic tubercular phthisis in all its stages, if devoid of
pyrexia, and provided strength and respiratory surface suffice for active exercise, is well suited on the Riviera, and such
patients lead a more enjoyable life there than in most
health resorts, having the combination of beautiful scenery,
abundant sunshine, and good living. In many cases the
slow process of cavity contraction appears to be somewhat
hastened by the dry stimulating climate, and arrest of the
disease gradually takes place, while old pneumonic consolidations certainly tend to diminish in extent, either by absorption or by undergoing fibrosis and contraction, in this dry
climate, and there is no lack of striking instances of arrest of
phthisis taking place in these regions, including many instances of large cavities and even of double cavities. Even
where the extent or degree of disease renders arrest, or even
great improvement, improbable the invalids may remain *in
statu quo* for long periods ; and I can remember patients whose
amount of pulmonary mischief has made me hesitate before
sanctioning their leaving England, passing not one, but seven
or eight winters in succession on the Riviera, with prolongation
of life and mitigation of suffering. The climate is well adapted
to pulmonary disease in advanced life, not only in reference
to phthisis, but bronchitis, asthma and emphysema ; and
many old people escape much suffering by wintering in this
region.

The last published meteorological observations [1] of the
different health resorts give a slightly higher winter mean
temperature to Hyères, Mentone, and San Remo than to
Cannes and Nice, and these two last accordingly share more of
the *bracing* climatic element.

Mentone, San Remo, Ospedaletti, Monte Carlo, and Alassio,
on the Western Riviera, and Nervi and Spezia on the Eastern,

---

[1] Marcet, *op. cit.*

are all admirably sheltered from the northerly winds and
bask in sunshine. While Cannes and Nice and St. Raphael
are more or less exposed to the mistral (N.W. wind) influence,
they enjoy one great advantage from their sheltering ranges
being neither so close to them nor so unbroken in line : for
the enclosed country offers a rich choice of sites for invalid
residences at varying distances from the sea and at different
elevations, and thus cases of irritable mucous membrane or
high nervous sensibility can be easily accommodated. This
want has been lately recognised at other places, and some
hotels are built as far as possible from the sea at Mentone
and San Remo. Hyères,[1] from its being the most southerly
station of all the Riviera, and its position on the side of a hill
four miles from the open sea (and three from the Rade
d'Hyères), possesses a climate of its own far less stimulating
than the rest, and consequently less bracing, but well suited
to cases of phthisis where there is a tendency to slight pyrexia
and great nervous excitability. It is the place where patients
sleep best on the Riviera, and twenty-four years' experience
of its effect on a large variety of consumptives has left a most
favourable impression on my mind.[2]

When rather more stimulating influence is required, its
suburb, Costabelle, with a charming hotel, completely sheltered
from cold winds and nestling nearer the sea, embosomed in
pines, had better be selected.

The cases in which the Riviera is specially contra-indicated
are hæmorrhagic phthisis, the so-called erethic forms, spe-

[1] It is only of late years that this remarkable climate, which may be
described as the most sedative of the Riviera, has been investigated according
to the methods of modern meteorology. This has been done by Dr. Biden,
one of the resident medical men, for a series of six years, and the results have
lately been published in the *Lancet*, from which I venture to extract the following averages for the six winter season months, November to April inclusive :
Mean temperature, 51·4° F. Relative humidity, 74 per cent. Rainfall, 14·61
inches. Days in which rain fell, 37. Days of sunshine, 135. Compared with
Cannes and Mentone it shows higher mean maxima and lower mean minima,
as might be expected from the distance from the Mediterranean, which consequently exercises less equalising influence.

[2] Though this is perhaps beside the mark, it may be mentioned that the
climate of Hyères is most beneficial in asthma, and I can count scores of cures
I have witnessed from its influence.

cially when accompanied by anorexia, and all cases of phthisis
in which there is regular pyrexia.

Where the Riviera climate is found to present too great
extremes and is somewhat too bracing, that of the Mediterranean Isles often suits better. Ajaccio (Corsica), Palermo,
Catania, Aci Reale, and Taormina (Sicily), offer a warmer and
more equable atmosphere, as well as fairly good accommodation and abundant food, and it is much to be regretted that
these fine climates, some of which I described some years
ago,[1] are not more utilised by English consumptives.

Malta is hotter and drier, and is a great contrast to the
other islands.

Algiers, with its larger amount of moisture but genial
warmth, its splendid choice of hill stations, Milianeh, Blidah,
Medeah, and Hammam R'Irha, for the early spring, and its
railway connection with Biskra and other oases in the great
Sahara Desert, is an admirable resort to fall back on.

Tangier is strictly outside the Mediterranean, but, having
a climate intermediate between that of Algiers and Madeira, is
here included.   The atmosphere is tempered by the Atlantic
breeze and far more equable than that of the Mediterranean
stations, and now that it is provided with a choice of good
hotels and boarding-houses, cheap and excellent food, with the
advantage of a resident English doctor, and is less than a week
by steamer from England, it will probably become more
and more popular.

Patients intending to winter at the Riviera and other
Mediterranean stations should not reach them till the end of
October or the first week of November, thereby escaping the
chance of diarrhœa, so common on that coast, and the plague
of mosquitoes, the numbers of which are soon reduced by the
approach of cool weather.

The return to England and the north when the Riviera
becomes too hot is always a matter requiring careful management, for consumptives appear to be rendered very susceptible
to cold and damp by the climate, and not hardened, as in the
Alpine stations.   Most invalids quit the Riviera too early, and

[1] 'Winter Stations of Sicily.'   *Lancet*, 1879.

it may be generally stated that the end of April or the beginning of May should be the earliest dates for starting, and England ought not to be reached before the last week in May or the first of June. The problem always is where to spend the intermediate period, and, according to my experience, the prudent and most successful courses are, (1) to spend fourteen days at Grasse, near Cannes (1,200 feet), or some similarly elevated place[1] away from the sea, then to move north, halting for ten or fourteen days at Fontainebleau, thirty-six miles south of Paris, where the spring is earlier than in England and the hotels are excellent, and the Forest provides charming walks and drives, or to stop at St. Germain, and after a few days in Paris in one of the well-sheltered hotels overlooking the Tuileries Gardens, home may be safely reached; (2) to leave the Riviera about the beginning of April and spend six weeks at Pau, where the spring climate is remarkably mild and still, or possibly the last fortnight of the period at Biarritz, and then return to England in June. Many halt for a month at Montreux, Vevey, Chexbres, Bex, or other places on or near the Lake of Geneva, and though I cannot say my experience of this method has been favourable, it is infinitely preferable to the more popular one of taking a tour to Rome, Pisa, Florence, and the North Italian cities, and after visiting the Italian lake district, always very damp at this season, returning home viâ the St. Gothard or the Cenis, generally in a worse condition than after quitting the Mediterranean coast.

Where a return to England is undesirable, patients may pass the summer at St. Martin Lantosque, St. Dalmas di Tende (3,000 feet), Certosa di Pesio (5,000 feet), good summer stations in the Maritime Alps.

The warm Pacific climates are suitable for the same class of cases as the Mediterranean coast, but appear to ensure greater equability of temperature, and patients are able to remain in them through the summer.

The moist and warm Atlantic climates were at one time

---

[1] The Grand Hotel de Valescure, two miles inland from St. Raphael, in the pine woods, is a good intermediate resting-place, but lacks the elevation of Grasse.

largely commended for all cases of phthisis, but the results
have proved that this reputation was undeserved in the case
of the disease generally. But for one form, viz., that of
catarrhal phthisis, the climate of Madeira is unsurpassed, the
irritable cough becomes softer and less frequent, the expectoration freer, and the whole respiratory tract seems soothed and
rendered less irritable by the mild atmosphere. The outward voyage of four to six days to Madeira or to Teneriffe, and
a short sojourn in one of these islands, is a good method to
get rid of obstinate bronchitic coughs which may have lasted
for months.

The cases best suited for sea voyages are (1) hæmorrhagic
phthisis; (2) scrofulous phthisis, especially where fistula has
been developed ; (3) cases of limited consolidation or cavity
where, without pyrexia, the cough is hard and obstinate,
probably from the bronchi being involved in the lesion; (4)
cases of phthisis and emphysema; and (5) cases where, in
addition to limited tubercular disease, the patients have been
overworked in mind or body. For these last the rest from
fatigue, the abundance of fresh air, and the entire novelty of
the scene acts most beneficially, and the influence is partly
stimulating, partly sedative; appetite returns, sleep is pro-
cured, strength and weight are gained, while the cough and
expectoration diminish and respiration grows easier. The
salt air seems to have a most favourable effect on the pulmonary vessels, and hæmoptysis at sea is very rare, and the
hæmorrhagic form of disease is specially benefited by its
influence.

The number of cases sent to the moist, temperate and
British climates must always be large, as consisting not only
of those for whom they are particularly indicated, but also of
those who are too ill or too poor to leave England, and we are
often constrained to recommend the British Channel stations,
as being near to the patient's home in case removal there may
become desirable. Nevertheless with all these drawbacks their
statistics yield fair results in chronic tubercular phthisis, and
specially in scrofulous phthisis, and the excellent food and
ood appetite to enjoy it which the English stations give, the

many comforts, and the easy removal to another climate if the selected one proves unfavourable, perhaps do something to make up for the blustrous weather, the smaller number of sunshiny days, and the occasional confinement to the house for long periods. Though the cough and expectoration may be more troublesome by day, the sleep at night is sounder, and gain of flesh is the rule where there is no pyrexia, and this last symptom is more easily controlled in England than elsewhere.

Among the British winter climates, St. Leonards, Brighton, and Worthing offer a choice of more or less bracing winter residences, while Ventnor and Torquay are more sheltered from cold winds, and Bournemouth makes up for by no means complete shelter from hills, by a remarkably dry soil and a certain degree of protection from the numerous pines.

CHAPTER XXIV.

PROPHYLACTIC AND ANTI-PHTHISICAL TREATMENT (CONCLUDED)— ANTIPYREXIAL TREATMENT.

Medicines - Cod-liver oil—Its mode of action and beneficial effects—Its solvent and penetrative powers—Various kinds of oil—Doses and times of exhibition — Emulsions—Glycerine, malt extracts—Oil sauces—Anti-phthisical tonics Arsenic—Quinine, strychnia, and mineral acids—Hypophosphites of soda, lime and iron.

Antipyrexial treatment: Salines—Various modes of counter-irritation—Antipyretics—Action dependent on cause of pyrexia—Quinine—Salicylates—Chairin Anti-pyrin Resorcin—Thalline—Antifebrin Hypodermic injection of carbolic acid Cold baths Sponging—Spinal ice bag and ice pack Treatment of night-sweats—Author's experience opposed to Niemeyer's conclusions—Mineral acids—Gallic acid—Oxide of zinc—Arseniate of iron—Belladonna and atropine—Picrotoxine and pilocarpine—Agaricin, muscarine and paracotoin.

Medicinal.—At the head of our prophylactic and anti-phthisical medicinal treatment must be placed cod-liver oil, which, from the advocacy of Dr. Hughes Bennett and Dr. C. J. B. Williams, and the remarkable confirmatory testimony of hospital and private practice, far surpasses in reputation any other article of medicine or diet used in the treatment of phthisis. The annual consumption of it at the Brompton Hospital amounts to nearly 1,500 gallons, and the quantity imported into Great Britain has reached colossal proportions.

Its exact mode of action in the system has not been completely ascertained, but the following remarks by Dr. C. J. B. Williams in the first edition of this work give us the best clue to it. 'That it is in itself a nutriment cannot be doubted; and that its nutritious properties go farther than to augment the fat in the body is proved by the well-ascertained fact that the muscles and strength also increase

C C

under its use. In fact, it has been proved to increase the proteinaceous constituents of the blood, except the fibrin, which is diminished.[1] In truth, the beneficial operation of cod-liver oil extends to every function and structure of the body. In cases most suitable for its use there is a progressive improvement in digestion, appetite, strength, and complexion; and various morbid conditions perceptibly diminish. Thus, purulent discharges are lessened, ulcers assume a healthier aspect, colliquative diarrhœa and sweats cease, the natural secretions become more copious, the pulse less frequent. It is difficult to comprehend how it can produce such marvellous and manifold salutary effects, but the extent to which it has been, and still is, administered pretty well prove that it has properties which render it congenial to the animal economy.

' Cod-liver oil forms an emulsion more readily than other oils, and leaves no greasy feeling in the mouth, and this corresponds with its easy digestibility and absorption from the alimentary canal. This may depend on its containing some biliary principles; it often has a marked effect in increasing the secretion of the liver, and, if this is sufficiently carried off by the several processes of combustion and elimination, no tendency to sickness results from its use. It is, therefore, not surprising that cod-liver oil can be administered in larger quantities and for a longer time in cold seasons than in hot; to persons who take exercise better than to the sedentary; and especially to those whose bowels act regularly and sufficiently. With many weakly persons it assists the digestive process by promoting the biliary secretion, and in not a few instances I have found it effectual in improving and rendering more fluid this secretion in persons liable to gall-stones or obstructions from inspissated bile. On the other hand, it is apt to disagree in cases of inflammatory dyspepsia, especially that affecting the duodenum; in those of hepatic congestion, with fulness and tenderness of the hypochondria; and in all states of high fever or inflammation. All such affections should be relieved by saline effervescing draughts, mild mercurial aper-

[1] *Simon's Animal Chemistry*, by Day, vol. i. p. 280.

ients, and such means, before the oil is given; and, in case
of persons prone to these disorders, they may be required
occasionally during its use.

'Thus we have an oily matter well borne by the stomach,
easily diffused by emulsion through the alimentary mass,
readily absorbed by the lacteals, where it contributes to form
a rich "molecular base" in the chyle; apt to saponify with
the basic salts of the blood; and, when diffused in this fluid
throughout the capillaries of the body, capable of penetrating
to all the textures, and of exercising its solvent and softening
action on the solid fats of old deposits, whilst it affords a rich
pabulum for the living sarcophytes[1] and bioplasm of the blood,
tissue-cells, and lymphatics.

'Its superior penetrative and suppling properties render
cod-liver oil valuable in the process of currying leather, and
previously to its introduction into medicine, this was its chief
commercial use. Its fluidity and divisibility enable it to per-
vade all tissues of the body, and to penetrate even into
caseous and imperfectly organised deposits, and so to dissolve
their solid fats and soften their concrete sarcophytes, as to
render permeable and supple their whole mass, and open
them to the immigration of new and active bioplasm, by the
operation of which their vitality and nutrition may be improved and maintained, or if incapable of such improvement,
their substance may be gradually dissolved and carried off.
A proof of its penetrative power in the living body is to be
found in the detection of the smell of the oil in the pus
of scrofulous abscesses when these have been opened in
patients who have been taking the oil for some time previously.

'If we call to mind also the large share which fatty transformation seems to have in the processes of resolution and
suppuration, which terminate inflammation and clear away
its products, we can see why the administration of cod-liver
oil proves so eminently beneficent in promoting these salutary results. For that very process of fatty transformation
which is so injurious and fatal when it affects vital organs, as
the heart, or when it spreads destruction in the caseation of

---

[1] Leucocytes.

a lung, is salutary and conservative when it helps to soften and carry off the obstructing and irritating products of inflammation. We can thus understand why under the use of the oil the cough becomes softened and the expectoration easy, being thereby fattened and ripened (*sputa concocta, crachats gras, cuits*).'

There is abundant proof that the oil can be taken in hot climates as well as cold, and many of our patients persevered with it in the East and West Indies and in Madeira, taking certain precautions, such as reducing the dose, and mingling it with food. The great points to be borne in mind are: *first*, that the patient must be able to take and digest solid food ; *second*, that the dietary must be adapted to the reception of so much oily matter, and all pastry, fat meat, rich sauces and stuffing, and the like, should be avoided and the amount of butter, cream and sweet things reduced; *third*, that the oil should be persevered with for months, if not for years, being only omitted if distinct symptoms of biliousness appear, and repeated again on their cessation.

To promote the prolonged use of this invaluable agent, certain precautions are necessary, such as the following :

The pale clear oil with little taste, such as is imported from Norway or Newfoundland, or prepared from fresh cod livers in England, is to be preferred to the dark-coloured and strong-smelling oils of De Jongh and others, it having been proved by Dr. Garrod[1] that the dark colour and high flavour are due to partial decomposition and putrefaction of the livers before and during the process of separating the oil from them. Cod-liver oil contains iodine, bromine, phosphorus and other elements, but in such small quantities as hardly to account for its remarkable effects, inasmuch as various combinations of these same elements have been tried without similar satisfactory results. The oil, and especially the pale varieties, contains a large amount of bile which renders the fatty matter easily assimilable, and it is therefore as an oil of remarkable fluidity and penetration that it probably produces such great beneficial results.

[1] *British and Foreign Medico-Chir. Review*, January 1856.

The specially prepared cod-liver oils of Bell, Allen and Hanbury and Peter Möller, are preferred by delicate palates, but the ordinary pale oil is now so carefully prepared as to perfectly suit most patients if judiciously combined with medicinal agents.

The oil should be taken either *during* or *immediately after* a meal, so as to mingle intimately with the food. Theoretically, the best period would be at the close of gastric digestion, when the food is passing into the duodenum to undergo chylification, as the oil undergoes no digestion in the stomach, but, practically, it is found to cause less eructation if introduced into the stomach with the food, and to leave that organ a longer period of rest before the next meal. The dose should begin with a tea- or dessert-spoonful and be increased to a tablespoonful twice and, in a few cases, three times a day. When two doses only are taken they should accompany the two first meals, as the stomach is, with the body, often fatigued later in the day. Some few exceptional individuals prefer to take it at bedtime. The above doses are generally as much as a patient can manage for any length of time, and even with these the oil can often be recognised in the fæces. Jaccoud, however, maintains that the dose should vary from 4 to 8 tablespoonfuls, and says he has known the exhibition of from 7 to 10 oz. with benefit.

'Many persons, especially children, take the oil alone without any difficulty, and in such a case it seems needless to recommend any adjunct. Yet even with these, if the remedy is to be continued for a long time, it is better to give some agreeably flavoured tonic with it, for this prevents the palate and stomach from being palled by repetition, which is very apt to occur when the oil is long taken alone, however well borne and even relished at first. To the great majority of patients it is more agreeable to disguise the taste of the oil ; and this may be done by giving it in another liquid, which may also act as an agreeable tonic to the stomach. Some physicians and chemists have endeavoured to render the oil more palatable by the addition of an essential oil or other flavouring matter; and in this way the taste may be nearly completely disguised. But the great objection to these "palatable oils" is that the essential oil, while it covers the taste at the time, considerably increases the tendency to unpleasant eructation afterwards. This effect is still more marked in the "etherised cod-liver oil," which is advocated on the ground of its stimulating the pancreatic secretion, and thus assisting the formation of an emulsion in the duodenum; but its pungent disagreeable taste, and the frequent eructations it gives rise to, have in our experience prevented private patients from continuing its use for any length of time.' Creasote has been used with great success to cover the taste of the oil, but it unfortunately requires disguising itself, which is done with an essential oil, thus leading to the above-mentioned inconvenience.' 'The best way is to take the oil floating on a well-flavoured tonic, such as the compound infusion of orange-peel, with the addition of a little diluted mineral acid, and either sweetened with syrup, or rendered more bitter by the addition of a little tincture of hop, calumba, quassia, or cascarilla, according to the fancy of the palate or the requirements of the stomach. The bulk of the whole dose should be small, so that it may be swallowed at a single draught. The mineral acid may be varied. The nitric generally suits best in inflammatory cases, and those attended with much lithic deposit in the urine, but its tendency to injure the teeth is an objection to its long continuance. The sulphuric is more eligible where there is a liability to hæmoptysis, profuse sweats, or diarrhœa. The nitro-muriatic acid suits patients best who are subject to liver disturbance. But in most cases, and for long continuance, there is reason to prefer the diluted phosphoric acid, which may be termed the most physiological of the acids, tending to derange the chemistry of the body less than the others. The chief advantage of exhibiting the oil in such a tonic as that now recommended is that, in addition to disguising the taste

---

' Jaccoud's form is:
  Cod-liver oil ℥ij.
  Creasote ♏ iij.
  Oil of Peppermint ♏ j.

of the oil, the tone of the stomach is also kept up, so that it bears the oil in full doses and for a long period ; and in this respect it is superior to orange or ginger wine, aromatic waters, lemon juice, coffee, milk, and other vehicles that are occasionally used.   In cases of peculiar weakness of stomach, with tendency to vomiting, a $\frac{1}{30}$ or $\frac{1}{40}$ of a grain of strychnia has been given in a solution with each dose with such success, that strychnia may be regarded as a specific against the retching of phthisis.'   Patients in this way are able to persevere with cod-liver oil for months and even for years without digestive derangement.   Some patients can only take the oil in the emulsion form, and this is easily done by combining it with yolk of egg, or mucilage, and a flavoured syrup, but in order to emulsify it completely an alkali is requisite, and the Mistura Olei Morrhuæ Præparata[1] of the Brompton Hospital Pharmacopœia, introduced by Dr. Reginald Thompson, is an excellent example, as the dose is small and the proportion of oil large.   Mackenzie's and Scott's emulsions are also useful forms, but the latter contains, in addition, hypophosphites, which are not always desirable.   The attempts to introduce the oil into capsules have invariably failed, from the large sizes of these necessitated by the dose of the oil.

The inunction of cod-liver oil is sometimes beneficial in the tabes mesenterica of children, but this process renders it so rapidly rancid and offensive-smelling, that it had best be, if possible, avoided, or other oils, as almond and olive oils, substituted.

Various substitutes have been proposed for cod-liver oil, but most of them have had a short-lived existence.   Glycerine was extensively tried in England and failed.   Jaccoud, however, maintains that when pure and taken in doses of from three to four tablespoonfuls daily, it promotes nutrition and leads to gain of weight, but in excess it gives rise to agitation, unaccustomed loquacity, and persistent insomnia.

[1] ℞ Olei Morrhuæ ʒvj.
    Liq. Ammoniæ Fortioris ♏ ij.
    Olei Cassiæ ♏ j
    Syrupi ʒij.
    ʒss = ʒiij of Cod Liver Oil.

Maltine and the various forms of malt extract are often mentioned as alternatives to the oil, but their composition and mode of action differ entirely from that of cod-liver oil, for instead of supplying an easily assimilable form of oil and fat, they act as a diastase, and assist the saliva in converting the different forms of starch into sugar, which may eventually assist in the formation of fat. Some of these are combined with cod-liver oil, others with pepsine, hypophosphites of iron and calcium, and form by no means unpalatable preparations, and I have often noted gain of weight from their use, though to a far less extent than when cod-liver oil and a bitter are taken. The objections to the various malt extracts are that they do not promote, but rather cloy, the appetite, and from their stimulating effect they sometimes cause an increase of cough.

In the employment of tonics in the early stage of phthisis, our first great aim is by judicious treatment of the stomach and liver to ensure steady persistence with the cod-liver oil. The administration of mineral acids and mild bitters, especially of strychnia, in combination with the oil, answers this purpose well, and forms the agreeable 'oil sauce' of Dr. C. J. B. Williams, a few of whose typical prescriptions are subjoined.[1]

[1] ℞ Acidi Nitrici dil. ℨiij.
    Tinct. Calumbæ,
    Syr. Zingiberis, ãã ℥j.
    Infusi Aurantii Comp. ad ℥viij.
A tablespoonful to be taken twice a day with a teaspoonful of pure cod-liver oil, gradually increasing the oil to a tablespoonful.

℞ Acidi Sulphurici dil. ℨiij.
    Tinct. Aurantii ℥jss.
    Salicini Əij.
    Syr. Zingiberis ℥j.
    Inf. Aurantii Comp. ad ℥viij.
A tablespoonful with each dose of the oil.

℞ Acidi Phosphorici dil. ℥ss.
    Sodii Hypophosphitis ℨj.–℈ij.
    Tinct. Quininæ Comp. ℥jss.
    Glycerini puri ℥j.
    Inf. Aurantii Comp. ad ℥viij.
A tablespoonful with each dose of the oil.

℞ Acidi Phosphorici dil. ℨiij.
    Acidi Hydrocyanici dil. ℈j.
    Tinct. Lupuli,
    Syr. Zingiberis, ãã ℥j.
    Inf. Aurantii Comp. ad ℥viij.
To be given with the oil as above, when the stomach is irritable.

℞ Acidi Phosphorici dil. ℥ss.
    Liquoris Strychninæ Hydrochlor. ℈j.
    Tinct. Aurantii,
    Syr. Zingiberis, ãã ℥j.
    Inf. Aurantii Comp. ad ℥viij.
To be given as the former ones.

The dyspepsia of phthisis has sometimes to be met by occasional alterative and purgative medicine, and a short course of alkali and bitter,[1] as the Mistura Gentianae Alkalina introduced by Dr. Quain into the Brompton Pharmacopœia, and the stomach thus prepared for the reception of the necessary dietary and the cod-liver oil. Where there is great weakness and pallor, and the cough is slight, iron in some form may be taken in small doses, but guardedly, on account of the cough, on which it seems to exercise a specially irritating influence. Quinine, in addition to its antipyretic effects, to be alluded to later, is often of great advantage in promoting the appetite and adding to the muscular power, but where the cough is severe, and the expectoration scanty, it is, like iron, contra-indicated.

Arsenic is, perhaps, the most useful tonic next to strychnia in the treatment of phthisis, by reason of its various effects on the animal economy. In the forms of the Liq. Arsenicalis, Liq. Sodii Arseniatis, or the Liq. Arsenici Hydrochloricus, in doses of 3 to 5 minims, after food, it aids nutrition most powerfully, checking slight pyrexia and night-sweats, improving the respiratory powers, and giving a more sleek appearance to the patient, with a considerable gain of weight. It appears to mix well with the oil, and be tolerated by the stomach and system generally for months and even years, without giving rise to any symptoms of accumulation, and, what is of great importance, it does not excite the vascular system or increase the cough. Arsenic is especially indicated in chronic phthisis, combined with asthma, and certainly exercises a marked tonic influence on the bronchial muscle and pulmonary plexuses of nerves, reducing the tendency to spasm, and rendering the respiration easier and deeper. In cases of psoriasis and chronic eczema accompanying phthisis, the skin, as well as the lungs, demonstrates its beneficial influence.

The hypophosphites of lime, soda, and iron have also

---

[1] ℞ Sodii Bicarb. gr.xv.
    Acidi Hydrocyanici dil. ♏ iij.
    Infus. Gentianæ Comp. ℥j.
    Ft. haust. bis terve die sumendus.

proved of great service as tonics. When administered alone, in doses of from 5 to 10 grains, they cause gain of weight and general increase of strength, but when combined with oil the tonic effect is even greater. They seem to act best in cases where phthisis is combined with great nervous and bodily prostration, and with atonic dyspepsia, but must be omitted where there is pyrexia, as they tend to raise the temperature. They agree well with the liver, and are tolerated by even irritable stomachs, and may be regarded as an easily assimilated form of phosphorus, which they resemble in being inflammable at a low temperature, and in their beneficial effects on the nervous system. According to Dr. C. J. B. Williams, they seem to increase the failing powers, perspiration and circulation, and he hazards the suggestion, ' Can this be by increasing the affinity of blood for oxygen, so that it can attract it, and maintain the blood-changes even under the increased difficulties and obstructions caused by disease ? '

Antipyrexial treatment.—A great part of our treatment in phthisis consists in attempts to subdue the inflammatory conditions accompanying it, and on their success or failure depends much of the patient's future prospects. We must bear in mind that these conditions, whether they be the result of direct bacillar irritation, or of the ptomaine or poison which the bacilli secrete, are inflammatory and must be dealt with accordingly, for if allowed to increase in strength and virulence they will lead to further crippling, and in time to further disorganisation, of the whole lung tissue.

The inflammation in phthisis is rarely of the sthenic and generally of the asthenic kind ; in some cases, however, for the most part where croupous pneumonia accompanies tuberculisation, the pulse is hard, as well as frequent, the skin hot and dry, the urine scanty and high-coloured, the temperature raised, the cough hard, with glairy, viscid or sanguinolent expectoration, the breathing distressed and quickened, accompanied by pain in the chest, and here great good can be done by an effervescing saline containing carbonate of ammonia and antimonial wine (ɱv. to xv.) given every four or six hours, by free linseed poulticing, and by keeping the liver in full activity with an occasional dose of blue pill or calomel. The addition to the saline of a few minims of tincture of aconite (iij. to v.) or of tincture of digitalis ($\mathfrak{m}$x. to xx.) will sometimes moderate the heat of the skin and steady the pulse, and when the skin becomes moist, counter-irritation in the form of blistering will give great relief.

The subacute and chronic forms of inflammation which generally accompany tuberculisation and excavation require a modification of the above treatment. Our chief aim should be to increase those secretions which have a natural tendency to relieve inflammation, as the expectoration, the urine and perspiration, and also to derive lymph and serum from the pulmonary and bronchial systems by some form of counter-irritation. The effervescing ammonia saline,[1] with or without antimonial wine in small doses and a moderate amount of sedative, is, in my experience, unequalled for these purposes, for it promotes secretion from the skin, bronchial membrane, liver and kidneys, without checking the action of the bowels or deranging in the least degree the appetite and digestion. By giving it three times a day, or, better still, twice, viz. at bedtime and in the early morning, we secure free expectoration and good sleep for our patient, and leave the rest of the day free for the administration of cod-liver oil, tonics and food, which also aid in reducing the low forms of inflammation.

Other expectorants, such as senega, squills, and ipecacuanha combined with carbonate of ammonia or bromide of ammonium are useful, and in case of expectoration being abundant but the power of raising it feeble, the Mistura

---

[1] The subjoined is a convenient formula :—

   $\mathbb{R}$ Potassii Bicarbonatis ʒss.
   Ammonii Carbonatis ϶ij.
   Vini Antimonialis ʒss.–ʒi.
   Syrupi Papaveris ʒss.
   Spir. Chloroformi ʒj.
   Aquæ ad ℥viij. Misce.

Signa : An eighth-part with a powder of Citric Acid (϶j) dissolved in water night and morning.

Ætheris cum Ammonia [1] of the Brompton Hospital is of great service.

Counter-irritation.—Of all the means at our disposal to reduce the local inflammations of phthisis counter-irritation is the most effective, and if continuously employed will often render cough mixture and linctus unnecessary, to the great relief of the patient's appetite and digestion.   In many cases fly-blisters from one to two inches square, applied over the seat of the inflammation, reduce the cough and thoracic pain; in others a larger blister kept on for eight hours, and replaced by a warm linseed poultice, is preferable, this producing a good size of blistered surface without fear of cantharides absorption, which a longer application of the plaster renders possible.   The Liquor Epispasticus acts more speedily, while the Solutio Cantharidis Acetica,[2] largely used by Dr. C. J. B. Williams, is more suitable for prolonged use.   Lint dipped in compound camphor liniment and applied to the chest wall till redness, or, if necessary, vesication is produced, is a good and easily regulated form of counter-irritation, and this as well as the Linimentum Chloroformi and Linimentum Ammoniæ may be used where cantharides is contra-indicated. The various preparations of iodine—tincture, liquor, linimentum, carry out a slow form of counter-irritation and are best adapted for prolonged use, but they never seem to so effectually relieve cough and pain and to reduce expectoration as the cantharides preparations, possibly by reason of their not causing so copious an amount of serous exudation.

Rigollot's mustard leaves, which also contain pepper, mustard poultices and compound mustard liniment are all useful where a slight degree of counter-irritation is desirable. In cases accompanied by chronic inflammation of the lungs

---

[1] ℞ Spir. Ætheris,
  Spir. Ammoniæ Aromatici, ãã ʒss.
  Tinct. Aurantii ʒjss.
  Aquæ Camphoræ ad ℥viij.
  Ft. mist. Dose: ʒss.–ʒj.
[2] ℞ Aceti Cantharidis ʒvj.
  Spiritus Camphoræ ʒij.
Ft. Linimentum : To be applied at night till redness is produced. *Brompton Hospital Pharmacopœia.*

and bronchi, where the cough is hard, convulsive and very obstinate, and the expectoration by no means free, pustulation is necessary and more lasting in its effects, which can be produced by the liniment and other preparations of croton oil, or by thapsia plasters. The French physicians, including Jaccoud, strongly recommend the actual cautery, in the form of needle points heated white-hot, or the galvano-cautery, applying these frequently to the chest wall. Jaccoud also recommends the application of Vienna paste as a caustic over very small surfaces (size of threepenny bit) of the chest wall, separated by wide intervals, and of a sufficient depth to induce inflammation and sloughing of the skin at those points. He maintains that in this way, the pulmonary lesion is prevented from extending, either entirely or for a long period.

The use of setons was formerly in vogue, and at the present time many physicians keep open the blistered surface with oil of savin or elemi. I have tried, or seen tried, all these methods, except the galvano-cautery, and unhesitatingly prefer to them some form of blistering as less annoying to the patient and equally effectual for purposes of counter-irritation. When we note the relief given to a tubercular lung by a discharging abscess, either in the axilla, neck or groin, or from a fistula, we see the desirability of a safety valve, but this result may be less painfully and more conveniently arrived at by occasional blisters or croton oil liniment.

Antipyretics.—In a large proportion of cases and specially in the acute forms of phthisis, or where active changes are proceeding in the lungs, the above methods of treatment, while relieving many of the symptoms, fail to exercise any reducing influence on the high temperature, which is the cause of most of the patient's misery and of the extreme wasting. We have recourse then to the medicines specially reputed to reduce and control the temperature of the body, and in selecting from these we must be guided by the character of the pyrexia, and especially by its causation. As was explained in Chapter XII., there are two principal forms of pyrexia in phthisis: (1) the pyrexia of tuberculisation,

as is seen in the active first-stage cases; and (2) the
pyrexia of suppuration, where excavation is proceeding and the
septic products are at the same time undergoing re-absorption. This last is the hectic type of many authors and of
which a characteristic chart may be seen on p. 176. It
is obvious that the first, being due to the irritation to the
pulmonary tissue which the presence of the tubercle bacilli
causes, should be more amenable to antipyretics and other
treatment than the second, which owes its origin to suppuration and the absorption of the matter, and tuberculosis
of fresh portions of the lung. This more or less approaches
to the type of pyæmia.

The first type generally yields to counter-irritation, and
salines, with aconite and digitalis. Even arsenic introduced
into the tonic three times a day will often succeed in reducing
it. Where these are ineffectual, quinine (gr. iv. to gr. vj.)
given in effervescence [1] shortly before the temperature rises,
and repeated once or twice later on in the day, will often control the fever, but salicylic acid or salicylate of soda, in gr. x.
to gr. xv. doses three or four times a day, may be tried with fair
prospect of success. Jaccoud recommends for this form of
pyrexia the hydrobromate of quinine in large doses, 15 to 30
grains for three successive nights, which, he states, always
succeeds in lowering the temperature for three or four days.
I have, however, carefully carried out this treatment according to Jaccoud's directions in two cases of this kind, and in
neither find any perceptible reduction, though cinchonism was
noticed in both patients.

Where pyrexia is due to the softening of tubercle, it does
not appear to be much influenced by antipyretics, but
generally ceases when free expectoration and the physical
signs show excavation to be complete. In the third stage,

---

[1] ℞ Quininæ Sulphatis ℈ij.
　　Acidi Citrici ℈iv.
　　Aquæ ad ℥iv.
Dose: A tablespoonful with two tablespoonfuls of the following:
　　℞ Potassii Bicarbonatis ℥jss.
　　Tinct. Aurantii ℥jss.
　　Aquæ ad ℥viij.

(active) pyrexia, I have tried a vast number of remedies, of which the following are a fair example. Quinine, different forms of arsenic, salicine, salicylic acid, salicylate of soda, iodoform, Heim's pill, chairin, chinoline, resorcin, antipyrin, thalline, antifebrin, carbolic acid (hypodermic injections of), cold baths and cold sponging, spinal ice bags, and the ice pack.

Quinine in large doses of gr. v. to gr. xx., dissolved in acid and given every four hours, will sometimes reduce the temperature considerably, but, on the withdrawal of the drug the temperature generally rises again; and I have found Heim's pill,[1] which combines opium and digitalis with the quinine, exercise a temporary effect on the fever, though not a permanent one.

Salicylic acid, and, better still, *salicylate of soda* as less nauseous, in gr. x. to gr .xx. combined with a bitter, to avoid the mawkish taste, will generally reduce the temperature, especially if given four or five hours before the usual rise, but it must not be continued long, on account of the symptoms it gives rise to, these being noises in the ears, deafness, and often a great depression of spirits. It may be injected hypodermically in 10 gr. doses dissolved in water, with even more decided effect on the pyrexia. *Salicin*, from which the above drugs are derived, also has an antiperiodic influence, and gives rise to no bad symptoms. *Iodoform* I have largely tried, with no success, and I was not much more fortunate with *chairin* or *chinoline*, which have the additional disadvantages of being exceedingly nauseous.

Antipyrin, or dimethyloxychinizin, is a most valuable antifebrile agent, and in many cases in a few doses of gr. xv. to gr. xxx. reduces the temperature to normal, acting apparently by inducing perspiration, which sometimes becomes excessive. On the eighth day of its use, an erythematous rash resembling measles appears, which ceases when it is discontinued. Antipyrin often gives rise to anorexia and vomiting,

---

[1] ℞ Quininæ Sulphatis,
Digitalis, āā gr.j.
Opii gr.ss.
Fiat pil. Iiis horis sumenda.

and cannot be persevered with for a long period, but is very valuable in its first effects on phthisical pyrexia, and its action is as rapid as it is powerful. *Resorcin* in doses of gr. x. to xxx. exercises a similar effect to antipyrin, though perhaps somewhat weaker.

Thalline (Tetrahydroparachinanisol) is even more powerful as an antiseptic, but, on account of its inducing collapse and shivering, has to be used cautiously. The sulphate or tartrate are generally employed, and it is better to give small and frequent doses, such as one or two grains every one or two hours, than a larger dose at longer intervals. In two cases of third-stage pyrexia in which I tried thalline, it reduced the temperature 2° and 3° soon after administration, and the effect did not at once disappear on its discontinuance. It is highly commended by Ehrlich and Lagner.[1]

Antifebrin, or phenyl-acetamide, is perhaps the most powerful and satisfactory antipyretic I have ever tried, and, as yet, I have seen no bad symptoms from its continued use, except exhaustion from the sweating it entails. Three to five grains given an hour before the expected rise of temperature will often prevent it, and this dose taken three times a day has completely controlled the temperature of a pyrexial case within normal limits. When the temperature has risen, a fall of 1 to 2 degrees may be predicted within an hour of the dose, and the pulse falls in frequency and rises in tension, while the reduction of heat appears, according to my experience, to be the result of the profuse and often long-continued perspiration. In one case, when the dose was taken at 3 P.M., the perspiration began at 4 P.M. and lasted till 6 A.M. on the following day. Being almost insoluble in cold water, it had best be given either in the form of powder, or dissolved in warm water or spirit. Unfortunately like other antipyretics in phthisis, its action only lasts during the period of its use, and not after.

In two cases I tried the hypodermic injection of a solution of carbolic acid, gradually increased in strength from 1 in 50

[1] *Berliner klinische Wochenschrift*, 1885, No. 51.

to 1 in 20, and here again, though the patients gave signs of carbolism, no effect was produced on the pyrexia. Three [1] patients in the third stage (active) I immersed in baths, lowering the temperature of the water from 90° F. to 60° F. by ice. In all three the temperature fell, and in two it never rose again, but in the third, an extending cavity case, after three baths the pyrexia was much the same as before, though the appetite had improved, and the action of the skin was greater. The ice pack similarly causes a temporary reduction of the temperature, which is acceptable to the patient, but on its removal the pyrexia returns.

Cold baths and ice packs are all somewhat dangerous to cases of advanced phthisis, and, with the risk of possible collapse, should be avoided, but great comfort to the patient may accrue from sponging with tepid water containing a small quantity of acetic acid or of toilet vinegar, and the high temperature may be also reduced by wearing Chapman's spinal ice bag during the periods of great heat.

The more the pyrexia of phthisis is studied, the more clearly it appears to arise from the various changes taking place in the lungs, and to cease when these pass from activity into quiescence, as seen in the chronic third-stage chart, and though we may modify such pyrexia by treatment, we cannot expect to abolish it, except by a removal of the causes.

Night-sweats naturally should be considered in connection with pyrexia, as they often accompany the morning defervescence, but they are not always due to the reduction of temperature, as they may occur profusely when there is scarcely any pyrexia. In this case they arise from the great weakness of the patient and partake of the nature of a flux, which, besides interrupting the needful slumber of the patient, exhausts him by the loss of fluid and saline material. In all cases they should be checked, and as a rule this is effected by the anti-phthisical measures we have already described, without recourse being had to any special remedies. Where they do not succeed, there is no lack of agents of which one or other is pretty sure to answer the purpose; and we cannot

[1] *Clinical Transactions*, vol. vi.

understand how Niemeyer[1] could have come to the conclusion, ' that there are no means of relief for this distressing
symptom ; ' for our experience is quite the reverse, and we
have found this symptom more amenable than many others
to rational treatment.

Where the perspirations are only slight, cold tea at night,
sponging the chest with toilet vinegar, or dilute sulphuric
acid and water, at bed-time is sufficient ; where more profuse,
a night draught of half a drachm of dilute nitro-hydrochloric, or sulphuric, acid in glycerine and water will often
answer the purpose. Gallic acid in ten-grain doses once or
twice a day is an excellent remedy, and sometimes the addition of the tincture of perchloride of iron, or sulphate of
quinine, to the daily tonic will have the desired effect ; but
these two last drugs are apt to increase the cough, and must,
therefore, be given with caution. The medicine which has
been found to act almost as a specific on night-sweats is the
oxide of zinc, in doses of two or five grains, in the form of a
pill at night, and the good results have generally been so
prompt and lasting, that in few cases has it been necessary
to continue its use for any lengthened period.

The preparations of arsenic are generally very efficacious
for reducing night-sweats, and the arseniate of iron (gr. ⅙ to ⅓),
strongly recommended by Dr. Pollock,[2] and taken as a pill at
bed-time, is excellent. Dr. Murrell[3] has made an extensive
trial of various medicines, and praises Dover's powder (Pulv.
Ipecacuanhæ Comp. gr. v. to x.) as superior to oxide of zinc
though inferior to atropine. Belladonna and atropia, as might
be expected, from their physiological action in the secretions,
powerfully control night-sweating, but as they provoke dryness of the throat, and in time dilatation of the pupil and
disordered vision, they are not advisable for long continuance.
The doses are, of the extract of belladonna gr. ¼ to ⅓ in a
pill at night, and of sulphate of atropine, one to four minims
of the Liquor Atropinæ Sulphatis given by mouth or injected
hypodermically at night. Dr. Lauder Brunton recommends

---

[1] *Text-Book of Practical Medicine.*   [2] *Medical Times and Gazette*, 1874.
[3] *Practitioner*, 1879.

strychnia, and in addition to the above I have had excellent results from picrotoxine (gr. $\frac{1}{60}$), nitrate of pilocarpine (gr. $\frac{1}{20}$) in the form of a pill at night, whilst agaricin (gr. $\frac{1}{20}$), muscarine (gr. $\frac{1}{5}$) and paracotoin (gr. $\frac{1}{2}$ to 3) have been strongly recommended by others.

Where the tendency to this flux is very strong, food ought to be taken during the night, solid being preferable to liquid, and combined with a small amount of alcohol, and the patient should sleep in flannel or some woollen material.

Before closing this chapter, a word on the general management of pyrexia in phthisis is not out of place. When the fever is high it is undoubtedly advisable to keep the patient in bed; but as the pyrexia is usually post-meridian, and only lasts for a certain number of hours, confinement to bed may be restricted to these, and for the rest of the day he may be up, and, if weather and climate permit, lie or sit in the open air. Even during the pyrexial period the patient's bed may be wheeled to the open window, or, as is practised in many foreign hospitals, on to a balcony opening out of his room, and thus he can be brought under the beneficial influence of the free play of fresh air on his fevered body. The diet during this period should include food, for the most part liquid, every two hours (night and day), and alcohol in moderate quantities with each supply of food, alcohol having been shown by Binz and others to limit the waste of tissue and to act as respiratory food.

CHAPTER XXV.

THE PALLIATIVE TREATMENT OF CONSUMPTION.

Palliative treatment subordinate to anti-phthisical and antiseptic—Treatment
of the varieties of cough—Effervescing cough mixtures—Linctus, lozenges,
sprays and inhalations—Objections to use of sedatives—Pain in the chest,
its causes and remedies—Strapping—Treatment of different forms of
hæmoptysis—Styptics—Leeches and dry cupping—Treatment by derivation—Induction of artificial pneumothorax—Intrapulmonary injection of
styptics—Dangers of high feeding—Treatment of the different kinds of
diarrhœa—Diet—Warm applications and blisters to the abdomen—Astringents—Opiate injections and suppositories—Linseed tea—Treatment of
constipation by diet and mild aperients—Treatment of dyspnœa, phlebitis,
and bed-sores—Treatment of laryngeal phthisis—Blisters—Insufflation—
Inhalations—Cocaine solution—Dysphagia relieved by rest and nutritive
injections—Treatment of albuminuria and dropsy—De-albuminised milk—
Treatment of pneumothorax—Aspiration—Treatment of meningitis.

In the preceding chapters the general and local treatment of
Consumption in their anti-phthisical and antipyrexial aspects
have been discussed, and we will now say a few words on
the *palliative* treatment of the disease, premising that we trust
the day will come when such measures will cease to be necessary, and the improvement of the general health and strength
of the patient and the destruction of the bacillus will be our
sole aim. Nevertheless, as certain symptoms, by the irritation they cause, often induce sleeplessness, feverishness,
vomiting, &c., and seriously interfere with the general progress of the patient, it is obvious that we must use all means
in our power to allay them, always taking care that our
palliative measures do not interfere with steady persistence
in the constitutional ones.

Cough is usually the most prominent and troublesome
symptom. If loose and slight in amount, it had best be left
alone, or expectoration promoted in every way; but when it

is hard and frequent, and interferes with the patient's rest at
night and cannot be reduced by other means, such as counter-
irritation, it should be allayed by the use of sedatives, the
choice of which and their combination with other drugs must
depend on the nature of the cough and the amount of expectoration. When the cough is hard and the expectoration,
though free, only slight in amount, linctuses, containing opium
or its salts, codeia or morphia, combined with such simple
expectorants as lemon-juice and chloric ether, are most useful.
When the cough is very violent, and ends in retching and
vomiting and not in expectoration, dilute hydrocyanic acid
may be added with advantage; and when the expectoration is
offensive, glycerine of carbolic acid goes far to correct the
fœtor, and at the same time assists the expectoration.[1] Sometimes the cough is convulsive, and accompanied by a good
deal of wheezy breathing or stridor; it may then be relieved
by belladonna, stramonium, or Indian hemp, in doses of a
quarter or half grain in a pill at night, or by extract of
hyoscyamus or of conium (gr. iij).

In cases of consumption marked by cough with difficult
expectoration, a combination of bromide of ammonium with
American cherry and henbane will be found of great use, on
account of its allaying the cough and promoting expectoration
without constipating the bowels.[2]

Strong expectorants, equally with the old-fashioned emetics,
are, as a rule, to be avoided, as tending to upset the stomach;
but when the expectoration is difficult, or if there be temporary

---

[1] A few formulæ are annexed:—

℞ Liq. Morphinæ Acetat. ʒjss.
   Spir. Chloroformi ʒj.
   Succi Limonis ℥ss.
   Mucilaginis Acaciæ ad ℥ij.
Dose: a teaspoonful.

℞ Liq. Opii Sedativi,
   Acidi Hydrocyanici dil., āā ʒss.
   Spiritus Chloroformi ʒj.
   Aquæ ad ℥ij.
Dose: a teaspoonful.

℞ Liq. Morphinæ Acetat. ʒij.
   Glycerini Acidi Carbolici ʒj.
   Oxymellis Scillæ ℥ss.
   Mucilaginis Acaciæ ad ℥ij.
Dose: a teaspoonful.

[2] ℞ Ammonii Bromidi ʒjss.
   Syr. Pruni Virginianæ ℥ss.
   Tinct. Scillæ ℥ss.
   Succi Hyoscyami ℥ss.
   Aquæ Menth. Pip. ad ℥viij.
An eighth part two or three times a
day.

bronchitis and increased bronchial secretion, small doses of squill or ipecacuanha are indicated. Ipecacuanha spray, as recommended by Drs. Ringer and Murrell, is useful. When the cough is very hard and troublesome at night, a few drops of laudanum, liquor opii sedativus, or bimeconate of morphia, in an effervescing saline containing ammonia (see p. 395), generally allay the irritation and induce sleep, without greatly deranging the digestion.

Lozenges of morphia, or morphia and ipecacuanha, or opium, should be used, like the cough mixtures, only at night, so as not to derange the stomach, and thus interfere with the anti-phthisical treatment pursued during the day. Apart from this objection to their use in the daytime, they are more required at night, their great use being to ensure to the patient a certain amount of refreshing slumber, and thus increase his strength and appetite.

Another way of reducing the cough is by inhaling carbolic acid combined with chloroform, diffused by means of steam or hot water; but in most cases of phthisis, and specially the advanced ones, our great aim must be to assist the feeble powers of expectoration, and this is best done by combinations of ammonia, senega, and spiritus ætheris, with some slight sedative and a certain amount of stimulant in the dietary.

Pain in the chest is another symptom which sometimes requires direct treatment. It is referred to various parts of the chest, often to the lower regions of the chest as much as to the upper. When it is of a dull aching kind and not markedly localised, belladonna or opium plasters generally give relief. When localised, it generally proceeds from dry pleurisy, secondary to the formation of the tubercle beneath the pleura. In this case Dr. Frederick Roberts' plan of securing immovability of the chest wall by strapping the side is admirable, and Leslie's or Ewen's plasters answer best for this purpose. If more poignant, mild counter-irritation with either tincture of iodine, or one of the turpentine liniments, of which the Linimentum Terebinthinæ Aceticum is to be preferred, is advisable. If more severe, and especially if physical examination detects any decided cause for it in the existence of dry pleurisy or pleuro-pneumonia, then let recourse be had to vesication, produced by one of the methods already described.

Hæmoptysis must be treated according to the amount expectorated, the stage of the disease, and intercurrent inflammation accompanying it. Where no cavity exists, our object should be to reduce any inflammation arising from the blood effused into the alveoli rather than to check the hæmorrhage, which will probably subside of itself : for, we must bear in mind that tubercle bacilli are often found in the blood of hæmoptysis, and are doubtless the carriers of infection to fresh portions of the lungs. A blister to the chest, accompanied by a powder of gallic acid (gr. x. to xv.), and acid tartrate of potash (gr. x.), taken every four hours, is generally sufficient, provided the ordinary precautions of rest and abstinence from alcohol and warm drinks be observed. When the bleeding arises from the rupture of a pulmonary aneurysm into a cavity, every effort must be directed to promote rapid coagulation of blood and to reduce the rate of the circulation. The former is attained by giving tannic acid (gr. ij. to iv.), alum (gr. x.), and acetate of lead, in five-grain doses, every few hours, followed in the case of the latter by morning draughts of sulphate of magnesium and sulphuric acid, to prevent lead accumulation. The latter is arrived at by digitalis and aconite, ice bags to the heart, and ergot of rye. An effective combination can be made of tannic (gr. ij.) and gallic acid (gr. x.), and in our experience acetate of lead amply deserves the praise bestowed on it by Dr. Paris, that as a styptic there was nothing either ' simile aut secundum ' to it.

Oil of turpentine (♏x. to xx.) is a powerful styptic and may be given either with mucilage and peppermint water, or in the form of capsule, the best of all being Clertan's ' perles,' which are small to swallow, and contain 10 minims each of the oil. The vapour of turpentine, as produced in a Siegle's steam spray apparatus, often checks hæmoptysis. Digitalis in large doses (of infusion, ℥ss.) will often stop hæmoptysis, but on account of its lowering effect on the pulse cannot be long continued. Ergot of rye is much relied on by many authorities, and is sometimes remarkably satisfactory in its action, but sometimes quite the reverse. To control the hæmorrhage it should be administered in doses of ʒj. to ʒij. of the liquid extract every two or four hours, or in hypodermic injection of the pharmacopœia preparation of ergotinum (♏iii. to x.) or of ergotinine (♏v. to x.). The patient's room should be kept cool ; alcohol should be avoided, and the food should be taken cold, and accompanied by ice for him to suck, and it is well to mix ice even with the medicine taken.

In severe cases, when the ruptured aneurysm is a large one, or if several burst, the rapid outpouring of blood necessitates recourse to dry or other cupping, often most helpful, and even to leeches, and the derivation of blood from the bleeding vessel by the use of Junod's boot, or the application of heat to the extremities, is sometimes advisable, and a blister on the overlying chest wall often reduces the bleeding. In one extreme case Dr. Cayley[1] punctured the pleura, and induced artificial pneumothorax, causing collapse (partial, there being some pleuritic adhesions,) of the bleeding lung, and thus arrested the hæmorrhage, but the patient died two days after of acute tuberculosis of the opposite lung. Another method, which I should advocate in preference, is the injection, through a fine-drawn pointed syringe, of a styptic into the lung itself or into the cavity containing the ruptured aneurysm. I accomplished this in one desperate case, injecting 20 grains of tannic acid into the lung, with the result of entirely arresting the hæmorrhage. The patient experienced an astringent taste in the mouth directly after injection, and a few minutes later tannic acid was detected in the saliva and urine, showing the rapidly absorbent action of the pulmonary vessels. Unfortunately the patient sank two days later from syncope, and the post-mortem examination showed the tubercular lesions to be so extensive that the arrest of the hæmorrhage was not sufficient to save the patient. I think, however, in smaller doses, and selecting cases of more

[1] *Clinical Transactions*, vol. xviii.

limited disease for operation, this procedure might prove of great use.

In dieting patients subject to hæmoptysis, and specially those in which there is reason to suspect the existence of aneurysm of the pulmonary artery, we should bear in mind the danger of high vascular tension, and on this account avoid a too generous dietary, and we shall thus keep down the volume of blood present in the system generally and in the lungs in particular.

Treatment of the diarrhœa.—We have three forms to deal with in phthisis, (1) from indigestion, (2) from intestinal ulceration, and (3) from lardaceous diseases of the intestines. The treatment of the first form of diarrhœa need not detain us long, as it consists of simply correcting the dietary and ordering a few doses of alterative and purgative medicine, with some alkali to reduce excessive acidity. The second form, that arising from ulceration, requires very careful attention. The great point to be kept in view is the healing of the ulcers, and this can only be attained by shielding them from all irritable substances, and by promoting a healthy granulating action. The treatment, in fact, resolves itself into three sets of measures.

1st. Rest in bed and the administration of only such food as can be quickly and easily assimilated without causing much pain or distension of the intestine through accumulation of flatus. Such are chicken-broth, beef and veal tea, milk gruel, blancmange, which should be combined with the liquor pancreaticus (Benger), or Fairchild's peptonised powders, and prepared after the admirable methods of Sir Wm. Roberts.[1]

2nd. Warm applications to the abdomen in the form of linseed poultices, turpentine stupes, or hot water fomentations, to reduce the pain and promote a certain degree of derivation to the skin. If the pain be severe, the application of a small blister over the area of tenderness on pressure, as recommended by Dr. J. E. Pollock, is often very advantageous. I have noticed in some obstinate cases that the rising of the

---

[1] 'The Digestive Ferments and Preparation and Use of Artificially Digested Food.'

blister has been accompanied by diminution of the diarrhœa and subsidence of the abdominal pain.

3rd. Internal medicines. When we have reason to believe that the ulceration is slight and confined to the small intestine, the diarrhœa may be treated by bismuth and opium, or by some astringents. The liquor bismuthi et ammonii citratis (B.P.) is a convenient form, but not always so effective as the carbonate or the nitrate of bismuth in powder, in ten to twenty-grain doses. Dover's powder combined with it in ten-grain doses is often effective, but the most powerful astringent is the sulphate of copper in quarter to half grain doses combined with half grain to a grain of solid opium given three times a day. Of the various vegetable astringents I have found tannic acid in four-grain doses to answer best, far better than rhatany and catechu, but in all cases I combine it with a certain amount of opium to reduce the irritability of the ulcers. Logwood is an admirable astringent, and Indian bael, especially a preparation of the fresh fruit, is often efficacious in checking the diarrhœa if the ulceration be limited. If, however, the ulceration attack the large intestine as well as the small, it is obvious that more local treatment is advisable, and recourse should be had to injections or suppositories. The enema opii (B.P.) administered twice a day is sometimes sufficient, and may be strengthened by the addition of acetate of lead or of tannic acid in five-grain doses. This is a small injection, and it is doubtful how far in the intestines its local effect would reach. Where the ulceration is very extensive, and involves the greater part of the large intestine, an attempt ought to be made to apply the remedies more fully to the mucous membrane : and for this purpose injections of larger amount—from a pint to a pint and a half—may be used, consisting of gruel or of starch, or, best of all, of linseed tea, and all containing a certain quantity of opium (tinct. opii ℳ xxx. to xl.). I would specially recommend the linseed-tea, as it appears to exercise the same beneficial effect on the ulcers of the large intestine as it does on follicular ulceration of the throat. One of the most obstinate cases of intestinal ulceration I ever witnessed yielded

to linseed-tea injections, after almost every other treatmen
had been tried in vain, the tubercular ulcers apparently healing, the diarrhœa ceasing, and the patient living for two
years afterwards, and dying of pulmonary lesions. In cases
where the stools are very fetid, I have added glycerine of
carbolic acid to the injection with advantage. In many
cases, however, it is desirable to give the large intestine as
much rest as possible, and not to stretch the ulcerated
mucous membrane by any distension with fluids; in these
cases suppositories of morphia (from half a grain to a grain)
of tannic acid, or the suppositorium plumbi co.(B.P.), which
contains one grain of opium, are indicated.

The treatment of the diarrhœa arising from lardaceous
degeneration of the intestine is not very hopeful. Where the
very channels of assimilation, viz. the villi, ' our roots,' as
they have been justly termed, have undergone degeneration,
as well as the various structures from which the succus ente-
ricus is poured out, it is difficult to see how treatment can
restore the lost tissues. Dr. Dickinson's researches[1] show
that the loss of alkali is the chief characteristic of the disease.
Dr. Marcet's analyses[2] show that the chief chemical feature is
deficiency of phosphoric acid and potash and excess of soda
and chlorine, and on this principle we should give phosphates
of potash. When, however, the disease has so far advanced
as to reach the intestine, it may be considered beyond any
effective general treatment. We must be content to restrain
the diarrhœa, if we can, by astringents, the more powerful the
better. Tannic acid, in from two to five grain doses, with
dilute sulphuric acid, sulphate of copper or sulphate of zinc,
are the most useful, and injections of these substances often
do much good.

Though diarrhœa is common in the later stages of consumption, the opposite state, i.e. constipation, often prevails
in the earlier ones, and should be counteracted, for two
reasons : firstly, because it is impossible for the cod-liver oil
to agree if the biliary and intestinal secretions are not pro-

[1] *Pathology and Treatment of Albuminuria.*
[2] *Pathological Transactions,* vol. xxii.

perly discharged, but allowed to accumulate in the intestines, causing flatus, loss of appetite, &c.; and secondly, because phthisical patients are more liable to hæmoptysis when the bowels are costive than when they act regularly.

The simplest method of correcting this state is by introducing into the diet a fair amount of fresh or cooked fruit, by the use of porridge, or by substituting brown bread for white, the bran in the former acting like the oat, as a simple irritant to the intestinal mucous membrane and a promoter of peristalsis.

Where dietetic means fail, recourse must be had to aperient medicine, the continuous action of a mild purgative being preferred to the prompt but less lasting results of a stronger one. The aperient mineral waters Hunyadi Janos, Victoria, Friedrichshall, Æsculap, and the Carlsbad Salts, are all excellent. The confectio sulphuris, or an electuary[1] composed of sulphate of potash and confection of senna, taken in a teaspoonful dose every night, is often sufficient; but pills[2] of Barbadoes aloes, containing $1\frac{1}{2}$ or 2 grains of the extract, combined, in the more obstinate cases, with $\frac{1}{2}$ grain of extract of nux vomica, and taken at bed-time, answer admirably, it being sometimes necessary to substitute gr.$\frac{1}{4}$ or gr.$\frac{1}{3}$ of podophyllin for the nux vomica in cases of great costiveness.

Dyspnœa is common in advanced cases, and may as a rule be relieved, if coming on suddenly, by spirits of ether and sal volatile, or by the inhalation of oxygen or iodide of ethyl; if of a more chronic kind, by a combination of chlorate of potash and nitric acid taken by the mouth, and the supply of plenty of fresh air.

Phlebitis.—In the last stages of the disease phlebitis of various veins of the upper and lower extremities is not uncommon, the most usual seat being in the internal saphenic vein. Whether this be always septic or not is uncertain, but

[1] ℞ Potassii Sulphatis.
Syrupi Zingiberis, āā ʒss.
Confectionis Sennæ ʒj.
Dose: a teaspoonful every night.

[2] ℞ Extracti Aloes Barbadensis gr.jss
—— Nucis Vomicæ gr.ss.
Potassii Tartratis Acidæ gr.j.
Mastich. gr.jss.
Spiritus rectificati q.s.
To be taken every night.

the appearance is more that of inflammation in a portion of
the vein than of phlebitis of the whole vessel. In this latter
case rest in the horizontal position and painting with tincture
of iodine will reduce and disperse the fibrinous thrombi,
which may partly arise from a hyper-fibrinous condition of
the blood present in phthisis. Where phlebitis exists, it must
be treated by fomentations of flannel wrung out in hot water,
sprinkled with laudanum and applied to the limbs under oil-
silk. Painting the track of the inflamed veins with extract
of belladonna and glycerine often greatly relieves the pain.

Bed-sores should be prevented by carefully watching the
state of the skin of the back and sacrum, and by placing the
patient on a spring- or a water-bed : the half or three-quarter
length sizes of the latter are generally preferable to the full-
length ones, for while rendering the pressure more equable
they do not make the patient so hot. At the Brompton Hospital light spring beds have quite taken the place of water
beds. When the skin, though red, is still unbroken, a wash
of brandy- or of spirit-and-water, one part in four, has a fortifying effect, but if it be broken, collodion flexile may be
applied with a view to form a protecting film, and thus afford
the skin an opportunity of healing. This is, however, of little
use if means be not taken to remove pressure from the part,
which should be left free for dressings. A circular air-cushion
will do the former, but at the Brompton Hospital it has long
been found that these soon get out of order, and cushions of
down [1] of similar shape have been substituted. Another
method employed is to apply a piece of thick felt plaster or
of the material known as ' buffalo hide,' perforated with a hole
the size of the wound, and to dust iodoform over the ulcer,
which will then quickly heal.

The *treatment of laryngeal phthisis* consists, besides the
usual constitutional measures, of such local ones as will
reduce the inflammation and promote the healing of the
ulcerated surface. This would seem at first sight a reasonable object to expect to attain, considering how easily medicaments are applied to the interior of the larynx, but the

[1] To be obtained at Nightingale's, Wardour Street.

difficulty arises from the larynx lying in the track of the bacilli-laden sputum ejected from the pulmonary cavities, and often, to a certain extent, retained in that organ by its irregular surface and the various movements of the vocal cords.

As long as the ciliated epithelium remains perfect, the bacilli cannot penetrate, but if any break of surface occur, which they may themselves produce, fresh centres of infection are formed, and the area of ulceration is rapidly extended; and under these conditions it is not likely that, even under the most assiduous local treatment, healing of the ulcers will take place.

When the appearances in the larynx are for the most part catarrhal or congestive, counter-irritation in the form of blisters the size of half-a-crown, applied successively to one or other side of the larynx, gives relief to the pain, and will sometimes cause a return of the voice, and even when ulceration has commenced, this treatment renders the suffering less acute. The ulcers should be brushed with a solution of sulphate of copper (gr. xx. to xxx. to the ℥j.), or else the larynx be insufflated [1] with a powder such as Dr. Semon's [2] once or twice a day to promote healing. The cough and pain may be mitigated by warm moist inhalations of carbolic acid or creasote combined with a sedative, of which the Vapor Chloroformi Co.[3] introduced by Dr. C. J. B. Williams is a good example. The dyspnœa is sometimes sufficiently severe to require tracheotomy, but the great trouble is the dysphagia, which is often most painful to witness, and this may be rendered at first less severe by sucking and swallowing ice, or painting the pharynx and epiglottis with a solution (20 per cent.) of cocaine before deglutition is attempted, or by a gargle of tannic acid (gr. x.

[1] Various forms of insufflators can be obtained of Messrs. Krohne & Sesemann, Meyer & Melzer, and other makers.

[2] ℞ Iodoformi,
Acidi Boracici, ãã gr.j.
Morphinæ Acetatis gr.⅛.   Misce.

[3] ℞ Chloroformi ♏x.
Succi Conii ℥j.
Glycerini Acidi Carbolici ℥ij.
Aquæ bullientis ℥viij.
Vapour to be inhaled night and morning. *Brompton Hospital Pharmacopœia.*

to the ʒj.) if the patient is able to use it, or by the application
to the parts of glycerine of tannic acid. All these will make
matters easier. The food should be free from salt in any form,
which invariably causes much pain, and it should be soft and
not too liquid, consisting of jelly, arrowroot, gruel, quenelles
and purées, light custards and such materials as can be easily
swallowed, and at the same time have no pungent flavour or
smell. Where, however, the ulceration is very extensive, and
specially if the epiglottis be greatly involved, the best way
is to give the pharynx and œsophagus complete rest, and to
feed entirely with nutritive injections. This will also render
more easy any local applications to the larynx.

Albuminuria and *Dropsy*, the accompaniments of a limited
number of advanced cases of phthisis, require appropriate
treatment, though the exceedingly weak state of the patients
precludes the use of the same active measures which can be
applied in non-tubercular cases. To reduce the œdema of the
lower extremities, and the ascites, if it exists, a nightly pill of
blue pill, squill and digitalis, for a short period, and a diuretic
day mixture containing benzoate of ammonium, acetate or
iodide of potassium, with scoparius, juniperus, caffeine, or
strophanthus, is the best measure to be adopted under the circumstances, for stronger measures would provoke diarrhœa,
which only too frequently accompanies the albuminuria, and
is often very difficult to control. It generally occurs in the
case of lardaceous kidney, and sometimes reduces the dropsy,
but more frequently weakens the patient and leaves the
albuminuria unaffected.

The albuminuria can be diminished by diet regulations ;
and the restriction of the patient to a purely milk dietary
with a very small amount of beef tea, has in my hands caused
a considerable decrease in the amount of albumen, with an
increase in the amount of urea and in the quantity of urine
passed. This has also been accompanied by a reduction of the
dropsy and diarrhœa. To carry out this principle more completely, I requested the Aylesbury Dairy Company to prepare
me specimens of milk from which all the albumen except
1 per cent. was removed, and this they succeeded in doing,

and supplied me with a very palatable fluid, which was really a mixture of whey and cream, and has been used by my colleagues and myself for some time.   Patients with albuminuria have taken this for long periods with fair results, the amount of albumen diminishing and that of urea increasing.

The *treatment of pneumothorax* depends on its extent, on the air-pressure within the pleura, on the amount of irritation it sets up, and on the degree of dyspnœa present.

If the pneumothorax be localised and do not embarrass the breathing, it will not be desirable to interfere with it ; but if there be no pleuritic adhesions, and the air which has escaped through a valvular opening in the pulmonary pleura accumulate to such an extent as to cause complete collapse of the lung and displacement of the heart to the left or right of the sternum, as the case may be, and give rise to a certain amount of cyanosis in the patient, then the removal of some of the air by aspiration is clearly indicated, and is often followed by considerable relief to the breathing and a return of the heart to its normal position.

The first effect of pleural perforation is that of shock, often following closely on the sharp pains, and for this some authorities recommend opium in moderate doses.   I prefer the administration of some diffusible stimulant, such as spiritus ætheris, the inhalation of iodide of ethyl,[1] or, if the collapse be great, of nitrite of amyl,[1] following a certain amount of alcohol and food.

If the effusion of air be followed by that of serum and pus, the case becomes one of pyo-pneumothorax, and a rapid accumulation of the fluid will probably call for paracentesis, the insertion of a drainage tube and free discharge through it into antiseptic dressings.   Many of these cases do well for a time, but eventually die from lardaceous disease of various organs.   Counter-irritation does not seem at first sight to be indicated, but the pain of perforation can often be assuaged by a blister over the region of perforation, or occasionally

[1] Both these are sold in convenient capsules, containing from ♏ iii. to v., by Martindale, 10 New Cavendish Street, and other chemists.

by strapping the affected side of the thorax with Leslie's or Ewen's plaster.

The great point, however, is to maintain the patient's strength with food and stimulant, of which latter such patients are wonderfully tolerant, until the shock is over, and then to direct measures towards the re-absorption of the air and the expansion of the collapsed lung. The general treatment of pneumothorax is that of an acute cavity case, and abundant liquid food easily assimilable should be the diet, with a fair amount of stimulants.

Meningitis.—The treatment of the tubercular complications of the brain and its meninges is not a very hopeful subject, as in spite of all measures effusion takes place, which is not easily removed. I have seen some temporary relief from leeches applied to the nape of the neck, followed by blistering, and combined with large doses of bromide and iodide of potassium. Sometimes purgatives relieve by derivation, but all improvement is temporary, and tubercular meningitis generally closes an often protracted consumptive history only too rapidly.

CHAPTER XXVI.

ANTISEPTIC OR BACILLICIDE TREATMENT.

Objects of bacillicide treatment—Tubercle bacilli endowed with great vitality—
Observations of Schill and Fischer, and of Malassez and Vignal—Author's
experiments on life and growth of tubercle bacillus—Its rapid multiplication in beef juice—Influence of quinine, arsenious acid, boracic acid, iodine
and distilled water—Klein and Lingard's inoculations of guinea-pigs with
tubercle mixed with Sulphocarbolate of sodium, phenylacetic and phenylpropionic acids—Inhibitory action—Methods of applying bacillicides—
Uselessness of antiseptic respirators—Medicated hot water or steam inhalations—Antiseptic chambers—Bacterium termo spray treatment—Bacillicide
medicines—Author's experience—Intrapulmonary injections of antiseptics
—Drainage of pulmonary cavities—Author's method of cavity injection—
Meunier's treatment by hypodermic injection of antiseptics—Bergeon's
gaseous rectal injections.

To destroy the tubercle bacillus in its first halting place in the
body and thus to prevent its multiplication and spread either
to adjacent parts or to distant organs is clearly our first
object, and for this purpose we have to determine two important
portant questions :—

1st. What are the agents, medicinal or other, capable of
destroying the bacillus and yet fit to introduce into the
human body ?

2nd. What channel is it desirable to select for such purpose ?

If we were to trust some of the numerous writers on the
antiseptic treatment of phthisis, the process is a very simple
one, and consists in adding a few drops of carbolic acid,
eucalyptol or thymol, to an oral or ori-nasal respirator, which
the patient is directed to wear for several hours a day. It is
claimed for this practice that the cough is reduced in intensity
and the expectoration diminished in amount, and undoubtedly

the effect of carbolic acid vapour, whether used in a hot moist inhalation, or as a dry vapour, does reduce the sensibility of the bronchial mucous membrane, and by allaying the irritation it probably diminishes the amount of secretion, but where are the experiments to show that this method of treatment lessens the number of tubercle bacilli in the sputum, or interferes materially with the retrograde processes which they set up in the lungs?

It has been too hastily assumed by many that the same medicinal agents which are found to be destructive of putrefactive and other bacilli, will be equally destructive of tubercle bacilli, whereas all experiments with sputum and tubercular organs point to the conclusion that tubercle bacilli are both difficult to destroy and remarkably tenacious of life.

Schill[1] and Fischer showed that tubercular sputum might be dried and kept for months without losing its bacillar virus, and Malassez and Vignal[2] detected tubercle bacilli still unchanged in specimens of lungs which had been kept for nine months and had undergone various stages of decomposition.

In 1883 I carried out a series of experiments[3] on the cultivation of tubercle bacilli and on the various reagents which prevented or retarded this process. Tubercular sputum was selected for the purpose, and cultivation was conducted in a Page's Incubator, kept at a temperature of 38° C. (100·4° Fahr.), for periods varying from forty-eight hours to eight days. The cultivation fluids in use were syrup, hay infusion, Pasteur's solution (without sugar), beef solution and pork broth. No decided increase was noted in any but the beef solution, where an enormous multiplication of the bacilli took place at the end of three days; and up to the seventh or eighth day, even when decomposition commenced, the bacilli were seen increasing. The contrast between the numbers in the original sputum and in this cultivation is well shown in Plate 1.

[1] *Mittheilungen aus dem K. Gesundheitsamte*, 1884.
[2] *British Medical Journal*, August 1883.
[3] *Proceedings of Royal Society*, No. 231, 1884.

Solutions of quinine in strengths varying from 2 to 10 grains to the ounce; of arsenious acid ($\frac{1}{2}$ grain and 1 grain to the ounce); of boracic acid (strengths, 1 in 15 and 1 in 30); of iodine (1 in 12), and distilled water, were mixed with sputum in equal proportions. Multiplication was observed in the bacilli mixed with distilled water; arsenious acid and boracic acid exercised no destructive or retarding influence on them, and they increased in these solutions; but iodine very considerably reduced their numbers and in some specimens it became very difficult to find any tubercle bacilli. Sulphate of quinine caused the numbers to decrease rapidly, and this destructive influence was in direct ratio to the strength of the solution.

Schill and Fischer [1] tried the effect of mixing various drugs with sputum and inoculating animals. The drugs were the following: absolute alcohol, creasote, carbolic acid, thymol, arsenious acid, naphtha, iodide of potassium, chloride of sodium, bromide of potassium, liquor ammoniæ, iodoform vapour, sal ammoniac, turpentine, salicylic acid, solution of aniline, and others. In the case of the following the inoculations were entirely unsuccessful, showing that they exercised an inhibitory action on the tubercular virus; carbolic acid (5 per cent. solution), salicylic acid (saturated aqueous solution of), liquor ammoniæ, caustic and absolute alcohol (in proportion of 3 to 1 of sputum). These were all mixed with sputum for at least twenty hours before inoculation, it being found that a less time was insufficient.

On the other hand, steam effectually destroyed the virus of fresh tubercular sputum in fifteen minutes, and of sputum fourteen days old in from thirty to sixty minutes. All experiments with the other above-mentioned drugs failed to inhibit tuberculisation in the animal.

Dr. Klein and Mr. Lingard [2] also tried the use of certain disinfectants on the virus of tuberculosis, mixing these with the caseous matter from the pulmonary lesions of human and bovine tuberculosis. Guinea-pigs were inoculated with these, but in all instances control experiments were carried out, with pure tubercular matter. The three agents used were phenyl-

---

[1] *Op. cit.*   [2] *Local Government Med. Off. Report*, 1884-85, p. 190.

acetic acid, phenylpropionic acid, and sulphocarbolate of sodium. It was found that both in bovine and human tubercle, when the two first acids were used in strengths of 1 in 400, and the times of exposure were forty-eight, seventy-two, and ninety-six hours, slight and incomplete inhibition of the development of the virus took place, the animals dying later than those inoculated with the pure virus. But when stronger solutions (1 in 200) were used for the same period, and even for shorter periods, inhibition of the tubercular virus, bovine and human, was complete. Sulphocarbolate of sodium was also tried in the same manner, and it was found that when a 10 per cent. solution was mixed with either bovine or human tubercle, for twenty-four, forty-eight, or seventy-two hours, no inhibitory action took place, the animals dying of general tuberculosis in the usual time; when, however, a 15 per cent. solution was used, for even a shorter time of exposure, complete inhibition of the virus was the result.

In these and others we have a basis of scientific experiments to guide us in the selection of agents suitable for the destruction of the bacillus, or, as we may term them, *bacillicides*; and the next question to settle is how to apply them. The channels open to us are (1) through the air passages and lungs in the form of dry or moist vaporous sprays; (2) through the blood, either by medicines taken into the stomach and alimentary canal or injected under the skin; (3) or more directly into the lungs by injections through the thoracic walls.

The treatment of phthisis by vapours or inhalations is very old, and has never been very successful. The form most in fashion at present is that of an ori-nasal respirator, containing a few drops of carbolic acid, creasote, thymol, eucalyptol, with or without some sedative to prevent irritation, a common adjunct being spirits of chloroform. The shapes of the respirators vary greatly, from the simple perforated iron oval one, covering the mouth, of Sir William Roberts, to the more elaborate ori-nasal form of Curschman.[1] Coghill's, Words-

---

[1] To be had of Krohne & Sesemann, 10 Duke Street, Manchester Square.

422   ANTISEPTIC TREATMENT OF CONSUMPTION

worth's and Burney Yeo's are very portable and convenient respirators, but inferior in comfort to Curschman's on account of
the larger size of the latter, and of its being fitted with a ring
of air cushion to prevent undue pressure against the face. They
are worn for periods of from one hour to six or seven hours,
taking them off occasionally when they are irksome to the
patient. During the last six years I have made trial of every
form which seemed to offer any advantage in phthisis, using
in them a great variety of medicaments. I have specially
noted their influence on cases where tuberculisation was commencing. Sometimes, though rarely, the cough has somewhat
lessened, and the patients have felt soothed by their use ; but
I have never found them to have the slightest effect in diminishing the local disease, or in permanently reducing the cough.
I have noted the extension of crepitation from one apex downwards, or its replacement by the signs of excavation, in spite
of continued use of this form of treatment, and though I have
often perceived a diminution of the crepitation in early cases,
I have never been able to connect it with the use of these
respirators. In the case of cavities with large purulent
discharge, some good might be expected from the introduction of antiseptics and the purifying of their walls by the continued action of these.

An objection to respirators is that they seriously impede
the freedom of respiration, partly by limiting the movements of the jaws, and partly by the obstruction to the
exit and entrance of air, which is caused by the wire gauze.
They thus more or less muzzle the patient, and prevent that
entire freedom of respiration which is so essential in phthisis.
Some experiments made by Dr. Hassall[1] on the validity of
various antiseptics when used in respirators, go to show that
creasote, carbolic acid, and thymol are scarcely capable of
volatilisation at the temperature generally used in these
appliances, and that the amount inhaled when the respirators
are used for one or two hours daily is so small as to be
practically useless. Iodine, on the other hand, is easily
volatilised ; but, on reaching the mucous membrane of the

[1] *Lancet*, May 5, 1883.

mouth, appears to be converted into an iodide by combination
with the saliva, and it is doubtful how much of the pure
iodine reaches the respiratory surface, though some probably
may. When carbolic acid and iodine are inhaled together,
according to Dr. Hassall, a strong chemical action is set up
between them, whereby probably the antiseptic properties of
both are impaired.

Another mode of antiseptic treatment is by hot-water or
steam inhalations of various kinds, some form of inhaler being
used, and the different drugs kept at a temperature suitable
to promote their vaporisation. In this way we do get a
certain amount of the drug inhaled. The best form of this
treatment seems to be jets of steam spray charged by means
of capillary tubes with necessary medicaments, such as may
be seen in use in the inhalation rooms of the New Wing of
the Brompton Hospital. The patients receive a good deal of
the drug into their bronchial tubes and lungs in a short time,
owing to the force of the steam current ; but the objection lies
in the damp and hot atmosphere which it causes and the
inexpediency of subjecting the patient for any length of time
to such strong measures.

I have applied a number of medicinal agents by this
method to the lungs and air-tubes, such as tar, creasote,
carbolic acid, eucalyptol, combined with conium, henbane and
other sedatives, but I have never noted anything more than
temporary beneficial results from this method of treatment,
such as decrease of the cough, and no reduction in the number
of tubercle bacilli in the sputum. Siegle's and Oertel's steam
sprays are the best form for warm inhalations, but like all this
class can only be used for limited periods, say two or three
times a day.

The third method is by diffusing through the air of a
chamber medicated vapours. In this way consumptive patients
can be kept under the influence of special drugs for long
periods. In this method we only imitate some varieties of
climate, such as those of the sea coast, of pine woods, and of
sulphur springs, or in the neighbourhood of volcanoes. Three
rooms are set apart for this purpose in the New Wing of the

Brompton Hospital, and for some time two of my wards, containing three beds each, were kept specially impregnated with the vapour of iodine and chlorine respectively, and selected cases of consumption were submitted to this mode of treatment. The results were merely negative in each case, and the treatment was discontinued.

A novel mode of inhalation treatment has been initiated by Professor Cantani of Naples,[1] who, taking for his principle the struggle of existence among micro-organisms, has attempted to destroy a pathogenic organism in the form of the tubercle bacillus by means of a non-pathogenic one in the shape of the bacterium termo, an organism present in decaying animal matter and putrefactive fluids. A pure cultivation of the termo mixed with broth was reported to have been sprayed into the air-passages of a woman aged 42, who had a large cavity of the left lung, and whose sputum contained elastic fibres and tubercle bacilli. According to Cantani, one month of this treatment caused disappearance of the tubercle bacilli from the expectoration and their replacement by the bacterium termo, accompanied by reduction of pyrexia and general improvement and diminution of cough and expectoration. Salama of Pisa confirmed Cantani's conclusion by reporting a similar successful case; but, on the other hand, Filopovitch[2] of Odessa treated six cases of advanced pulmonary phthisis by this method for periods varying from seven to fifty-two days. In one case the experiment was given up at the end of a week, the patients growing worse from the beginning, the fever steadily rising, bronchitis increasing, and hæmoptysis appearing. Three patients died under the treatment; one after fifteen, another after seventeen, and a third after twenty-five days of treatment. The remaining two left the hospital in no way improved. In none was there any diminution of the sputum, and in one fatal case it became more profuse, more liquid, and with the characteristic bad odour of a pure cultivation of the bacterium termo, and after death numerous excavations showed a culture of this organism. In the non-

[1] *London Medical Record*, November 16, 1885.
[2] *British Medical Journal*, October 2, 1886.

fatal cases nothing like a diminution of the tubercle bacilli was noted, and Filopovitch came to the conclusion that no good whatever may be expected from the treatment of tuberculosis by the inoculation of the bacterium termo, and that some of his cases indicated that infection of the human system by this organism was not quite so harmless as had been alleged. Dr. Theodore Acland, in December 1886, visited Professor Cantani's wards and laboratory, and after careful investigation came to the conclusion that no pure cultivations of bacterium termo had been employed, but a spray of mixed bacilli had been used. Nor was he able to discover any of the cases which had been so successfully treated on this method.

It is by no means certain that out of the body the bacterium termo is capable of surviving the tubercle bacillus, and I have often noted that in decaying tubercular lungs the tubercle bacilli are to be detected after the bacterium termo and other putrefactive bacilli have long disappeared. These latter are to be found in abundance in extending cavities, where the tubercle bacilli are also abundant, but they do not appear to interfere with the process of softening and excavation. Dr. Theodore Acland favoured me with a fact strongly bearing on the matter. In one experiment he found the two organisms growing side by side at the point of inoculation. The animal died of tuberculosis, and in the glands nearest the point of inoculation the two kinds of bacilli were present, while in the organs the tubercle bacilli alone were detected, showing clearly that the lymphatic glands stop the advance of the bacterium termo and not of the tubercle bacillus.

It is by no means impossible that the destroyer of the tubercle bacillus may be found in some other and stronger organism, but if this is to be, it will probably be in the form of the class that is capable of penetrating the tissues deeply, and of dispensing, which the bacterium termo cannot do, with the presence of oxygen.

Bacillicide medicines.—Our knowledge of agents capable of destroying the tubercle bacillus is too recent to admit yet of decisive results in this line, but we may fairly expect excellent ones, as experiment widens our list of bacillicides.

Taking my stand on Klein and Lingard's researches, I have lately been trying sulphocarbolate of sodium, phenylacetic acid, and phenylpropionic acid. The first-named salt was highly spoken of by Dr. Sansom as a good method of conveying carbolic acid to the tissues, and was given in doses of from seven to twenty grains three times a day to fifteen patients for periods varying from twenty-eight to eighty days, the average number of days of this treatment being fifty-two.

The patients who took sulphocarbolate of sodium were six males and nine females, of ages varying from fifteen to forty-four, of whom five had a cavity in one lung and one had one in both lungs. Nine had tubercular consolidations only. One lung, right or left, was alone attacked in twelve, and both lungs in three cases. In two instances there was ulceration of the larynx, and in three, considerable pyrexia. The patients were a fair sample of the ordinary consumptive in-patients, omitting perhaps the very acute and the very advanced cases, and the diet was the ordinary one of the hospital. The results were as follows: twelve showed general improvement, three, none at all; ten gained weight, three lost weight, and in two it was stationary. In one the gain amounted to 19 lbs. The cough and expectoration decreased very markedly in eleven, and increased in four. The influence of the sulpho-carbolate on the temperature was carefully watched, but it appeared to exercise no influence over it. The gain of weight showed that the appetite was not reduced. The *local* condition of the lungs showed increase of disease in four patients, a stationary condition in four, and a distinct diminution in ten. The observations on the comparative number of the tubercle bacilli in the sputum were too few to warrant inferences. The urine never gave any sign of carbolism.

The general impression left on my mind from these cases was, that though the effect on the local disease was not striking, the influence on the general condition of the patient, as seen in the weight and in the decrease of the cough and expectoration, was sufficiently good to justify a further trial. Phenylacetic acid and phenylpropionic acid are also being administered by the mouth, and have the merit of being singularly agreeable medicines, resembling honey in taste and smell. The dose varies from ten to twenty grains, and they are apparently well tolerated by the system for long periods.

In twelve patients (9 males and 3 females) treated with phenylacetic acid for periods varying from twenty-two to eighty-nine days, these cases being somewhat more advanced than those treated with sulphocarbolate of sodium, gain of weight was noted in nine, amounting in one case to $9\frac{1}{2}$ lbs., loss of weight in two, and a stationary condition in one. The cough diminished in seven and the expectoration was reduced in six, but it increased in two and remained the same in the rest. The local condition of the lungs showed deterioration in the way of advance or extension in two patients, a stationary condition in two, and marked improvement in eight. Tubercle bacilli were detected in the expectoration of all except one of these patients, but as their numbers were at no time large, comparative observations before and after taking the medicine gave no certain results. The general conclusion which these cases point to is not unfavourable, but it would be well to try the medicine on a larger scale before deciding definitely on its antiseptic properties. The results of phenylpropionic acid on twelve patients are somewhat more favourable than those of phenylacetic acid.

Other bacillicides are to be found in the preparations of quinine and iodine, though it does not as yet appear that their administration by the mouth or under the skin have exercised any specific effect on the bacilli. Drs. Filleau and Léon Petit [1] claim a reduction of tubercle bacilli in the sputum under hypodermic injections of carbolic acid, but their observations are too few in number to draw any deduction from. The two cases in which I gave this treatment a full trial I did not succeed in obtaining a like result.

Bacillicide intrapulmonary injections.—The direct introduction of antiseptics into tubercular consolidations and cavities has been practised by Mosler, Pepper, Berkart,

[1] *Bulletin du Laboratoire des Recherches expérimentales sur le traitement de la Phtisie pulmonaire*, October 1886.

Lepine, True and others, and deserves a further trial at the
hands of the medical profession than it has hitherto received,
partly on the ground of its being the only method by which
we secure the direct influence of medicinal agents on the
bacilli, and partly because all experience goes to show that
puncture of the lungs in such cases promotes fibrosis of the
lung and consequent obsolescence of the tubercle.

It will not be necessary here to discuss the whole question
of paracentesis of pulmonary cavities, which has been done
elsewhere[1] with regard to bronchiectatic and gangrenous
excavations. With regard to the tapping of tubercular
cavities for drainage, as was unquestionably done by Dr.
Hastings and Mr. Storks[2] and by Mosler of Griefswald,[3] it
appears to be only justifiable when the discharge is highly
offensive, and there is evidence that the channel is not sufficiently large to admit of efficient drainage. This may be the
case occasionally in basic cavities, but it is hardly possible
in, what are most common, cavities of the apex. Still, before
hastily condemning the operation, we should bear in mind,
(1) the great success which has attended the evacuation of
the caseous matter from strumous glands in various parts
of the body; (2) the many obstacles which a pulmonary
abscess, whether inflammatory or tubercular, encounters to
free discharge and to contraction of its walls, owing to its
being surrounded by the rigid thorax, to which it is probably
bound by a closely adherent pleura.

There are occasionally cases where the evacuation of a
caseous mass from the lung would probably remove the infective centre, and thereby preserve the remnant of that lung,
and possibly the whole of the opposite one, from infection.
Even Sir Spencer Wells'[4] proposed operation of pneumotomy,
or the removal of portions of diseased lung, may be justifiable

---

[1] *Vide Medico-Chirurgical Transactions*, vol. lxiii. and lxvii., for excellent
papers by Dr. D. Powell and Mr. Lyall, Dr. Biss, Dr. Cayley and Mr. Pearce
Gould ; also vol. lxix. for cases of bronchiectasis treated by paracentesis by
C. Theodore Williams, M.D. and Rickman J. Godlee, M.S., F.R.C.S.

[2] *London Medical Gazette*, 1815.

[3] *Berliner klinische Wochenschrift*, October 1873.

[4] *British Medical Journal*, June 7, 1881.

in certain cases, and there is no reason why such a proceeding should not be safely performed under antiseptic precautions.

The injection of antiseptic solutions into tubercular consolidations and cavities is by no means a formidable matter, and has never given rise to any alarming symptoms.

Mosler injected the cavities of three phthisical patients with various solutions: two with dilute permanganate of potassium, and one, first, with permanganate of potassium, and, later on, with a dilute solution of iodine and carbolic acid. In no instance did any harm result from the injections, and in the third case, where the latter were employed continuously, considerable improvement was noted in the pulmonary symptoms, though death took place from the general disease at a later period. Dr. Pepper [1] of Philadelphia injected six tubercular cavities in six patients through a fine Dieulafoy needle and with Lugel's solution of iodine (1 in 20) diluted in proportions varying from four minims up to twenty-four minims to the drachm of water, and claims to have benefited three patients considerably and very decidedly. He always used local anæsthesia to the chest wall before puncturing, and generally succeeded in detecting iodine in the saliva and sputum of the patients after the operation. A careful reading of Dr. Pepper's cases led me to believe that the iodine solution was hardly strong enough for bacillicide purposes, and that the contraction of the cavities noted by him was probably more the result of the frequent punctures, than of the iodine solution. His operations were proved, by post mortem examinations on two of the patients, to have done no harm, and, as I have also noted in more than one instance, the track of the needle could never be traced in the lung tissue.

Professor Lepine and M. True [2] of the Faculté de Lyon injected creasote dissolved in alcohol (in the proportion of 2 to 4 per cent.) into tubercular consolidations, not cavities, in twenty-five patients; but in most of these cases the number of injections were few, and though M. True claims improve-

[1] *American Journal of the American Sciences,* 1874.
[2] *Lyon Médical,* 1885.

ment in the form of diminution of cough and expectoration, with return of appetite and sleep, the results were only temporary, if not negative.

Dr. Singleton Smith[1] tried intrapulmonary injections of iodoform dissolved either in oil or in ether (1 part in 5) in five cases, three of which were tubercular. In one some improvement was noted, in the others the results were negative. Dr. Beverley Robinson made twenty-nine injections of various solutions in eighteen patients, with some improvements of the symptoms, specially of cough and dyspnœa.

Before commencing a trial of intrapulmonary injections, I devised a syringe capable of holding one drachm, with a fine aspirator needle tube about four inches in length, and ending in a solid double-edged harpoon point. The fine tube towards its end was perforated in three or four places, so that the liquid should issue not in the direction of the sharp point, which it would corrode, but behind and at right angles to it, in a series of lateral jets, with great force.

I hoped that after the lung had been punctured and the cavity reached, this free play of jets of antiseptic fluid on its walls would penetrate and soak into the cavity lining, which, it is well known, swarms with tubercle bacilli, and would thus destroy a large number of bacilli, and promote the evacuation of many more by free expectoration.

The syringe was first duly tested on the dead body, the fluid used being a solution of fuchsine, in order to mark the track of the needle through the tissues, and the injections were generally made in the first or second interspaces. In cases where cavities had been diagnosed during life in that region, the cavity was reached without difficulty. On removing the thoracic wall and opening the lung, the cavity was generally found full of the coloured fluid, the rest of the lung being quite free from any tint of the injection.

On the other hand, when the injection was made into tubercular consolidations devoid of excavations, the whole lung from apex to base was found permeated with the staining fluid.

[1] *British Medical Journal*, October 31, 1885.

In each case of injection into phthisical patients some innocuous staining fluid was mixed with the antiseptic solution, in order to trace it in the expectoration, and for this reason an aqueous solution of methylene blue was selected. A middle-aged man, with a cavity in the upper lobe of the right lung, with apparently an adherent pleura, was selected. There were signs of scattered tubercle below the cavity, but the lung was apparently healthy. There was no history of pyrexia or hæmoptysis, and the cough was by no means troublesome, and the case was chosen as an example of quiescent cavity. The surface of the skin was anæsthetised by a 20 per cent. solution of cocaine, and the needle, which hardly exceeded a hypodermic syringe needle in diameter, plunged into the first intercostal space overlying the cavity (as indicated by physical signs), to the depth of $2\frac{1}{2}$ inches, and when the end appeared to have reached a space in which it could be freely moved in all directions, the fluid was slowly injected. In this case it consisted of a drachm of a 15 per cent. solution of sulphocarbolate of sodium, with 10 minims of liquor opii sedativus, and sufficient methylene blue solution for colouring purposes. The patient did not complain of the pain of puncture but of a pain shooting down the right arm, which soon ceased after the operation. Coughing followed the injection, and the expectoration was immediately tinged with blue. The puncture wound was so small that the next day it could hardly be traced, and no blood flowed. No symptoms followed this injection, and it was repeated three times; but the patient complained of pain down the arm, lasting for some minutes. Other cases have been tried, and apparently without any bad results: but from the great variation in the number of tubercle bacilli present, it is difficult to determine how far they are reduced by the injection. It is, however, a somewhat irksome proceeding for both doctor and patient, and only to be attempted where the limitation of the disease is unquestionable, and the general symptoms favourable. Trial is being made of a number of these antiseptic solutions in this way, but in a matter requiring much caution and perseverance, it is not likely that we shall arrive at a

satisfactory conclusion without long and patient investigations
requiring not months but years.

Hypodermic injections of bacillicides.—Dr. Meunier of
Lyons recommends the hypodermic injection of iodoform,
eucalyptol, turpentine, and other reputed antiseptics, dissolved
in vaseline of great liquidity and purity. By Dr. Percy Kidd's
kindness I have been able to procure some of Dr. Meunier's
preparations and to test the process on two patients, one a
case of chronic cavity with some tuberculisation, and the
other a chronic first-stage patient. The preparation tried
was iodoform (5 centigrammes) and eucalyptol (18 milli-
grammes), dissolved in vaseline, and five cubic centimetres
(ʒjss) were injected under the skin of the abdomen once daily
for a month. To my surprise the whole of this large quantity
was completely absorbed in about ten minutes, leaving the
mark of the needle point and no more. The urine and saliva
were tested for iodine, but with negative results, and the introduction into the system of so much iodoform seemed to have
no effect on the temperature, pulse, respiration, or on the
cough and expectoration. The preparations certainly did no
harm, but as yet I have failed to find any distinct benefit
arising from their use.

Gaseous rectal injections.—Dr. Bergeon of Lyons has lately
introduced a novel method of treating consumption, asthma,
and some other forms of lung disease by the injection of
carbonic acid and sulphuretted hydrogen gases in large
quantities per rectum. The system is based on experiments
of Claude Bernard, who showed that certain gases, such as
carbonic acid and sulphuretted hydrogen, which were toxic
when inhaled, were absorbed by the large intestine in large
quantities, without any bad effects, then passed into the portal
circulation, and, reaching the heart and pulmonary circulation, were eliminated from the system through the lungs.
Dr. Bergeon from a simple apparatus gradually introduces
per rectum a mixture of carbonic acid and sulphuretted hydrogen, or of bisulphide of carbon, in quantities of four litres
at a time, the operation occupying about twenty minutes,
great care being taken to avoid mixture with atmospheric air,

and in a short time, according to his statement, the odour of sulphuretted hydrogen can be detected in the patient's breath. The abdomen becomes distended, often largely, but subsides without the distension giving rise to any pain or material discomfort. Dr. Bergeon states, after an experience of this treatment in hundreds of cases, that in early phthisis, even in acute general phthisis, in two or three weeks there is generally an arrest and in a few months a cure, and even in advanced incurable phthisis, including laryngeal phthisis, great amelioration is obtained. He claims that the pulse is lowered, the temperature falls, night-sweats cease, cough and expectoration diminish, the appetite returns and weight is rapidly gained, under this treatment, which consists of two injections daily. A certain amount of confirmation of Dr. Bergeon's results have been afforded by Professor Cornil, Drs. Chantemesse, Morel and Bardet, and also by Dr. Henry Bennet [1] and Dr. Burney Yeo [2] and some American physicians, but the treatment appears to exercise no reducing influence on the number of tubercular bacilli in the sputum, and, to judge by some of Drs. Bergeon and Morel's cases, [3] there is a recurrence of unfavourable symptoms on the cessation of the injections.

The treatment has been tried for too short a period to enable a fair judgment to be passed on it, and though somewhat opposed to preconceived notions, and savouring of carrying coals to Newcastle, as a colleague of mine described the injection of sulphuretted hydrogen into the rectum to be, the idea of supplying the lungs with medicinal agents through the portal system is an ingenious one, and worth a fair trial and application. My own experience of the Bergeon treatment is limited to six cases treated at the Brompton Hospital, which certainly do not bear out the results of the author of the system. The patients complained of occasional pain and distension after the injections, which, in one case, had to be

[1] *British Medical Journal*, December 18, 1886.
[2] *Lancet*, April 16, 1887.
[3] Petit, *Études expérimentales et cliniques sur la Tuberculose.* Premier fascicule, 1887.

omitted, on account of the aperient effect on the intestines. To test whether sulphuretted hydrogen was, or was not, exhaled in the breath, the patients were directed to breathe into solutions of acetate of lead. These, after several hours, were found to contain some carbonate, but no sulphide, of lead, showing that the H_2S is not exhaled as such from the lungs, but probably undergoes changes in the liver. Silver coins placed in the mouth also showed no change. The effect on the system was sedative, but temporary in duration. In all the cases the temperature fell, in some the pulse and respiration were reduced in frequency, and in one the cough and expectoration were lessened. The patients objected, in most cases strongly, to the treatment, and unless more brilliant results than these, which can be attained by other and less troublesome methods, can be arrived at, it is not likely that a system so opposed to people's feelings will be long persevered with.

INDEX.

PHY

Physical signs of pneumonia passing into phthisis, 100
of contracting cavity with emphysema, 106
of arrested phthisis, 107
of pneumothorax, 211
Pleura, adhesions of, common in chronic phthisis, 23, 54
gelatinous thickening of, 24
Pleurisy, origin of fibroid phthisis from, 271
Pneumonia, consumption originating in, 332
catarrhal, 10
croupous, 12
interstitial, 271
Pneumothorax, tubercle bacilli in, 119
forms of, 204
as sequela of acute phthisis, 205
— frequency in consumption, 206
— relative liability of lungs to, 207
— number of openings in, 207
— analysis of air in, 208
— localised, 209
— cases of recovery from, 213
— double, 219
— diagnosis of, 220
— prognosis of, 221
— treatment of, 416
Pollock, Dr., cicatrisation of cavities, 53
— cardiac displacement, 56
— hereditary predisposition, 62, 69
— temperature in consumption, 184
— scrofulous phthisis, 262
— duration of phthisis, 310
— treatment of night-sweats, 402
Poncha springs, 363
Ponchet, monads in sputum, 93
Ponfick, tuberculosis of lymphatic duct, 45
Pork broth, growth of tubercle bacillus in, 419
Portal, duration of phthisis, 310
Post-mortems. See CASES.
Powell, Dr. R. D., 277
— gelatinous infiltration of pleura, 24
— pulmonary aneurysm, 138
— intestinal ulceration, 191
— pneumothorax, 206
Predisposing causes of consumption, 58
Predisposition, family, 58

RIN

Predisposition, hereditary. See FAMILY PREDISPOSITION.
Prescriptions, 390, &c. 105, &c.
Prognosis of phthisis, 328
— — of different forms, 249, 250, 263, 270, 278, 285, 287
— — utility of tubercle bacillus in, 127
Prophylactic treatment, 337
Ptomaines, 46
Pulmonary aneurysm. See ANEURYSM.
Pyæmia after hæmoptysis, 157
Pyo-pneumothorax, 204
Pyrexia in consumption, laws of, 181, 184
— — treatment of, 394

QUAIN, Dr., 393
— disease of bronchial glands, 23
— fatty degeneration of heart, 24
— pulmonary aneurysm, 138
— phthisis and albuminuria, 237
Quinine as an antipyretic, 398
— — a bacillicide, 420
Quito, 363

RAGATZ, 376
Ransome, Dr., tubercle bacillus in air, 39
— infection of consumption, 87
Rapallo, 362
Rarefied air, influence of, in consumption, 356
Rasmüssen, aneurysm of pulmonary artery, 138
Raw meat, use of, 337
Rectal gaseous injections, 432
Red corpuscles, marked decrease of, in albuminuria, 236
Redtel, Dr., 344
Registrar-General for Scotland on damp soil causing consumption, 81
Renal disease, death from, 232, 297
Report of Registrar-General on deaths from phthisis, 309
Resorcin, 399
Respirators, antiseptic, different forms of, 422
Ribbert, tubercle in poultry, 42, 137
Rindfleisch, Prof., changes in terminal bronchi, 21
— tubercle bacillus stain, 34
— perforation of intestine, 192
— large cells in scrofulosis, 261

SCIENTIFIC WORKS.

NATURAL PHILOSOPHY for General Readers and Young Persons; a Course of Physics divested of Mathematical Formulæ, expressed in the Language of daily life, and illustrated with Explanatory Figures, elucidating the Principles and Facts brought before the reader. By Professor GANOT. Translated and Edited by E. ATKINSON, Ph.D. F.C.S. The Fifth Edition, with 20 pages of new matter, 2 Plates, and 495 Woodcuts. Crown 8vo. 7s. 6d.

ELEMENTARY TREATISE on PHYSICS, Experimental and Applied, for the Use of Colleges and Schools. By Professor GANOT. Translated and Edited by E. ATKINSON, Ph.D. F.C.S. Twelfth Edition, revised and enlarged; with 5 coloured Plates and 923 Woodcuts. Large crown 8vo. 15s.

CELESTIAL OBJECTS for COMMON TELESCOPES. By the Rev. T. W. WEBB, M.A. Map, Plate, Woodcuts. Crown 8vo 9s.

OUTLINES of ASTRONOMY. By Sir J. F. W. HERSCHEL, Bart. M.A. With Plates and Diagrams. Square crown 8vo. 12s.

HANDBOOK of PRACTICAL TELEGRAPHY. By R. S. CULLEY, Memb. Inst. C.E. Seventh Edition. Plates and Woodcuts. 8vo. 16s.

The PRINCIPLES and PRACTICE of ELECTRIC LIGHTING. By ALAN A. CAMPBELL SWINTON. With 54 Illustrations engraved on Wood. Crown 8vo. 5s.

WORKS BY JOHN TYNDALL, F.R.S.

FRAGMENTS of SCIENCE. 2 vols. crown 8vo. 16s.

HEAT a MODE of MOTION. Crown 8vo. 12s.

SOUND. With 204 Woodcuts. Crown 8vo. 10s. 6d.

ESSAYS on the FLOATING MATTER of the AIR in RELATION to PUTREFACTION and INFECTION. With 24 Woodcuts. Crown 8vo. 7s. 6d.

LECTURES on LIGHT, delivered in America in 1872 and 1873. With Portrait, Plate, and Diagrams. Crown 8vo. 5s.

LESSONS in ELECTRICITY at the Royal Institution, 1875–6. With 58 Woodcuts. Crown 8vo. 2s. 6d.

NOTES of a COURSE of SEVEN LECTURES on ELECTRICAL PHENOMENA and THEORIES, delivered at the Royal Institution. Crown 8vo. 1s. sewed; 1s. 6d. cloth.

NOTES of a COURSE of NINE LECTURES on LIGHT, delivered at the Royal Institution. Crown 8vo. 1s. sewed; 1s. 6d. cloth.

FARADAY as a DISCOVERER. Fcp. 8vo. 3s. 6d.

London: LONGMANS, GREEN, & CO.

www.ingramcontent.com/pod-product-compliance
Lightning Source LLC
Chambersburg PA
CBHW020900210326
41598CB00018B/1730